Transit management in the Northwest Passage: problems and prospects

Studies in Polar Research

This series of publications reflects the growth of research activity in and about the polar regions, and provides a means of disseminating the results. Coverage is international and interdisciplinary. Most books will be surveys of the present state of knowledge in a given subject rather than research reports, conference proceedings or collected papers. The scope of the series is wide and will include studies in all the biological, physical and social sciences.

Other titles in this series:

The Antarctic Circumpolar Ocean
Sir George Deacon

The Living Tundra
Yu. I. Chernov

Transit management in the Northwest Passage

PROBLEMS AND PROSPECTS

Editors

CYNTHIA LAMSON

DAVID L. VANDERZWAAG

Dalhousie Ocean Studies Programme

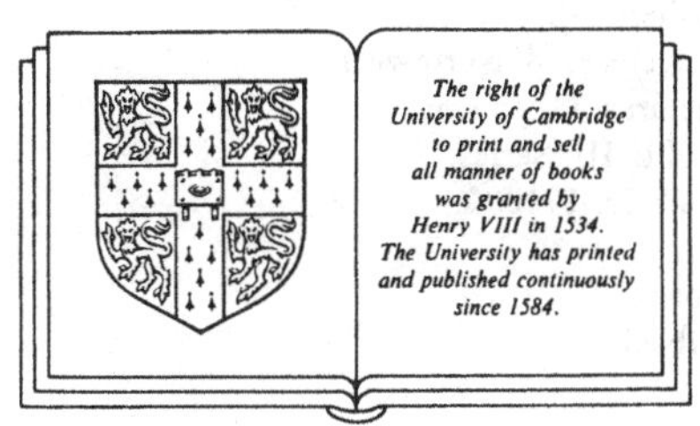

CAMBRIDGE UNIVERSITY PRESS

Cambridge

New York New Rochelle

Melbourne Sydney

CAMBRIDGE UNIVERSITY PRESS
Cambridge, New York, Melbourne, Madrid, Cape Town, Singapore, São Paulo, Delhi

Cambridge University Press
The Edinburgh Building, Cambridge CB2 8RU, UK

Published in the United States of America by Cambridge University Press, New York

www.cambridge.org
Information on this title: www.cambridge.org/9780521320658

First published 1988
This digitally printed version 2008

A catalogue record for this publication is available from the British Library

Library of Congress Cataloguing in Publication data

Transit management in the Northwest Passage:
(Studies in polar research)
Includes bibliographies.
1. Shipping—Arctic regions—Addresses, essays,
lectures. 2. Navigation—Arctic regions—Addresses,
essays, lectures. 3. Canada, Northern—Economic
conditions—Addresses, essays, lectures. 4. Northwest
Passage—Addresses, essays, lectures. I. Lamson,
Cynthia. II. VanderZwaag, David L. III. Series.
HE935.T7 1987 387.5′09163′27 85–171000

ISBN 978-0-521-32065-8 hardback
ISBN 978-0-521-09337-8 paperback

Contents

Acknowledgments xi

PART I: PERSPECTIVES ON THE PROBLEM

1 **The Northwest Passage: a contrast of visions** 3

Post scriptum 7

2 **The environment of the Northwest Passage** *Hal Mills* 8

Environmental perspectives 9

The international Arctic ocean 9

Canada's Arctic environmental initiatives 12

The conservation viewpoint 13

The physical setting 14

Bathymetry and currents 14

Ice conditions 22

Recurring polynyas 31

The biological setting 33

Arctic marine productivity 33

Marine and anadromous fish 34

Marine-associated birds 36

Marine mammals 38

Biologically significant areas 42

The human setting 49

The circumpolar Inuit 49

Renewable resource harvesting 54

Environmental issues 55

Introduction 55

Resource development and environmental issues 56

Transportation environmental issues 58

The policy muddle 59

Notes 60

3 **The development of Northern ocean industries** *Carlyle Mitchell* 65
The problem outlined 65
An analytical framework for Northern ocean industries development 66
The structure of the Northern economy 66
Theory and concept of Northern development 68
The role of Northern ocean industries 70
The Northern economy: a general perspective 71
Economic growth in the North, 1971–1980 71
Employment and incomes in the Northern economy 74
Northern waters industrial development in the 1970s 79
The renewable resource industries 79
Non-renewable resource industries 83
Theoretical evaluation 91
The future of Northern ocean industries 95
Notes 97

4 **Arctic marine transport and ancillary technologies** *Ernst Frankel* 100
The challenge of Arctic marine transportation 100
Introduction 100
Environmental problems 101
Navigation problems 104
Arctic transportation as a systems problem 106
Operational requirements 107
Modelling and systems analysis 107
Arctic shipping technology 108
Icebreaking ships 109
Submarine tankers 110
Icebreakers 111
Arctic tug-barge operations 113
Technological evaluation 114
Arctic interface technology 114
Terminal requirements 115
Submarine pipelines 115
Island terminals 116
Submerged terminals 116
Sensors and controls for interfacing 117
Cargo storage 117
Transfer and cargo handling 118
Arctic navigation, communications and control technology 119
Dead reckoning navigation systems 119
Radio-based navigation systems 121
Satellite communications 123
Navaids 125

Sensors and submerged markers 125
Traffic routing 126
Future requirements 127
Notes and additional sources 129

5 **Canadian arctic marine transportation: present status and future requirements** *William J. H. Stuart & Cynthia Lamson* 130
Introduction 130
Administration 131
Icebreaking vessels 134
Hydrography 138
Operational support services 140
Navigation and communications aids 140
Vessel traffic management 142
Search and rescue/salvage 144
Manpower and training 146
Shipbuilding 147
Summary 148
Notes 151

6 **Northern decision making: a drifting net in a restless sea** *David L. VanderZwaag & Cynthia Lamson* 153
Introduction 153
Government policies: the floats 154
Administrative mesh 157
Knitting an administrative net 158
Mending the administrative net 162
Finding the policy floats 164
Northern decision making: set- versus drift-netting 182
Legislative mesh 186
Land-use controls 187
Water-use controls 189
Special use provisions 197
Public review mechanisms 205
The decision-making net in action 210
Project reviews 212
Reasons for variation 227
Conclusions 229
Notes 230

7 **Constitutional development in the Northwest Territories** *Fielding Sherwood* 251
Introduction 251
Background 252
The 1960s to the present 256
The Nunavut proposal 260

The Federal–Nunavut relationship 262
Offshore resources and revenues 265
Other issues and conclusion 269
Notes 270

PART II: PARADIGMS AND PROSPECTS 277

8 **The designing of a transit management system** *Douglas M. Johnston* 279
The concept 279
Function 279
Scope 280
Alternative approaches, frameworks and theories 281
Introduction 281
Systems theory 282
Planning theory 283
Organizational theory 289
Impact assessment studies 290
Management system design 293
The question of values 294
Challenge to a nation 294
The values at stake 295
Some preliminary conclusions 299
Notes 302

APPENDIX: Statement on Canadian sovereignty 309

CONTRIBUTORS 315

List of figures

2.1	Bathymetry of the Beaufort and Chukchi Seas.	15
2.2	Mean general summer circulation of surface waters in the Beaufort and Chukchi Seas.	17
2.3	Bathymetry of the Northwest Passage and Arctic Islands.	18
2.4	Mean near-surface circulation in the Northwest Passage.	19
2.5	Main routes of the Northwest Passage.	20
2.6	Mean near-surface circulation of the Baffin Bay–Davis Strait Region.	23
2.7	Generalized winter ice zones of the Southern Beaufort Sea.	25
2.8	Major recurring polynyas and shore leads.	27
2.9	Average September ice conditions in the Central Arctic (Parry Channel) and preferred tanker route.	29
2.10	Average ice conditions during March in the Eastern Arctic.	30
2.11	Preliminary summary of food web relationships in Arctic offshore waters.	35
2.12	Seabird concentrations along the Canadian and West Greenland coasts.	37
2.13	Migration routes of marine-associated birds in the Western Arctic.	39
2.14	Walrus migration routes, wintering areas and known coastal summering areas in Parry Channel, Baffin Bay and Davis Strait.	41
2.15	White whale migration routes, wintering areas and summer concentrations in the Eastern Arctic.	43
2.16	Migration routes and summer ranges of bowhead and white whales in the Beaufort Sea.	44
2.17	Bowhead whale migration routes, probable winter range and known summer concentration areas.	45
2.18	Narwhal migration routes, wintering areas and summer concentrations in the Eastern Arctic.	46

2.19	Inuit use of sea ice and adjacent land areas for hunting.	52
3.1	The structure of the Northern economy.	68
3.2	Communities North of 60°.	76
3.3	World oil prices, 1960–1983.	84
3.4	Location of wells drilled for the calendar year 1982.	85
3.5	Wells drilled, Yukon and Northwest Territories, 1969–1980.	88
3.6	Mineral exploration and mining, Northwest Territories, 1981.	92
4.1	Conical drilling platform.	101
4.2	Movable drilling platform.	102
4.3	Icebreaking LNG carrier.	109
4.4	Submarine tanker.	110
4.5	*MV Robert LeMeur*	111
4.6	The Arctic Ocean, sea routes.	127
5.1	Canadian Coast Guard administrative regions.	132
5.2	Class 4 icebreaker/ice management vessel.	136
5.3	Class 4 supply vessel.	137
5.4	Status of surveys (1981).	139
5.5	Rescue Coordination Centres (RCC) and Search and Rescue Emergency Centres (SAREC).	144
6.1	Typology of policy instruments.	164
6.2	Evolution of the Task Force on Northern Oil Development.	167
6.3	Canada Oil and Gas Lands Administration (COGLA).	176
6.4	Total direct Northern expenditures.	180
6.5	Shipping safety control zones.	196
6.6	Typical COGLA approvals process for a fixed hydrocarbon system.	200
6.7	Major Northern energy projects.	211
7.1	Evolution of the Northwest Territories.	254
7.2	Inuvialuit land claims settlement area.	266
	Canadian Jurisdictional zones in the Arctic	314

Acknowledgements

Transit Management in the Northwest Passage is the first product of a four-year study by the Dalhousie Ocean Studies Programme entitled the "Canadian Northern Waters Project". The Project has been funded through the generous support of the Donner Canadian Foundation. Transport Canada has also provided valuable assistance, and particular appreciation is extended to Coast Guard Northern for keeping their doors open and providing information when requested.

Several individuals deserve special commendation for their yeoman efforts in bringing this book to fruition: Ena Morris (for word processing and typesetting beyond the call of duty), Lynda Corkum (for word processing multiple drafts), Susan Rolston (for the tedious task of assisting in the editing process), Lester Foster (for technical assistance in typesetting), and most of all, Professor Douglas M. Johnston, the original conceptualizer of the project whose continuous and constructive support was invaluable.

Cynthia Lamson and David L. VanderZwaag, editors

Part I: Perspectives on the problem

1

The Northwest Passage: a contrast of visions

The Northwest Passage – a marine short cut to the Orient for European trade. That was the historic vision of many. The history of the search for a Northwest Passage dates back to 1497 when John Cabot was sent by King Henry VII of England to find a northerly route to the Orient. Cabot's mission was taken up by later explorers, including Jacques Cartier, Sir Francis Drake, Sir Martin Frobisher, William Baffin, Captain James Cook, Sir John Ross and Sir William Parry, but none were successful and a number of voyages culminated in tragedy. In 1845 Sir John Franklin and the crews of the *Erebus* and *Terror* vanished while attempting to find a navigable route across the Arctic Ocean, and from 1847 to 1859 the vast Northern seas were explored from the east and west in search of the lost expedition. The first successful transit of the Passage by sea was achieved by the Norwegian explorer, Roald Amundsen, whose converted herring boat *Gjoa* completed the trip over a three-year span from 1903 to 1906.[1] In the 1940s a Royal Canadian Mounted Police Sergeant, Henry A. Larsen, was the first person to transit the Passage in a single season in the schooner *St Roch*. The *Labrador*, a Canadian Coast Guard icebreaker, navigated the Passage in 1954 and as of September 1984, 41 surface vessels have completed transits of the Northwest Passage.[2]

The Northwest Passage – a marine highway for transporting valuable cargoes such as oil, gas and hard minerals to world ports. That has been the more recent vision of many. In 1969 the *SS Manhattan*, a 43000 HP ice-adapted oil tanker, became the first commercial ship to negotiate the Passage and demonstrate the feasibility of transporting Prudhoe Bay oil from Alaska by the marine mode. Dome Petroleum Ltd, Esso Resources Canada Ltd and Gulf Canada Resources Inc. are proposing to develop and transport hydrocarbons from the Beaufort Sea–Mackenzie Delta region, perhaps by a combination of pipelines and tankers. Panarctic Oils

Ltd of Calgary is proposing a "pilot project" to transport oil from its Bent Horn field on Cameron Island in the Arctic Islands to southern markets by an ice-strengthened tanker during the summer months. If successful, Panarctic could also develop its larger oil discovery at Cisco off the coast of Lougheed Island, estimated to contain 300 million barrels of oil, which could involve year-round transport of the Passage by ice-strengthened tankers.[3]

Located between the mainland of Canada and the Arctic Islands, the approximately 900–mile long Passage offers tremendous transportation advantages by shortening the shipping distance between some of the major seaports of Europe, Japan, the USSR and North America. For example, the distance between London and Yokohama could be reduced from approximately 14,650 nautical miles[4] to less than 8000 nautical miles. The Hibernia oilfield off Newfoundland is approximately 9500 nautical miles from Yokohama via the Panama Canal[5] and only 6300 nautical miles via the Northwest Passage.[6]

The Northwest Passage also offers a contrasting vision for many, that of a bountiful reserve of renewable resources and a fragile natural heritage. Many Northerners depend on harvests of marine species such as Arctic char and ringed seals for economic well-being and cultural survival. Many individuals perceive the sparsely populated Arctic as a lifestyle alternative, a frontier largely free from the complexities of modern industrial society.

Such visions tend to be mutually exclusive. The image of a cold, pristine Arctic wilderness suddenly melts into a picture of a groaning, oil-laden tanker rending the Northern tranquillity. Persons who look at the North from an aesthetic perspective may even more strongly oppose what a tanker symbolizes – an uncontrolled twentieth century industrial thirst for economic growth and technological expansion.

Given the real world of competing demands, where the extraction of non-renewable resources may conflict with other resource uses, governments must act as regulatory, mediating or planning agencies trying to reconcile the multifarious visions of the ideal. The challenge is to be a creative welder of visions, sensitive to the past and the future, sensitive to all values, including biological, psychological, cultural, political and aesthetic. In the North, the choice for governments need *not* be between renewable and non-renewable resource development. Both may proceed in a socially and environmentally acceptable manner if government assures the choice of sound technologies and an integrated environmental management regime.

To facilitate the integration of visions concerning the uses of Canada's

Northwest Passage, Dalhousie Ocean Studies Programme, funded by the Donner Canadian Foundation, has initiated a four year-study entitled the "Canadian Northern Waters Project". The purpose of the Project is to identify and analyse problems associated with designing and implementing an effective system of "transit management" in the Northwest Passage. Analysis will extend beyond technical and operational questions and will address issues of Arctic Ocean policy making.

This book represents the first step towards developing a transit management system for Canada's Northern waters. The first six chapters summarize major problem areas which must be considered in designing an integrated ocean management regime for the Northwest Passage. Chapter 2 highlights important environmental and human concerns at stake in Arctic resource development. Chapter 3 reviews the economic difficulties in developing Northern ocean industries. Chapter 4 surveys Arctic marine technologies such as icebreaking vessels, submarine tankers and terminals, artificial islands and navigation-communication systems. Chapter 5 sets forth Canadian Arctic marine transportation capabilities and probable future needs. Chapter 6 provides an overview of the administrative and legislative framework for Canadian Northern decision making. Chapter 7 explores the evolving political development of the Northwest Territories. The final chapter examines alternative conceptual models for designing a transit management system and advocates an integrated research strategy.

The design of a transit management system for the Northwest Passage will not occur by divine fiat, in the twinkling of an administrative eye. The final form of legal and administrative structures will ultimately depend on a gradual evolution of mechanisms created by the synergistic interactions among the federal government, Territorial governments, community groups, industry leaders, academics and public interest groups.

Dalhousie Ocean Studies Programme will contribute to such a process through the publication of a Northwest Passage series, of which this volume represents the introductory overview. Subsequent volumes will examine such topics as the status of the waters of the Canadian Arctic Archipelago in international law, the political economy of the Northwest Passage, the environmental assessment process for Arctic shipping proposals, and options for achieving co-operative ocean management among Arctic states. Whether the kaleidoscope of ideas, offered by such publications, will crystallize into an appropriate framework for managing the Northwest Passage depends ultimately on the commitment of Canadians to transform Northern visions into creative decisions.

Notes

1. *See* "Arctic Tankers through the Northwest Passage", *Beaufort* (December 1982), pp. 21–27; and Donat Pharand, "Quel Sera l'avenir du passage du Nord-Ouest?", *North/Nord* (Summer 1980), pp. 2–9, and (Autumn 1980), pp. 2–7.
2. *See* Donat Pharand, *The Waters of the Canadian Arctic Archipelago in International Law* (in press). In September, 1984 the *MS Lindblad Explorer* became the first cruise ship to transit the Northwest Passage on a 7600 nautical mile journey from St John's, Newfoundland to Yokohama, Japan.
3. "Panarctic Eager to lift Arctic Islands Oil" (Toronto), *Globe and Mail* (7 June 1984), p. 36.
4. The distance between London and Yokohama is 14650 nautical miles via the Cape of Good Hope, 12683 nautical miles via the Panama Canal and 12111 nautical miles via the Suez Canal. *Lloyd's Maritime Atlas*, Distance Table (eleventh ed, 1977).
5. *Canadian Ports and Seaway Directory* (1983), p. 51.
6. T. C. Pullen, "Arctic Marine Transportation: A View from the Bridge", *Ocean Policy and Management in the Arctic*, Canadian Arctic Resources Committee (1984), p. 129.

Post scriptum

"Time waits for no man." In the field of academic research the saying should be modified to "Time waits for no book", for human events often march quickly beyond the facts captured by any manuscript.

In the fast-changing world of Arctic affairs with the rapid evolutions in technologies, legal regimes and political arrangements, the saying is particularly true and certainly applies to the present volume. When the manuscript was in preparation during the years 1982–1984, two of Canada's Arctic policies were particularly unclear – the extent of national jurisdiction over Arctic waters and a national commitment to expand icebreaking capabilities. However, on September 10, 1985, External Affairs Minister, Joe Clark, reacting to public sentiment aroused by the transit of the U.S. Coast Guard icebreaker *Polar Sea* through the Northwest Passage in August 1985, clarified the political waters by announcing in the House of Commons the drawing of straight baselines around Canada's Arctic Archipelago, in order to formalize Canadian sovereignty over the enclosed waters, and the intention to construct a Polar Class 8 icebreaker.

Because the statement demonstrated a new national commitment to develop and control Northern waters, the statement is reproduced in its entirety as an Appendix. The statement challenges Canadians to address questions about transit management including the designation and protection of environmentally significant areas, formulation of a sound Northern economic strategy, development of appropriate marine technologies, implementation of mandatory vessel management services traffic, and creation of an explicit and fair decision-making process for reviewing industrial proposals having a shipping component.

Cynthia Lamson
David VanderZwaag

2

The environment of the Northwest Passage

HAL MILLS

To describe the environment of the Northwest Passage fully in a single chapter is perhaps an impossible task. The Passage, broadly defined, cuts across five time zones and includes at least seven distinct ocean regions – the Beaufort Sea, Amundsen Gulf, Prince of Wales Strait, Viscount Melville Sound, Barrow Strait, Lancaster Sound, Baffin Bay and Davis Strait. Volumes would be required to describe in any detail the environmental characteristics of these diverse regions.

By necessity, then, this chapter is much like an impressionistic painting. Only the broadest strokes may be applied to the canvas. Section I provides an overall framework for viewing the marine environment of the Northwest Passage by looking at the Arctic Ocean as an international concern, by examining Canada's major environmental protection initiatives, and by raising the need for a Northern conservation viewpoint. Section II explores the physical environment of the Passage through three lenses, bathymetry and currents, ice distribution and characteristics, and the location of recurring polynyas (open water areas). Section III highlights the biological environment from five perspectives: Arctic marine productivity, marine and anadromous fish, seabirds, marine mammals, and biologically significant areas. Section IV reviews the human setting by describing the circumpolar Inuit and the importance of renewable resource harvesting from marine waters. Section V concludes by summarizing the environmental issues raised by Arctic resource development and marine cargo transportation.

I Environmental perspectives

A The international Arctic Ocean

The Arctic is regarded by Russian scientists as "the weather kitchen of the Earth" and rightly so.[1] On a global scale the Arctic serves as a heat sink whereby the earth's atmosphere maintains itself in equilibrium through a net poleward transfer of heat,[2] to compensate for the lower levels of solar radiation and the higher levels of surface albedo from snow and ice. The heat transfer takes place in three ways: by the transport of air from lower latitudes; by the transfer of water vapour to snow or rain which releases heat; and by oceanic transport. Sea ice affects the heat exchange between the ocean and the atmosphere by acting as an insulator.

The temperature balance is so delicate that small increases in seawater temperature could lead to the melting of the polar pack ice and the Greenland ice sheet,[3] which could lead to a dramatic increase in sea level throughout the world. However, scientists disagree about what sorts of human activities could result in an alteration of the Arctic heat balance. Possible threats include oilspills, atmospheric pollution from the burning of hydrocarbons, and the release of cooling water from nuclear power plants along the Arctic coast. Obviously, it is important to avoid any environmental damage that could set off a chain reaction which would result in a cataclysmic change in the earth's climate.

Arctic marine ecosystems are international in character, with five states – Canada, the United States, the Union of Soviet Socialist Republics, Denmark/Greenland and Norway – bordering on the circumpolar Arctic Ocean, which has an area of more than 5000000 square miles. Much of the region is permanently covered with pack ice and the level of primary productivity in the ecosystems generally is quite low in comparison to Atlantic or Pacific sub-arctic ecosystems. Despite the low species diversity and low primary productivity, arctic marine ecosystems do produce pockets of intense biological richness at certain times of the year, often with large concentrations of particular species. These include migratory species of marine mammals and sea birds which breed within Canadian arctic waters, but which are international over their range. Therefore, both the ecosystems and species within them may come under the jurisdiction of more than one nation, and their effective protection and management may require international co-operation. One example of this is the Agreement on the Conservation of Polar Bears, wherein the governments of Canada, Denmark, Norway, the Soviet Union and the United States pledged to protect polar bear habitats, to manage polar bear populations according to sound conservation practices and, based on the best available scientific data, to coordinate polar bear research.

The application of international law to the Arctic marine environment has been complicated by overlapping jurisdictional issues, and by political, strategic and economic considerations. In the Canadian sector of the Arctic, the extent of Canada's sovereignty over the Arctic Islands is not disputed, but the status of the waters is somewhat less certain. Canada takes the position that the waters between the Arctic Islands are internal waters, and has formally asserted that claim by drawing territorial baselines around the Islands. However, there is continuing uncertainty about the status of the Northwest Passage in international law, and specifically whether it should be characterized as an "international" strait.[4]

In terms of the use, and potential misuse, of the international Arctic Ocean, the most significant economic resources are the hydrocarbon reserves of the circumpolar continental shelves, where substantial reserves or favourable prospects occur in the Barents Sea, Kara Sea, Laptev Sea, East Siberian Sea, Chukchi Sea, Beaufort Sea, the High Arctic Islands and Baffin Bay–Davis Strait. Potential hydrocarbon production and the associated transportation of hydrocarbons to markets pose the most immediate threats to the Arctic marine environment. Projected transportation of Arctic hydrocarbons by supertanker has sparked interest in the use of the Northwest Passage as a route that would link the Beaufort Sea with Japan, Europe and the eastern seaboard of North America. In a 1970 diplomatic note to the United States, Canada reiterated "its determination to open up the Northwest Passage to safe navigation for the shipping of all nations, subject, however, to necessary conditions required to protect the delicate ecological balance of the Canadian Arctic".[5]

With the exception of Article 234 (the ice-covered waters provision), the recently completed United Nations Convention on the Law of the Sea (UNCLOS III) does not specifically deal with the Arctic Ocean. UNCLOS III apparently avoided considering Arctic issues because of special Arctic circumstances, such as military strategic importance and the limited number of states bordering the region.[6] Some doubt thus exists as to the applicability of the Convention to the Arctic Ocean, but there is every reason to believe that it eventually will apply. The Treaty will place general obligations on circumpolar states with respect to land-based sources of pollution, pollution from continental shelf activities, dumping, vessel source pollution, monitoring, and environmental assessment. Article 194(5) deals directly with marine habitats and fragile ecosystems:

The measures taken in accordance with this part shall include

> those necessary to protect and preserve rare or fragile ecosystems as well as the habitat of depleted, threatened or endangered species and other forms of marine life.

The International Union for the Conservation of Nature and Natural Resources (IUCN) has developed a World Conservation Strategy, listing objectives for conservation and setting priorities for national and international actions. The Strategy has designated the Arctic as a priority area:

> [B]ecause the Arctic environment takes so long to recover from damage, the Arctic shall be considered a priority sea. Within their Arctic territories the Arctic nations should systematically map critical ecological areas (terrestrial as well as marine), draw up guidelines for their long term management, and establish a network of protected areas to safeguard representative, unique and critical ecosystems.[7]

States which officially adopt the World Conservation Strategy take on obligations for implementation within their country and to co-operate with other states on a regional basis.

The need for regional co-operation may be particularly acute for the Arctic Ocean in that: it is almost totally enclosed by land; it is mostly covered by ice; it plays a significant role in controlling world climate; it has vast hydrocarbon reserves; and it contains unique ecosystems, highly susceptible to pollution or disturbance. The international obligation to co-operate on a regional basis is covered in the Convention on the Law of the Sea (Article 197) which says that states shall co-operate "for the protection and preservation of the marine environment, taking into account characteristic regional features". The World Conservation Strategy indicates that among the items of common concern for Arctic nations are:

> [M]easures (including joint research) to improve protection of migratory species breeding within the Arctic and wintering inside or outside the region; studies of the impact of fisheries and other economic activities in the northern seas on ecosystems and non-target species; [and] the possibility of developing agreements among the Arctic nations on the conservation of the region's vital biological resources, based on the principles and experience of the Agreement on Conservation of Polar Bears.[8]

Canada is a party to two major agreements pledging regional co-operation in the Arctic at the binational level. The Canada–Denmark Marine Environment Co-operation Agreement establishes joint contingency plans for shipping or seabed pollution incidents, sets forth the duty to consult on activities creating a significant risk of transboundary pollution, pledges co-operation in marine scientific research and mandates co-operation in vessel traffic management and in identifying appropriate vessel routing areas. The Canada–United States Joint Marine Pollution Contingency Plan for the Beaufort Sea establishes procedures for the US Coast Guard and the Canadian Coast Guard to respond jointly to any oil or noxious-substance pollution threatening the waters or coastal areas of both parties.

B *Canada's Arctic environmental initiatives*

The *Manhattan* incident undoubtedly stimulated Canada to begin taking an active role in the protection of the Arctic marine environment. During the mid-1960s oil was discovered on the North Slope of Alaska, and the Northwest Passage was proposed as one transportation route to East Coast markets. The proposed trial voyage of the *Manhattan* implicitly raised questions regarding the status of the Passage, with Canadian concerns focusing on sovereignty rather than the environment.[9] When the United States refused to recognize Canadian sovereignty, Canada decided to grant permission for the voyage (even though no permission had been requested) and to supply a Coast Guard icebreaker to support the *Manhattan*, thereby ensuring *de facto* sovereignty.[10]

During the 1969 voyage the *Manhattan* required assistance from a Canadian icebreaker on two occasions, and was unable to cope with the pack ice in M'Clure Strait, forcing her to transit the narrow Prince of Wales Strait through Canadian territorial waters. Publicity over the voyage, and the difficulties encountered by the *Manhattan*, began to raise concerns about the potential damage to Arctic ecosystems in case of an oilspill. Canada relied upon environmental issues as the basis for establishing jurisdiction over Arctic waters.[11]

Early in 1970 the tanker *Arrow* grounded on the coast of Nova Scotia and confirmed conservationist fears about the effects of oilspills in a cold coastal environment. With the *Manhattan* preparing for another transit of the Northwest Passage, intense political pressure pushed Canada to take action. On 8 April 1970 with the *Manhattan* steaming northward through Davis Strait, the Canadian government introduced legislation to extend its territorial sea to 12 nautical miles, thereby, ensuring that the *Manhattan* could not transit the Passage without passing through the territorial sea. In addition, the Arctic Waters Pollution Prevention Act (AWPPA) unilaterally established a 100-nautical-mile zone above 60° North which vessels would only be permitted to enter if they met specifications regarding vessel construction, manning, design, navigational aids, and cargo controls. Although passage of the AWPPA provoked a sharp response from the United States, Canada maintained that its justification was "based on the overriding right of self defence of coastal states to protect themselves against grave threats to their environment", and because of "the uniqueness of the Arctic environment, both in its exceptional susceptibility to pollution and its frozen state".[12]

Although the debate over the AWPPA was long and bitter, Canada eventually secured a diplomatic victory by drafting, and negotiating international support for, the ice-covered waters provision of the Law of the Sea Treaty. Article 234 reads:

> Coastal States have the right to adopt and enforce non-discriminatory laws and regulations for the prevention, reduction and control of marine pollution from vessels in ice-covered areas within the limits of the exclusive economic zone, where particularly severe climatic conditions and the presence of ice covering such areas for most of the year create obstructions or exceptional hazards to navigation, and pollution of the marine environment could cause major harm to or irreversible disturbance of the ecological balance. Such laws and regulations shall have due regard to navigation and the protection and preservation of the marine environment based on the best available scientific evidence.

Article 234 serves to legitimize the AWPPA, while extending the right to adopt and enforce pollution prevention regulations throughout the 200-nautical-mile exclusive economic zone (EEZ). Thus Canada has been exerting functional jurisdiction over Arctic waters in general, and the Northwest Passage in particular, without having to assert her claim to sovereignty. Paradoxically, this has been accomplished on the basis of concern about the Arctic marine environment, without actually doing much to protect it.

C *The conservation viewpoint*

The political debate over the *Manhattan* and the AWPPA fostered the common misconception that Arctic ecosystems are fragile. In point of fact, they are extremely hardy and adapted to survive in a hostile environment. The alleged fragility of Arctic ecosystems may just indicate that we are not using appropriate criteria.[13]

The ecological hazards for the Arctic marine environment do not arise from fragility, but from seasonal biological concentrations which are highly susceptible to pollution and disturbance. The Arctic Ocean has low productivity, low species diversity, short foodchains, and species have low growth and reproductive rates. Yet when species do concentrate in great numbers, usually in confined recurring polynyas and leads, they are extremely vulnerable to potential oilspills or other disturbance.

The conservation perspective, therefore, views these critical marine habitats as requiring protection through a comprehensive conservation strategy. These habitats of temporal productivity often serve as breeding areas for migrating fish, birds and marine mammals. Disturbance of these small areas at the wrong time of year could have a serious impact within the ecosystem. A comprehensive planning strategy will require international co-operation, national policies for balancing resource development and conservation priorities, research on critical habitats and interactions within ecosystems, and effective measures for protection. There is broad public support for the protection and conservation of the Arctic marine environment. Although Arctic ecosystems are not fragile, their special conditions and vulnerability demand special stewardship.

II The physical setting

A Bathymetry and currents

In the Western Arctic the Southern Beaufort Sea is fringed by flat coastal plains and a relatively shallow continental shelf, especially adjacent to the Mackenzie River Delta. Here the 10 m isobath extends as much as 25 miles offshore. The 200 m isobath averages about 93 miles offshore in the Southern Beaufort, but less than half that distance along the Alaskan coast. The shallow waters pose a major problem for oil and gas operations in the area, forcing companies to dredge either channels to harbours or locate ports farther away in deeper waters. North of the Tuktoyaktuk Peninsula there is an area of underwater pingos which form significant navigational hazards. Banks Island is characterized by higher sedimentary headlands and by deeper waters to the south in Amundsen Gulf and to the north in M'Clure Strait. The continental slope beyond 200 m drops off fairly rapidly, as illustrated on Figure 2.1.

Surface currents of the Western Arctic are dominated by the Beaufort Sea Gyre, which is driven by the atmospheric high pressure area centered over the region.[14] The mean drift speed of the Gyre is about 2–3 cm per second over the abyssal plain, intensifying to 5–10 cm per second at the southern rim of the Gyre.[15] Offshore currents from the Gyre enter the Arctic Island channels from the north and west. In the Southeast Beaufort, nearshore currents are affected by the outflow from the Mackenzie River and by the prevailing winds. One current moves northeast from the Delta

Figure 2.1. Bathymetry of the Beaufort and Chukchi Seas.

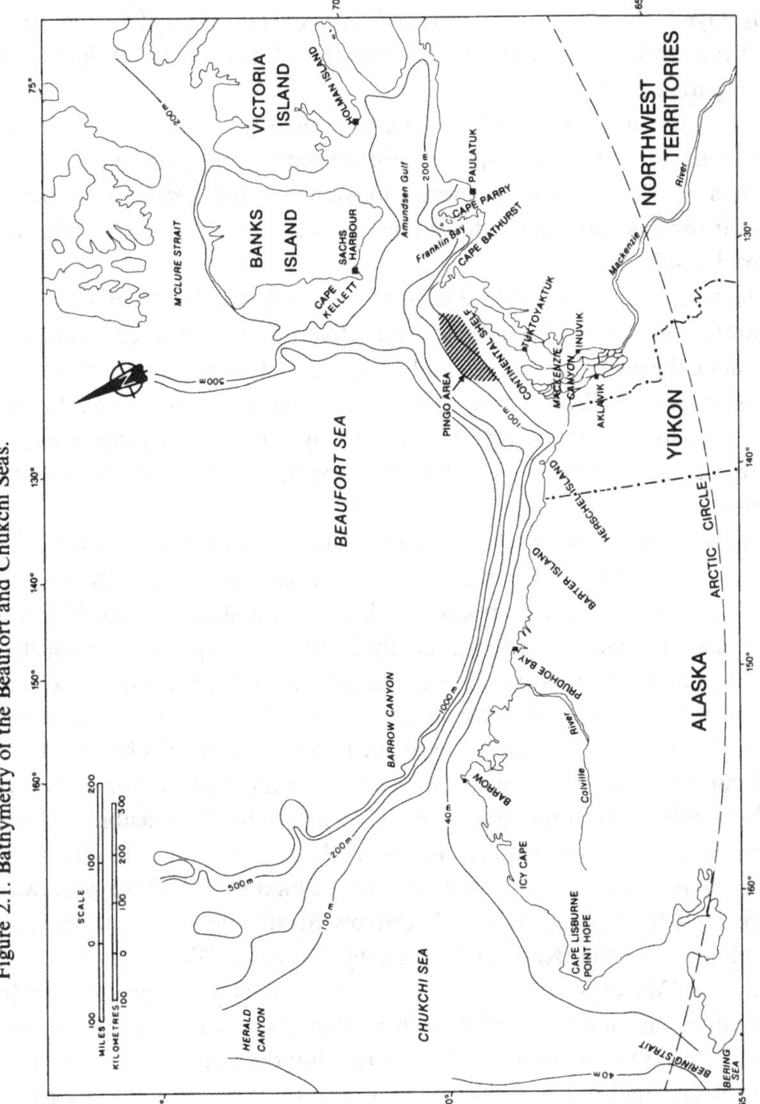

Source: Dome Petroleum Ltd, Esso Resources Canada Ltd, Gulf Canada Resources Inc., *Environmental Impact Statement for Hydrocarbon Development in the Beaufort Sea–Mackenzie Delta Region* (1982), V. 3A, p. 1.36.

area along the southern part of Amundsen Gulf, while another drifts northwest past Herschel Island. In the Western Beaufort, an influx of Pacific Ocean water passes through the Bering Strait. Part of the water moves eastward past Point Barrow in a strong nearshore countercurrent to the Gyre. The maximum easterly advance of this influx of Pacific water is believed to be in the vicinity of Barter Island, about 143° W longitude (see Figure 2.2).[16]

Nearshore currents in shallow waters are greatly affected by local wind conditions. The above descriptions give the generalized pattern of residual currents which determine net water movements, but nearshore currents may vary dramatically depending upon the direction, strength and duration of local winds.

The bathymetry and surface currents of the Arctic Island channels, as illustrated on Figures 2.3 and 2.4, respectively, along with ice conditions determine the preferred Northwest Passage route for proposed commercial shipping. Since at least seven alternative routes are possible (see Figure 2.5), the following discussion focuses on bathymetry and currents influencing the selection of the preferred route through the critical portion of the Passage.

The channel to the south of Victoria Island, which has been used for transits of the Passage by small craft in the past, has extremely shallow waters and is greatly encumbered with reefs and shoals in the Victoria Strait region. Given the approximately 20 m draft required by projected tankers, this route does not appear feasible as the Northwest Passage of the future. Similarly, the route south of Baffin Island through Fury and Hecla Strait appears unlikely due to the narrowness of the channel, swift tidal currents, and a shallow shoal area. However, hydrographic surveys of the shoals are continuing in an effort to determine the feasibility of the route. Bathymetry of the channels strongly suggests that the preferred route for the Northwest Passage is Parry Channel, the great waterway which includes Lancaster Sound, Barrow Strait and Viscount Melville Sound, and into the Beaufort Sea through Prince of Wales Strait.

Prince of Wales Strait is a 145-mile-long channel with a general width of 10 miles and a minimum width of 6.5 miles. There is a navigable channel of more than 50 m depth through the Strait, but the channel is very narrow at the Strait's mid-point. Although hydrographic surveys are being carried out to determine shoal areas and the width of the navigable channel, the route is certainly feasible as demonstrated by the transit of the SS *Manhattan*.[17] Weak currents move through the Strait in both directions, to the south along the Banks Island side, and to the north along the

Figure 2.2. Mean general summer circulation of surface waters in the Beaufort and Chukchi Seas.

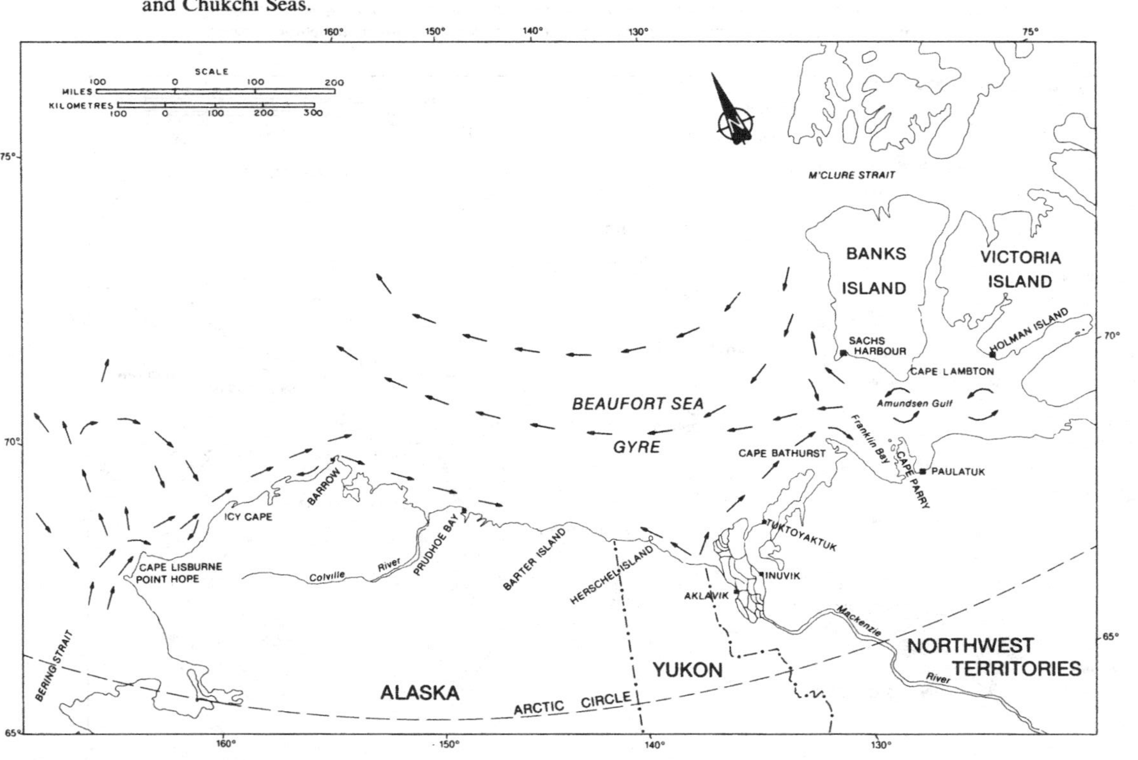

Source: Dome Petroleum Ltd, Esso Resources Canada Ltd, Gulf Canada Resources Inc., *Environmental Impact Statement for Hydrocarbon Development in the Beaufort Sea–Mackenzie Delta Region* (1982), V. 3A, p. 1.40.

Figure 2.3. Bathymetry of the Northwest Passage and Arctic Islands.

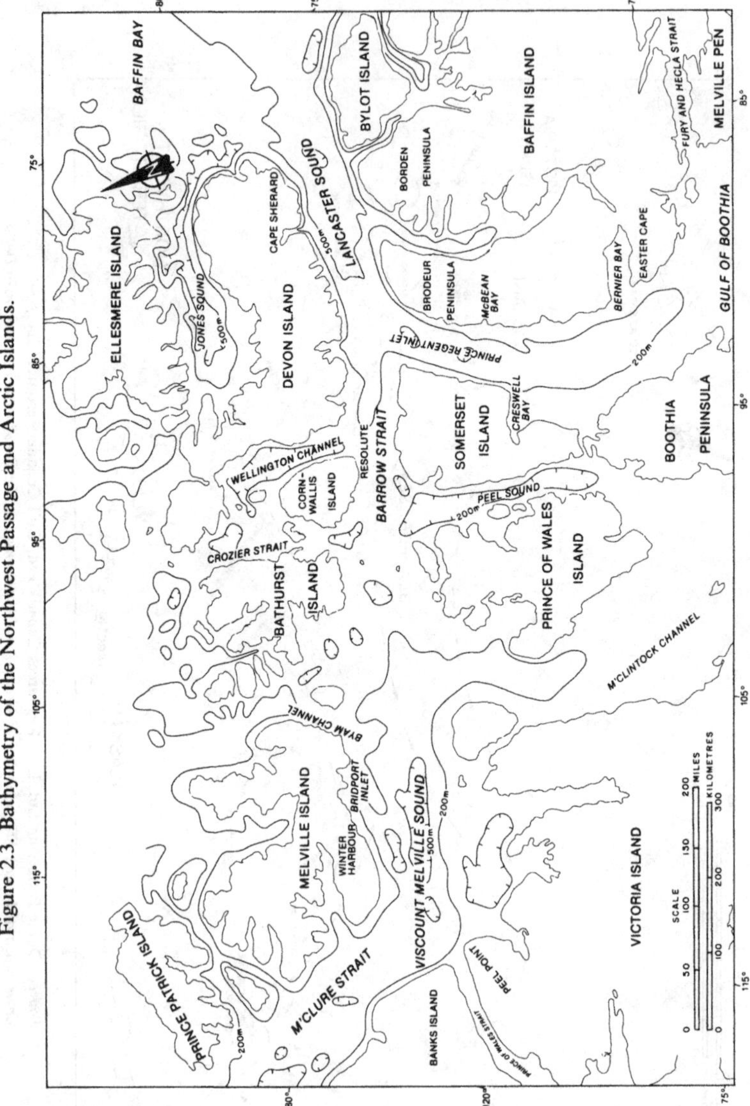

Source: Dome Petroleum Ltd, Esso Resources Canada Ltd, Gulf Canada Resources Inc., *Environmental Impact Statement for Hydrocarbon Development in the Beaufort Sea–Mackenzie Delta Region* (1982), V. 3B, p. 1.39.

Figure 2.4. Mean near-surface circulation in the Northwest Passage.

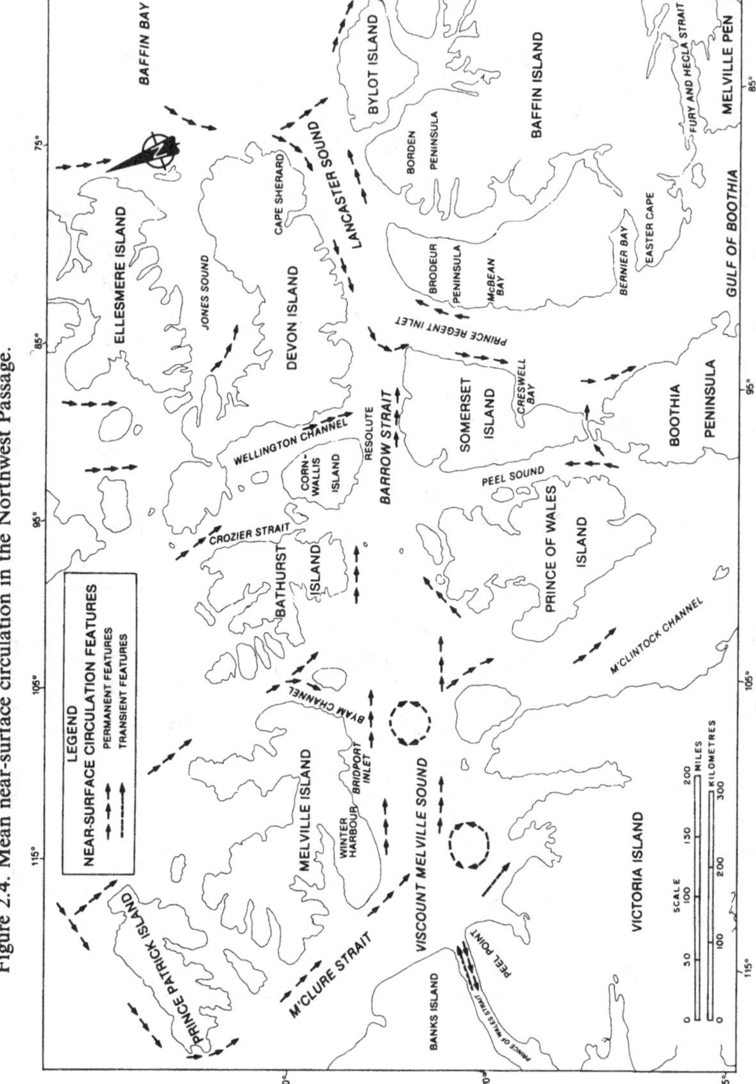

Source: Dome Petroleum Ltd, Esso Resources Canada Ltd, Gulf Canada Resources Inc., *Environmental Impact Statement for Hydrocarbon Development in the Beaufort Sea–Mackenzie Delta Region* (1982), V. 3B, p. 1.40.

Figure 2.5. Main routes of the Northwest Passage.

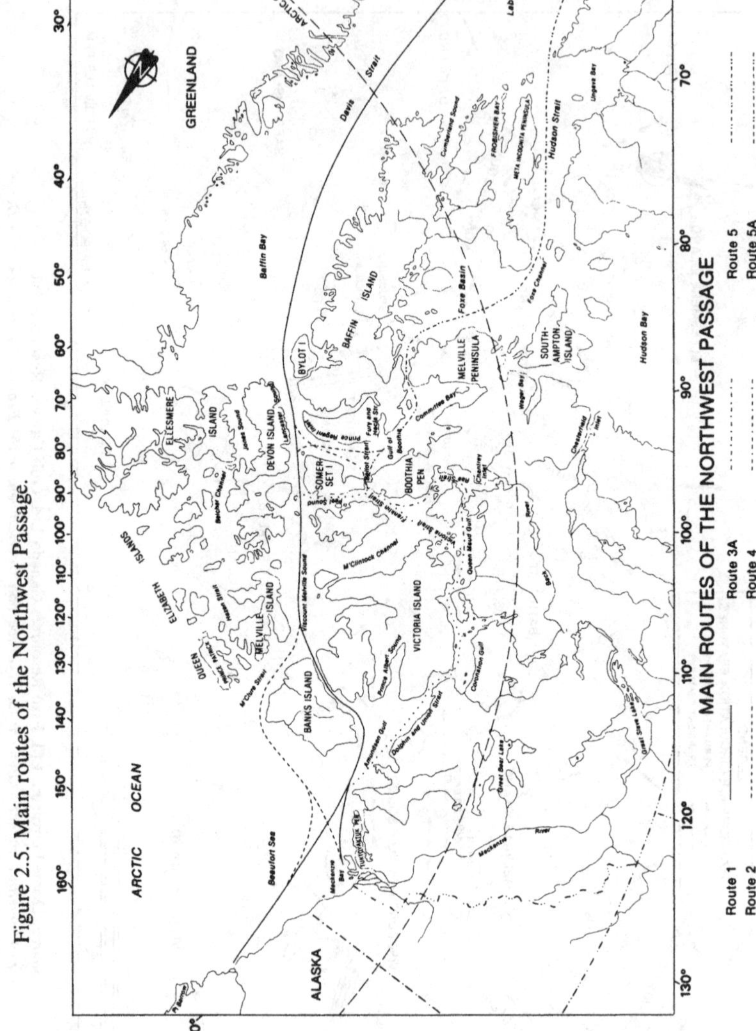

Source: Adapted from Donat Pharand, *The Waters of the Canadian Arctic Archipelago in International Law* (in press).

Victoria Island side.[18] The 50 m sill in the channel limits the exchange of water between Amundsen Gulf and Viscount Melville Sound.

M'Clure Strait has a navigational channel, about 56 miles in width with water depths of more than 200 m within a few miles off either coast. Most of the Strait has depths ranging from 300 to 500 m and, although the bathymetry has not been well surveyed, there do not appear to be any shoal areas.[19] An eastward current through M'Clure Strait transports a significant volume of Arctic Ocean water (and polar pack ice) into Viscount Melville Sound.

Viscount Melville Sound is the widest and longest body of water in Parry Channel. The central and western portions of the Sound have depths similar to M'Clure Strait through the middle of the Sound, with the 50 m isobath coming to within about 6 miles of the Melville Island shore. Water depths, although they appear more than adequate for navigation, gradually decrease towards the eastern end of the Sound and blend into shoal areas near the High Arctic Islands. Viscount Melville Sound also has a scarcity of bathymetric data and requires additional hydrographic surveys, particularly for the northern portion.[20] Surface currents (which only are a factor when the Sound is not covered by landfast ice) generally move to the east and the south.

Barrow Strait contains a number of islands and shoals. A sill at about 100 m limits the movement of Arctic Ocean water into Baffin Bay, while shallow sills also occur in the other channels. Alternative passages occur between the islands. All passages appear navigable according to the somewhat more detailed hydrographic surveys carried out for this area. Water depths increase steadily towards Lancaster Sound. Surface currents are from west to east.

Lancaster Sound, bordered by rugged and steeply sloping coasts, is 43 miles wide and 400 to 600 m in depth. No shoals or islands present navigational hazards.[21] The surface current is westerly along the Devon Island shoreline, but this current is deflected by the southerly current through Wellington Channel and returns as an easterly current along the Baffin coast.[22]

In the Eastern Arctic, bathymetry poses few problems for navigation. Most of the coastline is steeply sloping, so that the area of the adjacent shelf less than 200 m is small.[23] The northern portion of Baffin Bay, between Devon Island and Greenland, has comparatively shallow waters, but they still range from 250 m to 450 m in depth. The central part of Baffin Bay has water depths in excess of 2000 m, with the Davis Strait sill rising to about 600 m below sea level.[24]

The general pattern of surface currents in the Eastern Arctic, shown on

Figure 2.6, is dominated by the counter-clockwise flow of the West Greenland and the Baffin Currents. The former brings a strong flow of warm Atlantic water into the region, keeping the southwest portion of the Greenland coast open year-round. The main West Greenland Current heads north into Melville Bay before being deflected to the west. Throughout the Current's course, less well-defined currents branch off to the west across Davis Strait and Baffin Bay to join the southward drifting Baffin Current, which is suppplemented by currents from Nares Channel, Jones Sound and Lancaster Sound, all bringing Arctic Ocean water to Baffin Bay. The Baffin Current, considerably stronger than the West Greenland Current and assisted by currents from Hudson Strait, goes on to form the Labrador Current.[25]

B *Ice conditions*

1. *General*

The dominant feature of the Arctic marine environment is ice. Its presence, distribution, thickness, movement, seasonal patterns, physical characteristics, and variations from year to year, all contribute to making a marine environment that is significantly different from more southerly waters. A solid ice cover generally reduces productivity of the marine ecosystems and limits the access of seabirds and some marine mammals, while ice edges often have remarkably high productivity resulting in concentrations of all forms of life near recurring leads and polynyas. The presence of ice and the fact that pack ice moves and can trap or crush ships have placed severe restrictions on the use of Northern waters ever since the early days of exploration, and continue to pose a problem for proposed Arctic Class supertankers. On the other hand, the Inuit have used the sea ice as an extension of the land, which gives them greater access to their most important food source, marine mammals.

Sea ice, if defined in terms of mobility, occurs in two major forms. Pack ice, driven by wind and currents, moves rather freely. Fast ice, extending from a fixed base such as shorelines, grounded icebergs or shoals, is relatively immobile. A dynamic transition zone often occurs between pack ice and fast ice where ice is in a chaotic state due to the shearing of pack ice against the fast ice boundary and where shore leads continually open and close.

Ice is also sometimes described according to age. First-year ice, ranging in thickness from 30 cm to 2 m, does not have more than one winter's growth. Second-year ice, thicker than first-year ice, has survived one

Figure 2.6. Mean near-surface circulation of the Baffin Bay–Davis Strait Region.

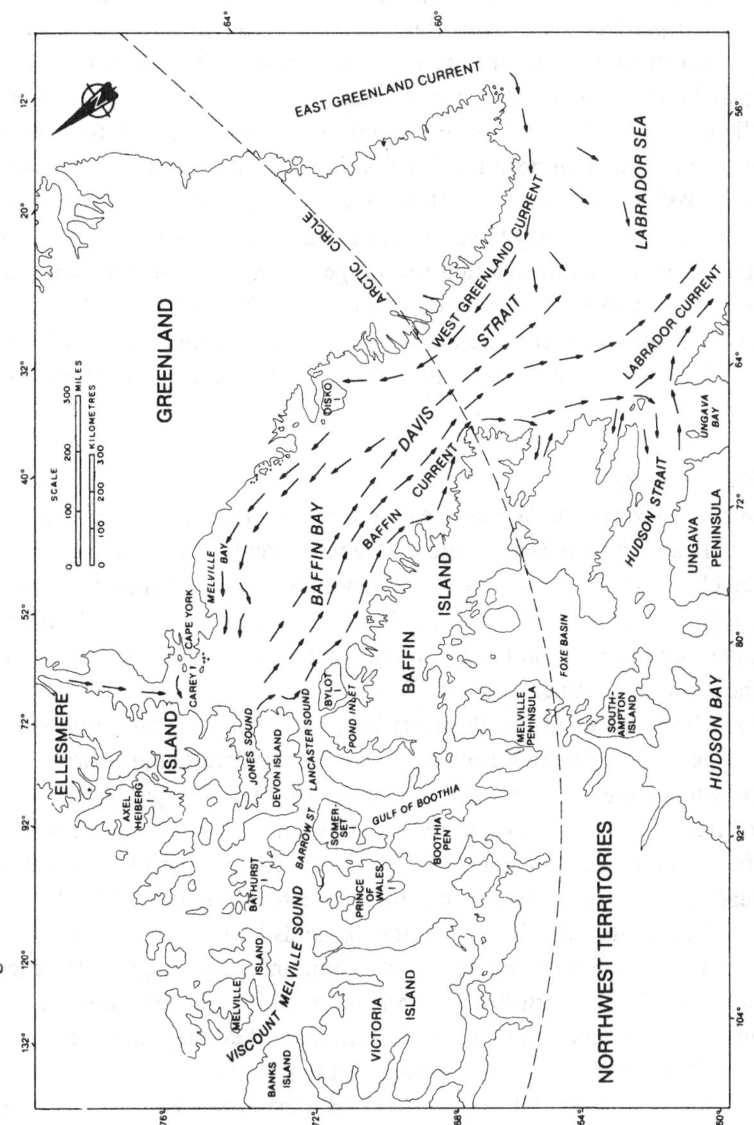

Source: Dome Petroleum Ltd, Esso Resources Canada Ltd, Gulf Canada Resources Inc., *Environmental Impact Statement for Hydrocarbon Development in the Beaufort Sea–Mackenzie Delta Region* (1982), V. 3B, p. 1.47.

summer's melt. Multi-year ice, up to 3 m or more in thickness, has survived at least two summers.[26]

Four types of ice formations may be particularly hazardous to navigation in the Arctic. Icebergs, an Eastern Arctic phenomenon, are calved from the foot of glaciers, drift with ocean currents, and often reach 70 m in height. Up to 90 per cent of their mass may travel below the sea surface. Ice islands, flat tabular structures broken off from the ice shelves north of Ellesmere Island, are massive, slow-moving (at a rate of 1 to 3 miles a day), and have often been used as floating airfields and scientific bases. Hummocked ice, mounds of broken ice forced upward by pressure, may span over a mile and may have ridges exceeding 20 m in height. Pressure ridges, linear masses of jumbled ice, range in height from less than 1 m to over 10 m, and keels to a depth of 47 m have been observed.[27]

Ice conditions vary considerably across the Canadian Arctic and may be discussed by geographical region – Western Arctic, Central Arctic and Eastern Arctic.

2. *Western Arctic*

Ice conditions in the Western Arctic are dominated by polar pack ice and the Beaufort Gyre (see Figure 2.7), which circulates the multi-year ice in a clockwise direction at a mean drift speed of about 1.2 miles per day.[28] Although the position of the nearshore edge of the polar pack varies considerably from season to season and year to year, it generally lies just off the physical continental shelf.

No icebergs occur in the Beaufort Sea, but ice islands and multi-year hummock fields can be of formidable size. The ice islands tend to drift with the Beaufort Gyre, but eventually break free to enter passages between the High Arctic Islands, to ground on the Alaskan coast, or to move across the North Pole and be carried out into the Atlantic east of Greenland. Grounding appears to play a major role in the break-up of ice islands into smaller fragments. Multi-year hummock fields form along the western edge of the archipelago, where the crushing and overriding of pack ice creates large fields of crushed ice and ridges parallel to the shore. If the hummock fields survive one or more summers, they may find their way into the Beaufort Gyre along the edge of the pack ice.[29]

As Captain George DeLong wrote in his diary, shortly before his ship, the *Jeannette*, was crushed by ice north of Siberia in 1881, "the pack is no place for a ship, and however beautiful it may be from an aesthetic point of view, I wish with all my heart that we were out of it." (DeLong and most of his crew died trying to return on foot.) For all surface ships, except the nuclear icebreakers of the USSR, that sentiment remains valid today.

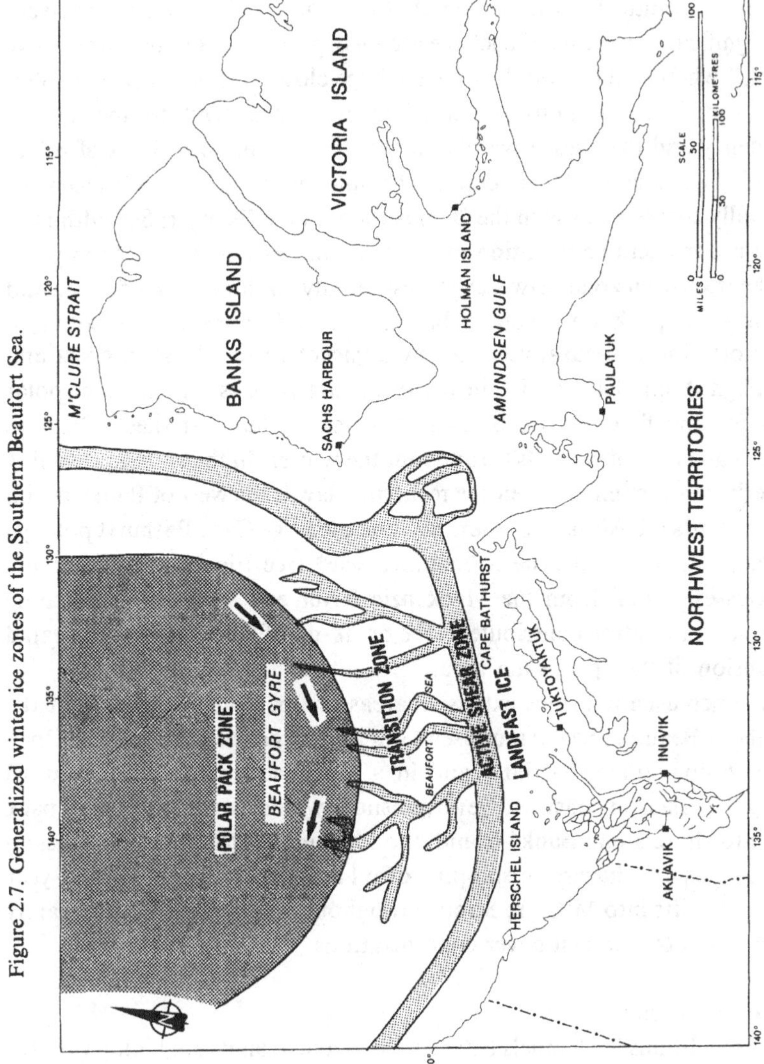

Figure 2.7. Generalized winter ice zones of the Southern Beaufort Sea.

Source: Dome Petroleum Ltd, Esso Resources Canada Ltd, Gulf Canada Resources Inc., *Environmental Impact Statement for Hydrocarbon Development in the Beaufort Sea–Mackenzie Delta Region* (1982), V. 3A, p. 1.2.

Even for Arctic Class 10 icebreakers, supposedly capable of operating in the polar pack, care will have to be taken to avoid pressure ridges, hummock fields, and ice islands within the pack.

Landfast ice in the Western Arctic gradually builds out from shore during the winter, with *in situ* thickness of about 2 m, reaching as far as 50 miles out into the Beaufort Sea.[30] Although this is primarily first-year ice, significant amounts of pack ice are incorporated about once every five years,[31] when onshore winds drive pack ice close to shore in early winter. This zone also contains pressure ridges, which tend to increase in frequency and thickness towards the transition zone, with keels of ridges reaching as much as 20 m in depth. The outer edge of the landfast ice zone generally corresponds with the 20 m isobath in the Beaufort Sea, although there is considerable variation in its maximum extent from year to year.

The transition zone betweeen the essentially stationary landfast ice and the moving pack ice tends to be quite a wide zone in the Southern Beaufort, and comparatively narrow adjacent to the Alaskan coast and the High Arctic Islands. Ice in the transition zone is a mixture of polar pack ice and first-year ice broken off from the landfast sheet. Offshore winds can create flaw leads throughout the winter. In the spring prevailing easterly winds open up extensive recurring flaw leads west of Banks Island and towards the Alaskan border. By June the large Cape Bathurst polynya develops in the southeastern Beaufort Sea[32] (see Figure 2.8). As spring progresses, runoff from the Mackenzie River and structural weakening from solar radiation contribute to the break-up of landfast ice and a rapid expansion of the open water area.

Summer usually brings extensive areas of open water throughout the Southern Beaufort Sea, Amundsen Gulf and Prince of Wales Strait. Open water occurs somewhat later and in a much narrower band westward along the Alaskan coast. However, onshore winds can quickly move pack ice onto shore along Banks Island, the Mackenzie Delta and the Alaskan coast – posing a hazard to shipping or oil and gas operations. Multi-year pack ice drifts into M'Clure Strait throughout the summer months, rarely in less than 60 per cent cover concentrations.

3. *Central Arctic*

Ice conditions in channels between the Arctic Islands are highly variable between good years and bad years, so that a description of typical ice conditions is of limited value for any specific time. Multi-year pack ice drifts into these channels from the northwest in accordance with the current patterns shown on Figure 2.4, and becomes entrapped in the landfast ice. The location and concentration of the multi-year ice in

Figure 2.8. Major recurring polynyas and shore leads.

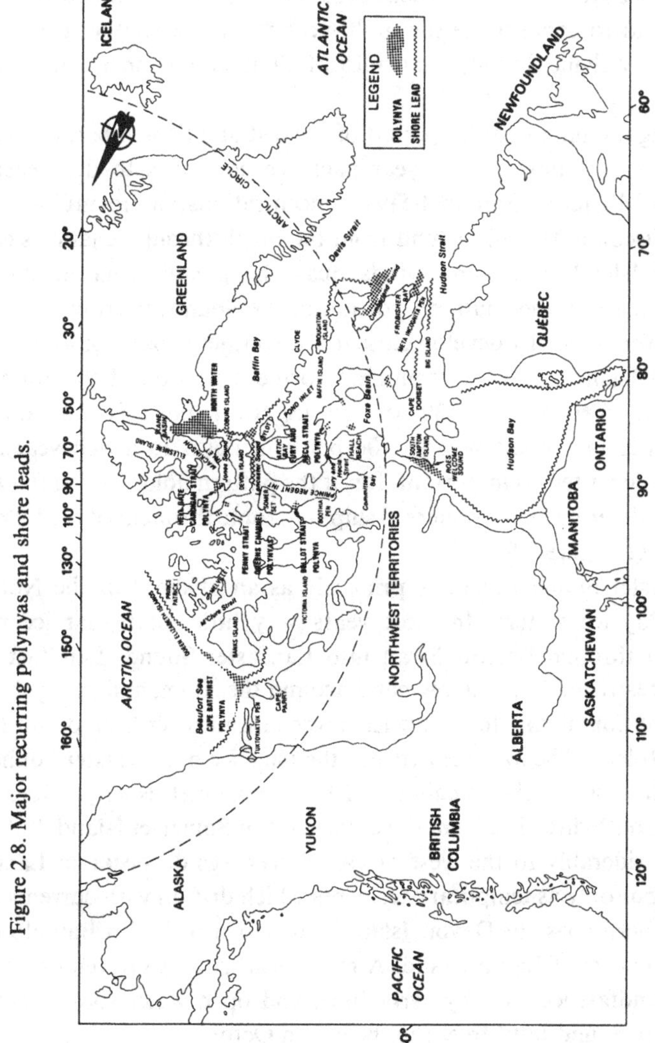

Source: Adapted from Ian Stirling and Holly Cleator, *Polynyas in the Canadian Arctic*, Canadian Wildlife Service Occasional Paper No. 45 (1981), p. 6.

potential shipping channels during the summer months is an important factor, along with bathymetry, in determining preferred shipping routes.

There is little motion of the multi-year ice during the winter months, when it is held in place by first-year landfast ice. During summer months, the ice is free to move through M'Clure Strait and, to a lesser extent, through the High Arctic channels. Predominantly north and northwesterly winds tend to concentrate the multi-year ice in the southern portion of Viscount Melville Sound,[33] and into M'Clintock Channel, where the ice stagnates.

During the minimum ice period of August and September, the general pattern is as follows: multi-year pack ice continues to drift eastward through M'Clure Strait in heavy concentrations; more multi-year ice enters Viscount Melville Sound from the north through channels east of Melville Island; prevailing winds push the greater concentrations of multi-year ice to the southern portion of the Sound; narrow shore leads of open water usually develop close to the southern coasts of Melville and Bathurst Islands; and the remaining northern portion of the Sound has something in the order of 30 per cent ice cover (but this can vary from open water in a good year to 100 per cent cover in a bad year). This geographic pattern is important, for, if the pattern holds during freeze-up, it largely determines the preferred shipping route throughout the following winter (see Figure 2.9).

Lancaster Sound functions primarily as an offshoot of the Northern Baffin Bay ice system. In most years very little multi-year ice moves eastward through Barrow Strait into Lancaster Sound. Landfast ice is usually restricted to a narrow band around the shore, and first-year pack ice, in motion, is the dominant ice cover in Lancaster Sound and Prince Regent Inlet.[34] The ice edge between the pack ice of Lancaster Sound and the landfast ice of the remainder of Parry Channel usually is located in a north-south direction from the eastern tip of Somerset Island, but it can vary considerably to the west or east.[35] Icebergs do occur in Lancaster Sound, but only as temporary residents which drift in with currents along the southern coast of Devon Island and return to Baffin Bay along the northern coast of Baffin Island. A major lead-polynya develops adjacent to the landfast ice edge by early June, and open water soon prevails in Lancaster Sound until freeze-up begins in October.

4. *Eastern Arctic*

Ice conditions in the Eastern Arctic (which includes Davis Strait, Baffin Bay, and Smith, Jones and Lancaster Sounds) are dominated by first-year pack ice, a well-defined counter-clockwise current pattern, icebergs, and

Figure 2.9. Average September ice conditions in the Central Arctic (Parry Channel) and preferred tanker route.

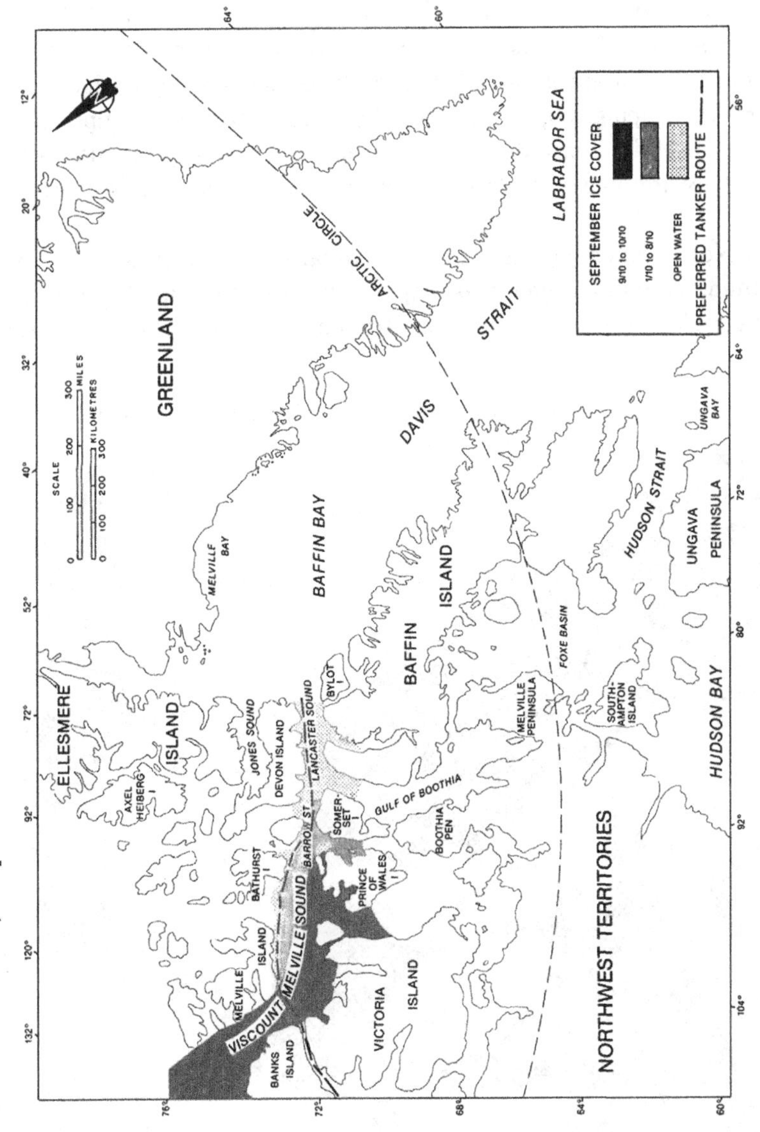

Source: Adapted from Dome Petroleum Ltd, Esso Resources Canada Ltd, Gulf Canada Resources Inc., *Environmental Impact Statement for Hydrocarbon Development in the Beaufort Sea–Mackenzie Delta Region* (1982), V. 3B, p. 1.15.

the North Water. The eastern portion tends to have less ice than the western portion due to the moderating influence of the West Greenland Current.[36] Some multi-year pack ice does enter the region through Nares Strait (north of Smith Sound) and through Lancaster Sound during summer months, but most of these old floes exit from the region by December.[37] Concentrations of multi-year ice can present navigational hazards to shipping. Winter ice cover, therefore, is primarily composed of first-year pack ice constantly in motion on the larger water bodies, with landfast ice forming close to shore and in the smaller bays (see Figure 2.10).

The North Water, a recurring polynya of fairly open water at about 78° N latitude in Smith Sound, is perhaps the most interesting phenomenon of this region. Although formerly attributed to an upwelling of warm

Figure 2.10. Average ice conditions during March in the Eastern Arctic.

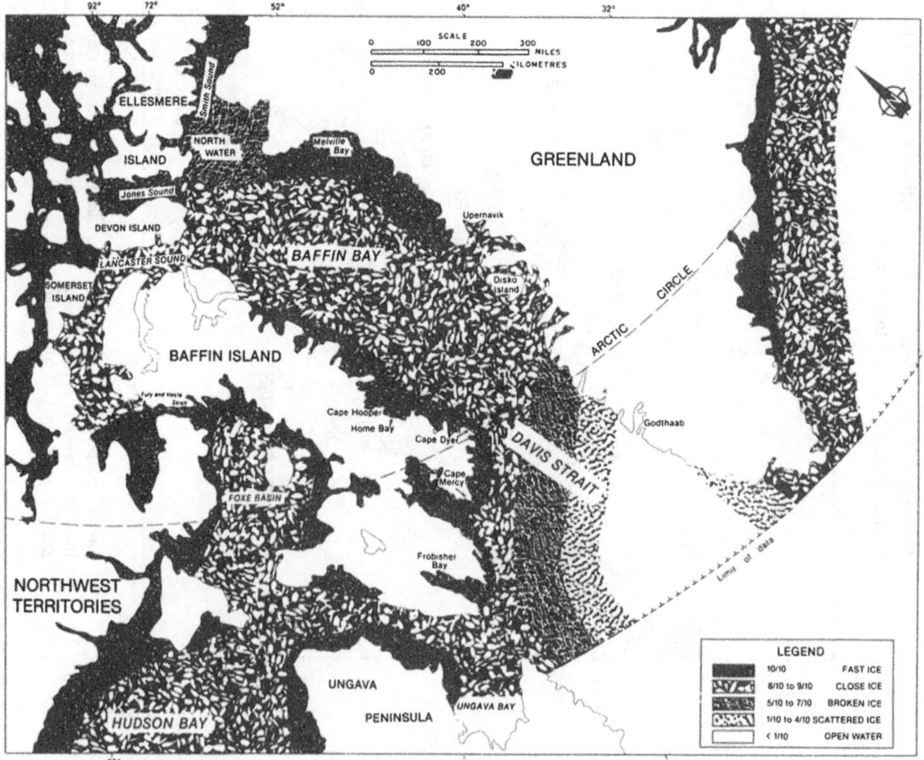

Source: Adapted from Dome Petroleum Ltd, Esso Resources Canada Ltd, Gulf Canada Resources Inc., *Environmental Impact Statement for Hydrocarbon Development in the Beaufort Sea–Mackenzie Delta Region* (1982), V. 3B, p. 1.15.

water, the phenomenon seems to depend on ice removal caused by prevailing winds and currents. The North Water is actually an ice factory which produces about 4.5 times as much ice as would otherwise be formed.[38] At almost all times the polynya displays significant areas of open water, and the polynya's size rapidly increases as spring and summer advance.

The pack ice moves with the surface currents, as described on Figure 2.6, and prevailing winds eventually drive the ice from the region via the Baffin and Labrador currents. The coastal boundary of the moving pack ice and landfast ice is a very dynamic zone. Pressure ridges do form, but ridge heights are much less than those within the Arctic Islands or the Beaufort Sea.

Break-up in Baffin Bay is stimulated by a rapid northward extension of the shore leads along the West Greenland coast, and by the southward extension of the North Water to join with open water from Lancaster Sound. A nearly ice-free channel from Southern Greenland to Lancaster Sound usually exists by early July, while more than 80 per cent pack ice remains in Central Baffin Bay.[39] Landfast ice rapidly breaks up along the coast, but may persist the longest in the central portion of Baffin Bay which is largely ice-free by early September.[40]

Icebergs are an extremely important feature of the marine environment in the Eastern Arctic. Although a few minor icebergs are produced from the glaciers of Ellesmere, Devon and Bylot Islands, the vast majority are produced along the West Greenland coast, particularly the area of Disko Bay (see Figure 2.10). Here most icebergs are calved in late summer, from approximately 25 very active glaciers.[41] Iceberg movement is predominantly controlled by currents, but wind and ice motion also have an influence. Large numbers of icebergs are carried northward to the Smith Sound-Lancaster Sound region, which can become quite congested with bergs stalling in the area for two or three years.[42] Some icebergs take a more direct route across Baffin Bay to the Baffin Current. Almost all icebergs eventually leave the area via the Baffin and Labrador Currents and terminate their existence in the North Atlantic.

C *Recurring polynyas*

A polynya (a word derived from the Russian "polyi" – open) is an area of open water surrounded by ice. Polynyas are created by one or a combination of the following forces: currents, tides, winds, upwellings and ice shearing. In Arctic waters many small polynyas form, but last for only short periods of time and probably do not have great biological importance. On the other hand, a number of recurring polynyas form at approximately the same location each year and tend to support large numbers of seabirds and marine mammals, thereby forming an important part of the Arctic marine environment. Although polynyas reappear annually at predictable locations, their size and shapes vary considerably. Some polynyas persist as open water (or at least partly open water) throughout the winter, whereas others open up at predictable times in the spring – long before break-up occurs in the surrounding ice cover.

Three types of polynyas recur in Canada's Northern waters: the unique North Water polynya in Smith Sound between Ellesmere Island and Greenland; the smaller non-linear polynyas scattered throughout Arctic waters; and the extensive systems of shore leads which form at the boundary area between pack ice and the outer edge of landfast ice (technically speaking shore leads are not considered to be polynyas, but shore leads are closely related features and perform similar functions).[43] The distribution of known recurring polynyas and shore leads is shown on Figure 2.8, but others exist. According to Dunbar, the most important polynyas are situated in: Smith Sound, Jones Sound and Lancaster Sound in Northwest Baffin Bay; the Eastern Beaufort Sea; and the shore leads of Hudson Bay.[44] Archaeological research shows a close locational relationship between settlements and polynyas throughout human occupancy of the Arctic. Such a relationship is presumably based upon the local abundance of marine mammals occurring in polynyas.[45]

The North Water is unique because it is the largest and the most northerly of the recurring polynyas. Relics of settlements dating back to AD 900 document the historic relationship between native peoples and this biologically productive phenomenon.[46] The North Water was most likely visited by Vikings from Eric the Red's colony during the thirteenth century, but it was not officially discovered until 1616 when William Baffin came upon "an open sea" at latitude 75° 40′ N. Vertical mixing, apparently caused by currents and prevailing winds, occurs within the North Water.[47] The large difference between surface water and air temperature in mid-winter has a spectacular effect on local climate by producing a heavy seafog and snow to feed the surrounding glaciers.[48]

The northern edge of the North Water is consistently covered by an ice bridge across the narrow head of Smith Sound. Some open water is always present along the Ellesmere or Greenland coasts, and consistent patches of open water occur near the ice bridge, at the entrance to Jones Sound, and within Lancaster Sound. The main area of open water begins to expand southward as early as mid-April and joins with open water from Lancaster Sound from late May to mid-June, and with West Greenland somewhat later.[49] The main importance of the North Water may not be in its own inherent biological productivity, but in the ecological relationship it plays with other open water areas in the Eastern Arctic, particularly those of Lancaster Sound and the Greenland coast.

Many of the other polynyas, such as the Cape Bathurst polynya (which is of vital importance to Western Arctic ecosystems), commonly result from shore leads, developing in mid-winter, and from prevailing winds which push the ice offshore. These processes are assisted by high tidal ranges in Cumberland Sound and Frobisher Bay. Other, usually smaller, polynyas form at the eastern end of channel constrictions as the result of strong currents – which are partly induced by tides piling up water on the westward side of land masses. Examples include Hell Gate, Penny Strait, Bellot Strait, and Fury and Hecla Strait. No sizeable polynyas occur in the vicinity of Viscount Melville Sound and adjacent waters, a fact which appears to have a profound effect on the distribution of marine mammals,[50] birds and people.

III The biological setting

A Arctic marine productivity

Marine primary productivity, the production of marine plant growth through photosynthesis, is extremely low in Arctic waters in comparison to Canada's Atlantic and Pacific waters. The amount of carbon produced under a square metre of sea surface in the North Pacific and the Atlantic is about 300 grams, while the comparative figure for the Beaufort Sea is about 10 grams.[51] Species diversity is also very low throughout Arctic ecosystems. Only about 25 species of fish live in Arctic waters, whereas the world has over 23000 species.[52] Some 40 species of bivalve molluscs live in the Arctic compared with approximately 120 species in the Northwest Atlantic.[53] The low level of species diversity is directly attributable to the low level of marine primary productivity, caused by cold water temperatures and extended periods of darkness and ice cover.

The paradox is that, despite the low productivity, the standing stock of certain species in specific Arctic locations can be very high. For example, during the early days of commercial whaling, herds of 1000 or more whales were encountered. A run of over 120000 Arctic char has recently been counted on a Victoria Island river.[54] The answer to the paradox seems to be that some Arctic marine habitats have high primary productivity for a short period of time, causing great congregations of migrating fish, seabirds and marine mammals. These species may spend most of the year at distant locations, but the Arctic marine habitat is crucial to their survival.

A key stimulus for Arctic marine primary productivity is the epontic biota, consisting primarily of diatoms which live on the underside of sea ice. These plant communities bloom for a short period of time as summer approaches, with the bloom peaking just prior to and during spring break-up.[55] This stimulus to primary productivity is especially effective at ice edges of polynyas and shore leads, where vertical mixing is fostered, and the short Arctic foodchains make use of the diatoms, which only are available in large numbers for about a month. Examples of the foodchain are: (i) diatoms – herbivorous zooplankton – carnivorous zooplankton – whales; and (ii) diatoms – herbivorous zooplankton – carnivorous zooplankton – Arctic cod – seals – polar bear.[56] Thus, the primary production of plant growth is quickly moved through the foodchain to marine mammals and birds (see Figure 2.11). Productivity is concentrated over space and time, while most Arctic waters remain quite unproductive.

The epontic communities are not the only source of energy for Arctic marine ecosystems. Other sources include: terrestrial vegetation, such as peat carried to the ocean by rivers, or coastal zone detritus; macrophytic algae (seaweeds), not abundant in the Arctic but locally important; benthic microalgae; and phytoplankton.[57]

B *Marine and anadromous fish*

Few species of fish exist in Arctic waters, and most of them are "small, difficult to catch and singularly unappetizing".[58] The sub-Arctic waters off West Greenland are an exception, containing some 86 species of fish. In Canada's Northern waters only two species are considered to have significant importance: Arctic cod as a critical foodchain link and Arctic char as a human food source.

Arctic cod, the most abundant fish in the Arctic, are distributed throughout the Arctic Ocean. They adapt to, and feed in, a variety of habitats: at the surface and under sea ice; at the edge of landfast ice; in shallow coastal waters during summer; and at depths of more than 900 m.[59] Arctic cod are widely dispersed during winter months, but gather into massive schools during the summer. The standing stock of Arctic cod in Lancaster Sound is estimated to be ten times that of the Beaufort Sea, with production estimated at 400000 tonnes per year.[60] They are an important source of food for the Arctic fox, polar bear, and numerous species of seabirds, which obtain over half of their summer energy consumption from cod.[61] Despite the abundance of cod, the Inuit seldom consume the fish.

Arctic char, trout-like fish with pink flesh, are anadromous, spending the first four to eight years of life in freshwater rivers and lakes, and

Figure 2.11. Preliminary summary of food web relationships in Arctic offshore waters.

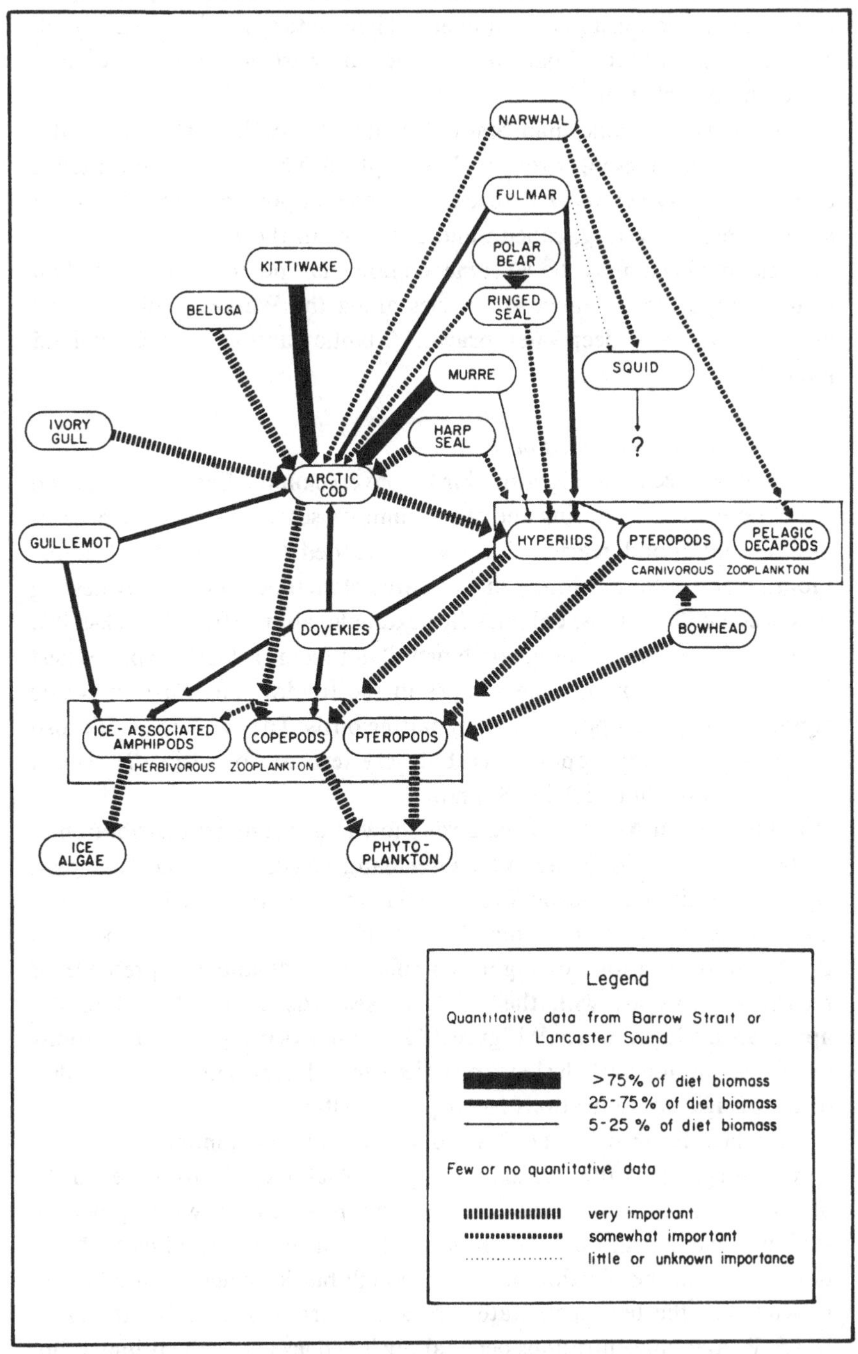

Source: Ralph A. Davis, Kerwin J. Finley and W. John Richardson, *The Present Status and Future Management of Arctic Marine Mammals in Canada*. Prepared for the Science Advisory Board, Northwest Territories (1980), p. 20.

thereafter migrating to the sea every summer for feeding purposes. Char are circumpolar in distribution, and occur fairly ubiquitously along the Canadian Arctic coast in the summer.[62] They mature slowly, attaining full size at about 20 years. Char are an important part of the Inuit diet and, therefore, heavily fished.

In the Western Arctic anadromous fish of biological importance include: Arctic cisco, least cisco, boreal smelt, humpback whitefish, broad whitefish and inconnu. Other marine species of lesser importance are: fourhorn sculpin, Pacific herring, capelin and snailfish. In the Eastern Arctic, the Greenland shark may have some commercial potential in Canadian waters. Important commercial species along the West Greenland coast are: Atlantic cod, deepwater prawn, Atlantic salmon, and Greenland halibut.

C *Marine associated birds*

About 80 species of migratory birds converge on the Eastern Arctic and the Western Arctic during the short summer season for the purpose of nesting and brood rearing. Nearly all are tied to the sea for feeding, moulting, or nesting.[63] Many of these are colonial nesters, usually nesting on coastal cliffs in large colonies (for example, about 800000 thick-billed murres nest in a single colony on North Bylot Island).[64] Other species nest in marshy areas or on mere scrapes in the tundra. The Eastern Arctic supports the highest populations of marine birds. The Western Arctic also supports quite high populations, but very few are found in the central vicinity of Viscount Melville Sound.

In the Eastern Arctic, marine birds show major summer distribution in nesting colonies along the West Greenland coast, along the Canadian coasts of Davis and Hudson Straits, and in the vicinity of the North Water and Lancaster Sound (see Figure 2.12). A strong correlation exists between the breeding locations of highly colonial seabirds and the presence of recurring polynyas, with the influx of seabirds closely linked to the appearance of open water.[65] Figure 2.12 also shows pelagic concentrations which tend to shift with the location of ice edges. The potential vulnerability of the pelagic concentrations to oilspills is self-evident.

Canada's Eastern Arctic seabird community is dominated by five species – Northern fulmar, glaucous gull, black-legged kittiwake, thick-billed murre, and black guillemot. These are pelagic birds which generally feed and live at sea, but come to shore to nest on protected cliffs. Most birds winter in the North Atlantic, although black guillemots are known to winter in the few open water areas of Northwest and North Baffin Bay.[66] Pelagic concentrations occur along ice edges where the primary diet

Figure 2.12. Seabird concentrations along the Canadian and West Greenland coasts.

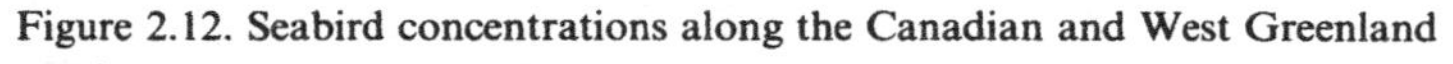

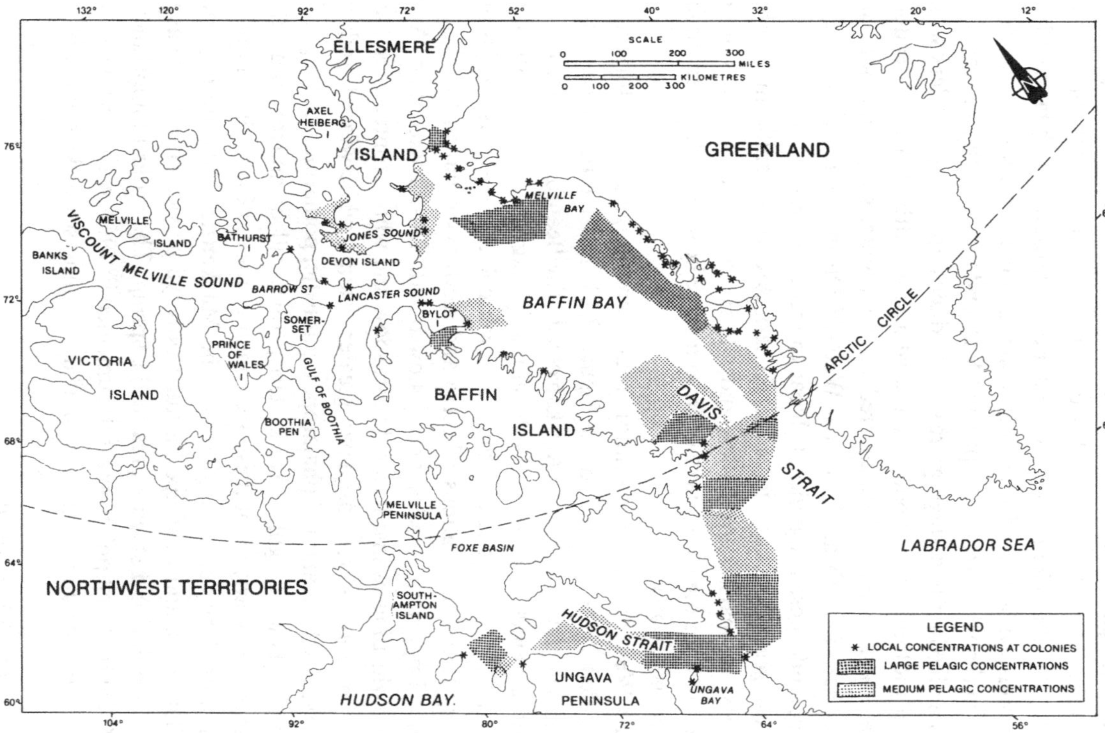

Source: Adapted from Dome Petroleum Ltd, Esso Resources Canada Ltd, Gulf Canada Resources Inc., *Environmental Impact Statement for Hydrocarbon Development in the Beaufort Sea–Mackenzie Delta Region* (1982), V. 3B, p. 2.33.

is Arctic cod, supplemented by amphipods, copepods, pteropods and sculpins. Most seabirds remain in their nesting-feeding area in the Arctic for about six months of the year.

Seaducks, particularly the common eider, king eider and oldsquaw, also migrate to the Eastern Arctic. The common eider nests colonially near salt water. The king eider is a solitary nester on well-drained habitat near lakes and ponds, while the oldsquaw nests on vegetated lowlands. The seaducks arrive in the High Arctic in April and May and use the polynyas and shore leads as feeding and staging areas. The ducks are marine feeders with a diet of molluscs, gastropods, and crustaceans. Although most seaducks winter to the south of ice-covered Arctic waters, some common eider winter as far north as the North Water.[67] Oil in polynyas during early spring could pose a significant threat to incoming seaducks.

Snow geese also migrate to the Eastern Arctic in significant numbers and nest on marshy coastal lowlands. They are not considered seabirds, although their life habits take them to marine beaches, salt water lagoons and coastal waters.[68]

The Western Arctic migration routes and staging areas for marine associated birds is depicted on Figure 2.13. Most birds arrive via the Mackenzie flyway or via the Alaskan coast. The Beaufort Sea area receives about 2000000 migratory birds annually, significantly less than the Eastern Arctic. In general terms, far fewer colonial seabirds but proportionally larger populations of geese and seaducks occur in the Western Arctic. Important species are: yellow-billed, red-throated and Arctic loons; whistling swans; Pacific brant; snow and white-fronted geese; king and common eider; oldsquaw; greater scaup; white-winged and surf scoters; northern and red phalaropes; parasitic, pomarine and long-tailed jaegers; glaucous gull; Arctic tern; thick-billed murre; and black guillemot.[69] Nearly all of these are bound to the sea for feeding, moulting or nesting. The coastal region and the Mackenzie Delta are important habitat areas, and the Cape Bathurst polynya plays a critical role as the major feeding and staging area in the spring. The critical importance of the Cape Bathurst polynya was documented in 1964, when open water did not occur until July, and over 100000 incoming seaducks starved.[70]

D Marine mammals

Marine mammals can be classified into two categories, permanent residents and seasonal migrants. The former category includes the ringed seal and the polar bear, which live among the pack ice, and the bearded seal and walrus, which select shallow water areas near open water for overwintering. The Arctic fox, although not really a marine mammal,

Figure 2.13. Migration routes of marine-associated birds in the Western Arctic.

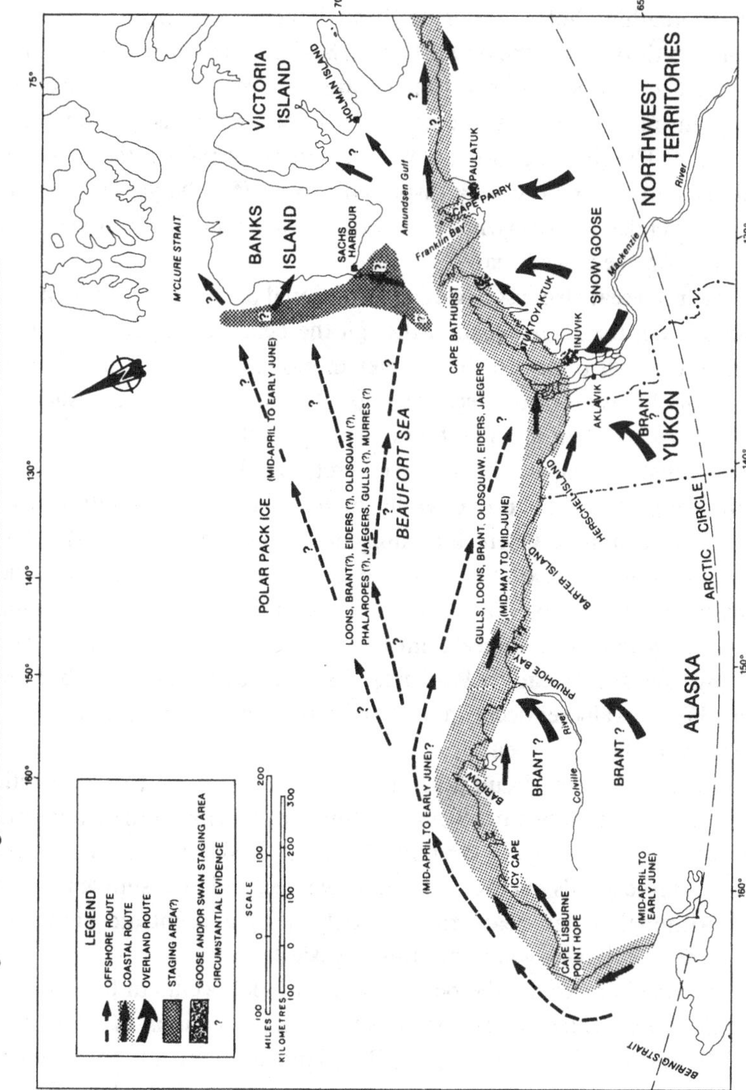

Source: Dome Petroleum Ltd, Esso Resources Canada Ltd, Gulf Canada Resources Inc., *Environmental Impact Statement for Hydrocarbon Development in the Beaufort Sea–Mackenzie Delta Region* (1982), V. 3A, p. 3.25.

sometimes functions as one during winter months when it moves out onto the landfast ice to feed on seal pups.[71] Seasonal migrants include the beluga, narwhal and bowhead whales, and the harp seal.

The ringed seal is well adapted to winter ice conditions in that it can maintain breathing holes by boring through ice with its front claws.[72] It can thus live throughout the Arctic Ocean. Seal pups are born in snow lairs on landfast ice. Food habits vary from area to area and from season to season, but include small Arctic cod, amphipods, euphausiids and other crustaceans. The ringed seal is the most common of the Arctic seals, but usually is solitary or in loosely dispersed groups.[73] Polynyas are mainly used by juveniles for overwintering. Polar bears and Arctic foxes prey extensively on the ringed seal.

The bearded seal also has a wide distributional pattern, but is much less abundant than the ringed seal. Although the bearded seal can maintain breathing holes, it prefers to overwinter in moving ice where cracks and leads occur.[74] It is a bottom feeder so distribution is restricted to shallow water areas where benthic invertebrates are plentiful.

Walrus distribution in the Eastern Arctic is shown on Figure 2.14. Atlantic walruses prefer to overwinter in shallow and open water areas, or in areas of thin ice where breathing holes may be maintained. They "haul out" onto the ice in winter and onto the beach in summer and usually huddle in groups. They apparently winter in the Davis Strait and the North Water, and migrate into Lancaster Sound in early spring following the shore lead.[75] Walruses, like bearded seals, are bottom feeders. Pacific walruses occur in the Bering and Chukchi Seas, but seldom range into the Beaufort Sea.

The polar bear is a circumpolar marine mammal found in all Arctic waters and along practically all coasts. It lives primarily on the sea ice but moves ashore for maternity denning. Its main food source is the ringed seal, but bearded seals, walruses, juvenile whales, and kelp supplement its diet.[76] Although polar bears range widely, they concentrate along ice edges or near polynyas where hunting is easiest.

In the Canadian Arctic, the beluga (white whale) occurs in five specific populations delineated according to summering locations in Lancaster Sound, Cumberland Sound, Ungava Bay, Hudson Bay and the Beaufort Sea. Most of the belugas from Lancaster Sound and Cumberland Sound probably overwinter along the edge of the pack ice in Davis Strait and in the Disko Bay region of West Greenland. Belugas from Northeastern Hudson Bay and Ungava Bay are also thought to overwinter in the Davis Strait area and off Northern Labrador (see Figure 2.15). Beaufort Sea belugas make winter migrations to the Bering Sea (see Figure 2.16). The

Figure 2.14. Walrus migration routes, wintering areas and known coastal summering areas in Parry Channel, Baffin Bay and Davis Strait.

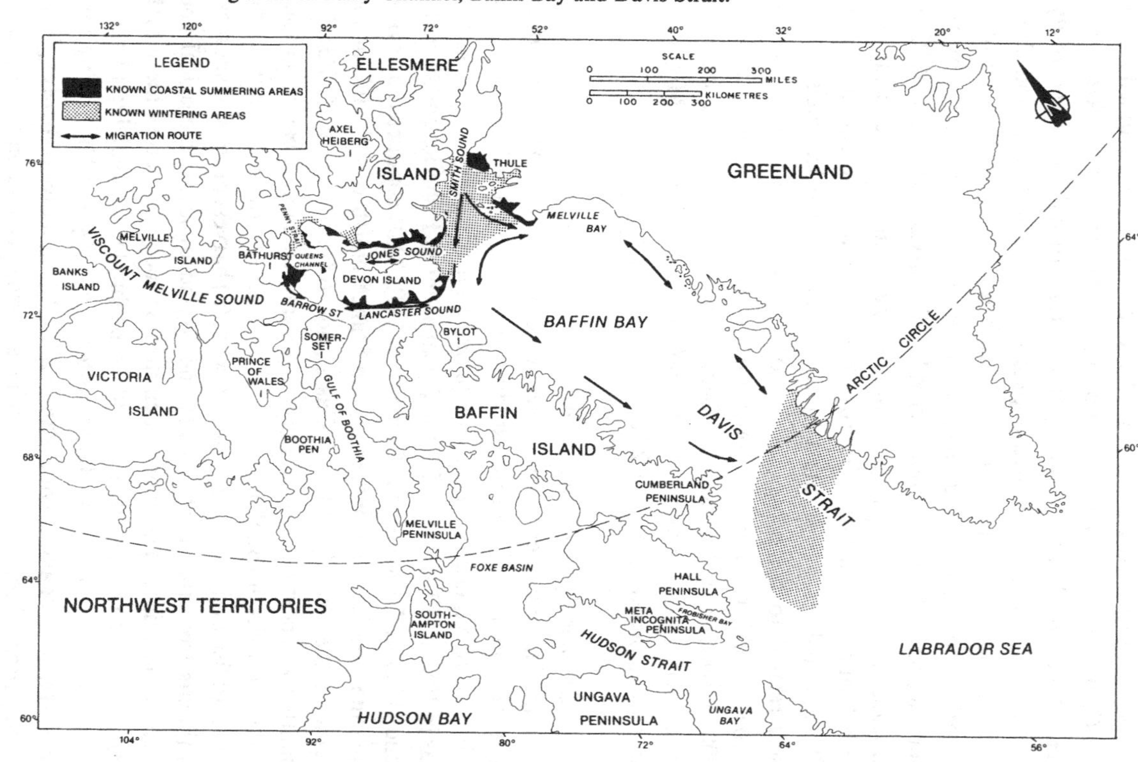

Source: Dome Petroleum Ltd, Esso Resources Canada Ltd, Gulf Canada Resources Inc., *Environmental Impact Statement for Hydrocarbon Development in the Beaufort Sea–Mackenzie Delta Region* (1982), V. 3B, p. 2.12.

total beluga population in the Canadian Arctic is estimated at approximately 24000–28000 individuals.[77]

Bowhead whales enter Canada's Northern waters from both the east and west, but they also are separated by an essentially unoccupied region in the Central Arctic.[78] The migration routes of the Western Arctic and Eastern Arctic bowheads are shown on Figures 2.16 and 2.17, respectively. The bowhead, with an estimated population of 300 in the east and about 2500 in the west, with the latter figure constituting about 75 per cent of the world's population, is designated as an endangered species by the Committee on the Status of Endangered Wildlife in Canada (COSEWIC).[79] Bowheads feed by skimming and straining large volumes of water through their baleen plates, feeding primarily on zooplankton and euphausiids.[80]

The narwhal, easily identified by its spirally twisted left incisor, and closely related to the beluga, is found only in the eastern portion of Canada's Northern waters. Population size is estimated to be at least 20000 individuals.[81] The narwhal, feeding on Arctic cod, squid, decapod crustaceans and Greenland halibut, prefers deep cold waters. Migration routes are depicted on Figure 2.18.

E Biologically significant areas

In Canada, a number of initiatives to provide special status for areas of natural significance are being studied, for example, the Minister of the Environment requested officials to prepare an inventory of significant areas in 1980 and 1981. In 1982, the Department of Fisheries and Oceans' Arctic Offshore Development Committee (ARCOD) conducted a review of marine areas of environmental significance North of 60° and, at the request of the Canada Oil and Gas Lands Administration (COGLA), submitted a report classifying these areas into categories based on four criteria: (i) environmental importance of the area to the maintenance and survival of fish and marine mammal stocks; (ii) harvest of these resources; (iii) threats from the environment; and (iv) adequacy of biological and oceanographic data. The report identified 38 locations across the North where hydrocarbon activities should be prohibited absolutely; 17 areas where exploration should not be permitted unless approved by an Environmental Assessment and Review Panel (EARP); eight areas where exploration may proceed following site-specific assessments, and only one area of limited importance where exploration could proceed without opposition from the Department on biological grounds.[82]

One of the difficulties in designating significant areas is the precondition of reaching consensus on a definition. Terms such as, "biologically significant areas", "critical marine habitat", "critical ecological areas",

Figure 2.15. White whale migration routes, wintering areas and summer concentrations in the Eastern Arctic.

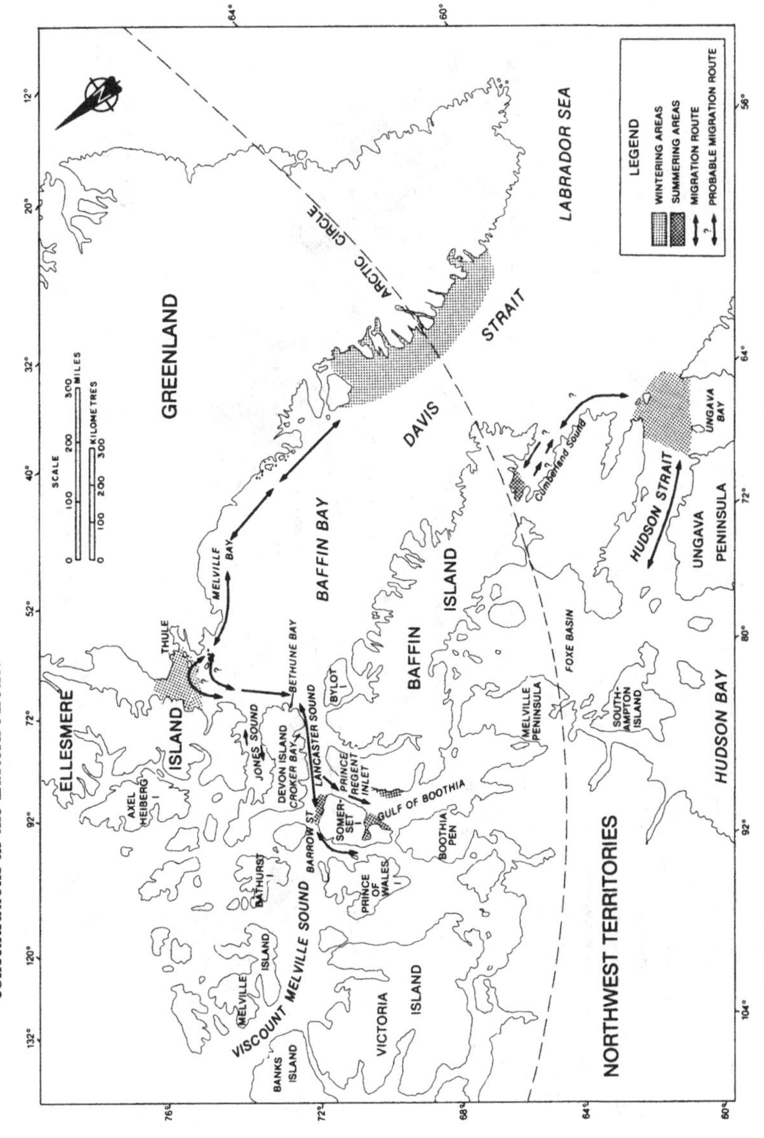

Source: Dome Petroleum Ltd, Esso Resources Canada Ltd, Gulf Canada Resources Inc., *Environmental Impact Statement for Hydrocarbon Development in the Beaufort Sea–Mackenzie Delta Region* (1982), V. 3B, p. 2.4.

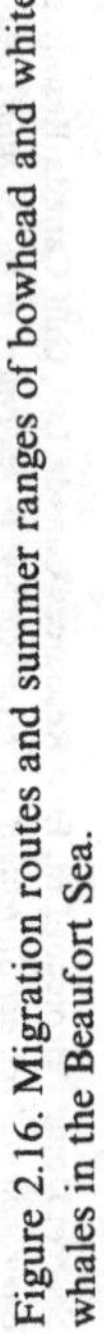

Figure 2.16. Migration routes and summer ranges of bowhead and white whales in the Beaufort Sea.

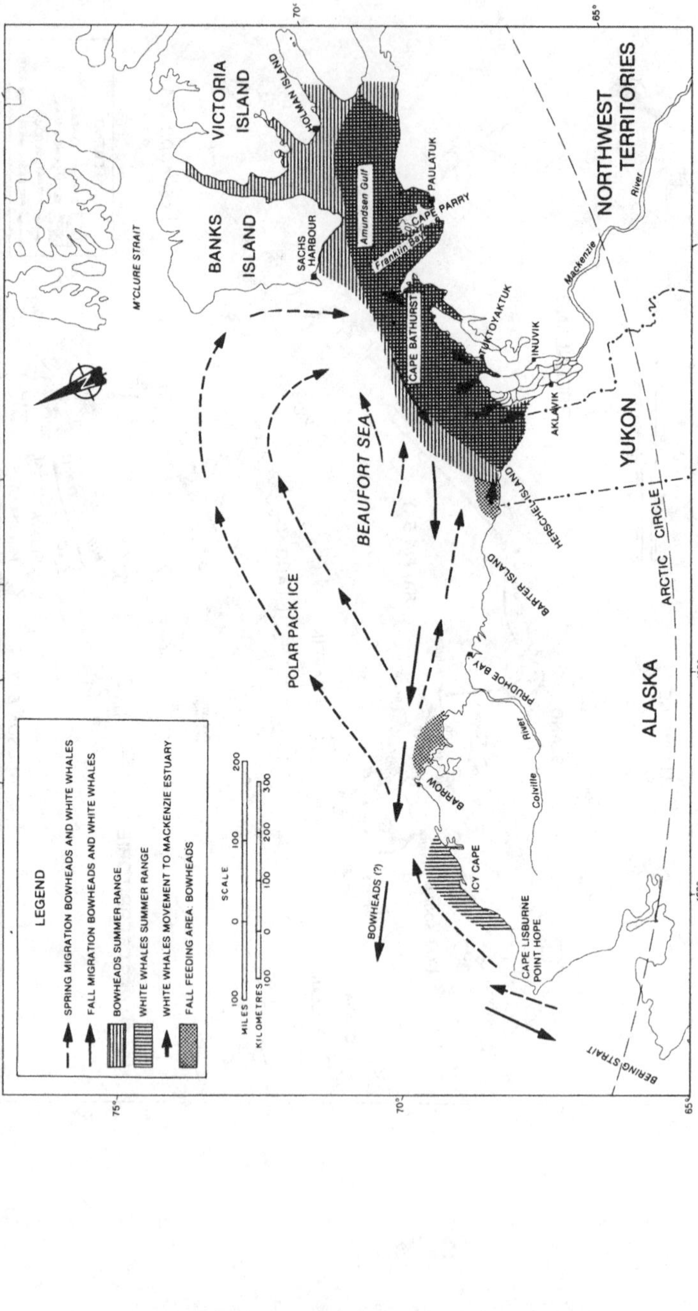

Source: Adapted from Dome Petroleum Ltd, Esso Resources Canada Ltd, Gulf Canada Resources Inc., *Environmental Impact Statement for Hydrocarbon Development in the Beaufort Sea–Mackenzie Delta Region* (1982), V. 3A, p. 3.10.

Figure 2.17. Bowhead whale migration routes, probable winter range and known summer concentration areas.

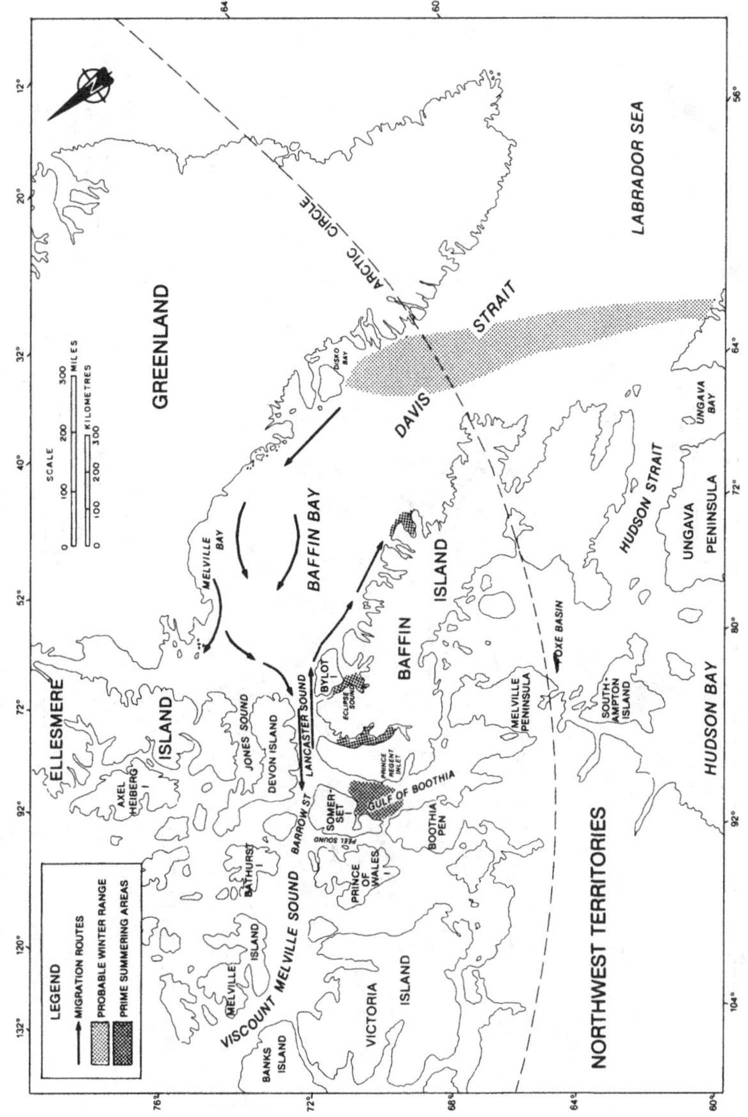

Source: Dome Petroleum Ltd, Esso Resources Canada Ltd, Gulf Canada Resources Inc., *Environmental Impact Statement for Hydrocarbon Development in the Beaufort Sea–Mackenzie Delta Region* (1982), V. 3B, p. 2.9.

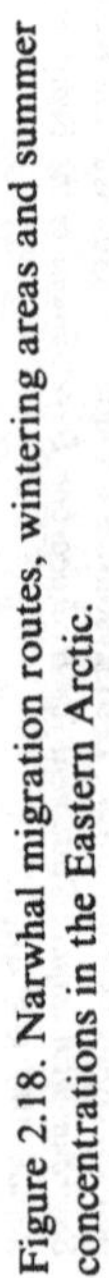

Figure 2.18. Narwhal migration routes, wintering areas and summer concentrations in the Eastern Arctic.

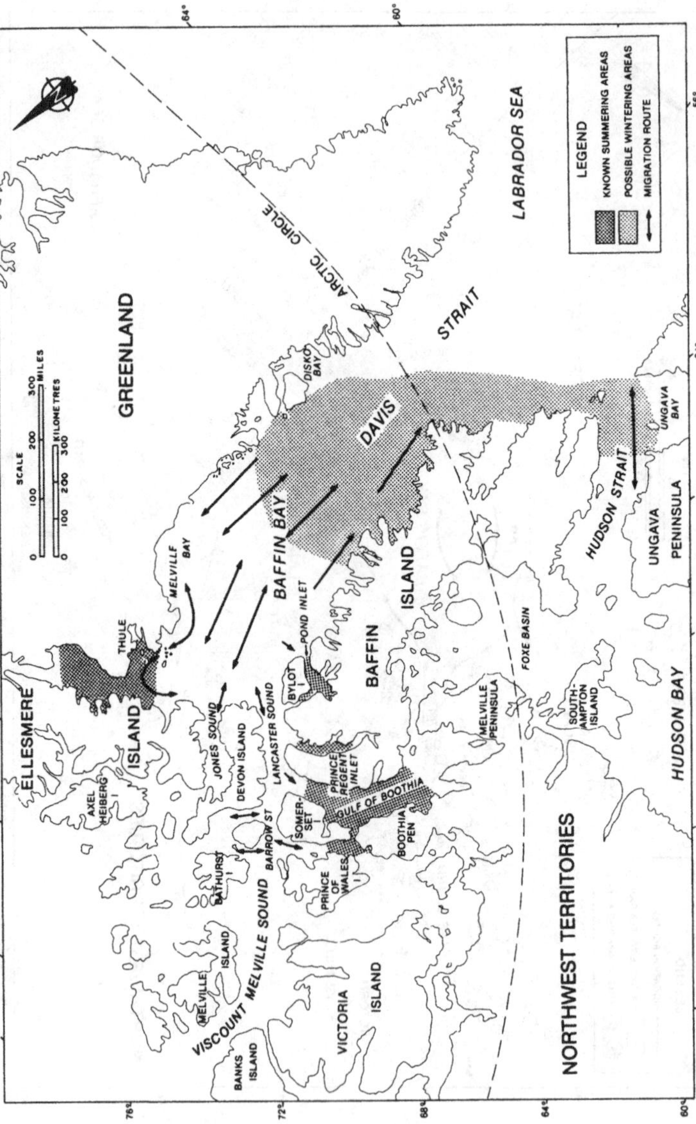

Source: Dome Petroleum Ltd, Esso Resources Canada Ltd, Gulf Canada Resources Inc., *Environmental Impact Statement for Hydrocarbon Development in the Beaufort Sea–Mackenzie Delta Region* (1982), V. 3B, p. 2.7.

"unique and critical ecosystems", and "natural areas of Canadian significance", all suffer from definitional problems. What does significant mean? What does critical mean? Can a significant area be critical in and of itself, or must it be threatened by environmental degradation from an outside source? What criteria should be used to determine significance and criticalness? Even if these questions could be answered, the problem of insufficient data for evaluating criteria would still remain.

Despite the difficulties in establishing a scientifically rigorous procedure for determining biologically significant areas, some structured format could be created. The selection of biologically significant areas should occur within the framework of a comprehensive planning process, as stated by Justice Thomas Berger in 1977:

> As part of comprehensive planning in Canada's North, the federal government should develop a northern conservation strategy to protect areas of natural or cultural significance. This strategy should comprise inventories of natural and cultural resources, identification of unique and representative areas, and withdrawal and protection of such areas under appropriate legislation.[83]

Although government has taken steps to develop a comprehensive planning strategy, for example, the Northern Land Use Planning Program, the Discussion Paper on a Comprehensive Conservation Policy and Strategy for the Northwest Territories and Yukon, and the Lancaster Sound Green Paper, Berger's recommendation has never been carried through to fruition. There is an indication, however, that a Northern conservation strategy could be developed in the near future. In September 1983 John Munro, Minister of the Department of Indian Affairs and Northern Development, established a Task Force on Northern Conservation. The Task Force released its report in December 1984 and recommended government initiatives in numerous areas, including:

(i) preparation of integrated resource management plans for conflict-prone areas such as the Mackenzie Delta-Beaufort Sea and Lancaster Sound;

(ii) establishment of a comprehensive network of protected areas;

(iii) implementation of land use planning agreements among the Governments of Canada, Yukon and the Northwest Territories;

(iv) establishment of a Conservation Advisory Board with board membership to facilitate the implementation of Task Force recommendations;

(v) establishment of a Cabinet Committee on Northern Natural Resources to coordinate federal response to conservation issues;
(vi) promotion of public and governmental awareness of Northern conservation issues; and
(vii) enactment of legislation for protecting marine areas.[84]

Regardless of the parameters of any future federal conservation policy, certain marine regions stand out as requiring special protection. For example, a green paper recently recognized Lancaster Sound as a biologically significant area:

> Lancaster Sound and the contiguous channels and inlets have been recognized by scientists as critical to the reproduction and survival of the seabird population in the Canadian High Arctic; the Sound is considered to be one of the most important marine mammal habitats in the Eastern Arctic as well. Several million seabirds nest, reproduce and feed in the Sound during the Summer months. Species concentrations are unrivalled anywhere in the High Arctic and include northern fulmar, glaucous gull, black-legged kittiwake, thick-billed murre and black guillemot. Marine mammals include 85 per cent of North America's narwhals and 40 per cent of the white (beluga) whales in addition to large populations of ringed, harp and bearded seals. There are small colonies of walrus, and the endangered bowhead whale still occurs in small numbers throughout these waters. Polar bears are found in large numbers on the ice.[85]

Major recurring polynyas and shore leads are also biologically sensitive areas. They are isolated pockets of primary productivity, they attract large concentrations of biological populations at certain times, thereby sustaining migrating species which are part of larger ecosystems, and they are particularly vulnerable to oilspills and disturbance. Two polynyas, in particular, should be designated as biologically sensitive areas requiring special protective measures:

1. *The North Water*

   The largest, most northerly, and possibly the most important polynya is the North Water, which is used by seabirds and marine mammals during migrations and overwintering. Its location between Canada and Greenland, and its ecosystem linkage with Lancaster Sound and the West Coast of Greenland, suggest that conservation and protection of the North Water should be a topic of bilateral discussions.

2. *Cape Bathurst polynya*

The recurring Cape Bathurst polynya, and the associated crack and lead system extending to beyond Point Barrow, Alaska, are critical to large numbers of seabirds and marine mammals in the Beaufort Sea and, therefore, should be a subject of bilateral discussions between the United States and Canada.

IV The human setting

A The circumpolar Inuit

All circumpolar states have native peoples who live by the shores of the Arctic Ocean and have a way of life which depends upon marine resources. The Lapps of Northern Scandinavia have a population of about 30000, which is divided among four countries: Norway, Sweden, Finland, and the USSR. This division results in a deep concern for the survival of the Lapp language and culture. More than half of the Lapp population depend upon the fish, seals and whales of the Barents Sea.[86] Approximately 140000 "minority people", speaking several different languages, inhabit the Northern USSR.[87] In Eastern Siberia these minorities include small groups of Inuit people, known as the Eskimosy. Other Inuit populations are found in Alaska, Canada and Greenland. The total world population of the Inuit is about 100000.

In Alaska, the Inuit people are known as Inupiat and Yupik, and they number about 30000 throughout the coastal regions of the state. The Inupiat constitute a majority of the permanent population of the North Slope Borough (9000), which covers the northern 15 per cent of the state, bordering on the Beaufort and Chukchi Seas. The discovery of oil at Prudhoe Bay provided the impetus for settling land claims in Alaska in 1971.[88] The Alaska Native Claims Settlement Act gave rights to 44 million acres of land, plus $962 million to the Indian, Aleut and Inupiat peoples, and confirmed additional rights for subsistence hunting and fishing, including exclusive rights to the harvesting of marine mammals.

Incorporation of the North Slope Borough in 1972 has given the Inupiat a degree of local control and responsibility for education, taxation and assessment, and land-use planning. The Borough has recently prepared a coastal zone management plan pursuant to the Coastal Zone Management Act.[89] Following public hearings and legal adoption of the coastal zone management plan, all state and federal agencies will be required to comply with the plan. The Borough encourages hydrocarbon developments onshore, but has opposed offshore exploration because of the potential impact on marine resources – particularly the bowhead whale. The Inupiat rely heavily on renewable resources for their subsistence, including the annual bowhead harvest, which is of historic and cultural significance to their way of life. However, the bowhead may be an endangered species and the Alaska Eskimo Whaling Commission and the International Whaling Commission agreed to a limit of 18 strikes for the 1983 season.[90]

Greenland, territorially a part of Denmark, is the world's largest island, and is predominantly covered by a thick ice cap. It has a population of some 50000, of whom about 85 per cent are native Greenlanders of Inuit ancestry. The population is concentrated along a narrow strip of the southwest coast facing Davis Strait, Baffin Bay and the Atlantic Ocean. About 70 per cent of the population live in the communities of Nuuk, Sisimiut, Ilulisatt, Maniitsoq, Aasiatt, and Qaqortoq, with much smaller communities farther north and a single community on the East Coast.

Following a referendum Greenland obtained a form of Home Rule from Denmark in 1979. Her degree of autonomy represents the political aspirations of other Inuit in many ways.[91] The Home Rule Assembly is fully elected and has responsibility for most internal powers formerly exercised by Denmark. Both Greenland and Denmark can exercise a veto over resource policies and projects, such as offshore oil and gas. The population desires to manage the affairs of Greenland on a more independent basis,[92] but economic and fiscal independence does not appear to be a practical goal in the short term. Greenlanders voted to withdraw from the European Economic Community primarily because of disagreements over fisheries policies and EEC allocations of cod quotas to its members without the consent of the Home Rule government.[93] A host of legal, economic and political issues complicate negotiations on withdrawal, not the least of which is the fact that Greenland presently enjoys free access to EEC markets and annually exports about 80 per cent of its fish catch to the EEC. Despite possible economic repercussions, Greenland withdrew from the EEC in January, 1985, and achieved "Overseas Countries and Territories" status within the EEC.

The economy is heavily dependent upon marine resources, and commercial fishing for cod, halibut and deepwater prawn is the main economic activity. In Southern Greenland, some native people do tend sheep, cows and horses, but marine mammals are harvested extensively for additional income and for subsistence food. Relative dependence on marine mammals increases with latitude along the West Coast of Greenland.

Canada has an Inuit population of about 25000 dispersed in small communities along its lengthy Northern coastline. The majority of the Inuit reside in the Northwest Territories, but about 4200 live in Quebec, and 2200 in Labrador. Several decades ago, the Inuit were widely dispersed in hunting camps, lived off the land and followed their traditional lifestyle. But government policies brought the Inuit into central communities where health, education and other services were available. However, the Inuit remain fairly scattered with no more than 2500 residing in any single community. Frobisher Bay, Pangnirtung, Eskimo Point and Baker Lake are the only communities containing more than 800 Inuit.

The division of the Inuit population among the Northwest Territories, Quebec and Labrador complicates Inuit aspirations for political and economic development within Canada. At present the Minister of Indian Affairs and Northern Development, and, to varying degrees, the governments of the Northwest Territories, Quebec, and Newfoundland and Labrador are responsible for administering Inuit affairs. In recent years, the Inuit have developed their own organizations and have become heavily involved in Northern political issues and local government. The Inuit Tapirisat of Canada (ITC) is an umbrella organization representing all Inuit in Canada. Regional entities include: the Labrador Inuit Association, the Makivik Corporation of Northern Quebec, the Baffin Region Inuit Association, the Keewatin Inuit Association, the Kitikmeot Inuit Association in the Central Arctic, and the Committee for the Original People's Entitlement (COPE) in the Western Arctic.

Aboriginal rights, based upon native use and occupancy of the land since time immemorial, have been the subject of negotiations between the government of Canada and native peoples since 1973. For aboriginal peoples whose rights have not been extinguished by previous treaties, these negotiations are referred to as "comprehensive land claims negotations." Government policy on native claims has been spelled out in the publication *In All Fairness.*[94] The policy requires that the negotiation process and settlement be fair and thorough so that the claim will not arise again, and provides for settlement rights and benefits which may include lands, limited subsurface rights, wildlife harvesting and management rights, and monetary compensation. The James Bay Agreement of 1975 covered most, but not necessarily all, claims of Quebec's Inuit, while negotiations with the Labrador Inuit have yet to commence.[95] On 28 March 1984 Cabinet approved a final land claims agreement with the Inuvialuit in the Western Arctic which promises to establish several new resource management mechanisms granting native Northerners new participation rights but not exclusive decision-making authority. Negotiations are continuing between the federal government and the Tungavik Federation of Nunavut (TFN), the land claims negotiating arm of the ITC, over areas of the Central and Eastern Arctic.

The Inuit are also insisting on rights to marine areas, primarily delineated by Inuit historic use of landfast ice to harvest marine mammals. The area involved, as depicted on Figure 2.19, is well documented in

Figure 2.19. Inuit use of sea ice and adjacent land areas for hunting.

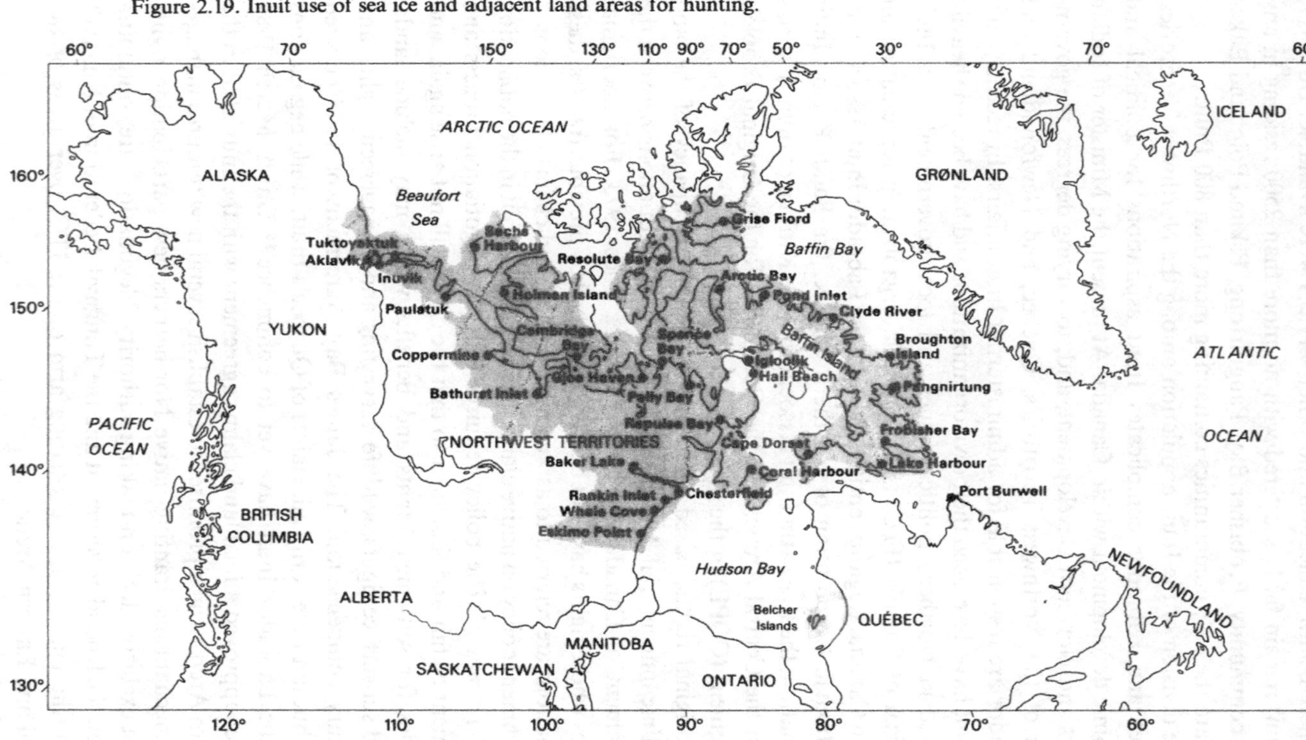

Source: Adapted from Milton R. Freeman (ed.), *Inuit Land Use and Occupancy Project*, V. 3 (1976), p. 153.

land-use and occupancy studies.[96] The Inuit view the landfast ice as a natural extension of the land, and make no differentiation between their use of the two. The argument has been advanced that Canada, by officially recognizing Inuit rights to these marine areas, could strengthen its claim that Arctic waters are historic waters in international law through which no right of innocent passage exists.[97] The Inuit have never used or occupied the waters in the northwestern portion of the High Arctic. They have, however, used the ice-covered areas of Barrow Strait and Lancaster Sound, which provides a possible basis for claiming at least part of the Northwest Passage as historic internal waters.

Separate from, but often erroneously linked to, comprehensive land claims negotiations is the question of political development in the North. The Carrothers Commission of 1966 documented the eventual need to divide the Northwest Territories (NWT) into separate Inuit and Dene jurisdictions, and political pressure from the Inuit for division has been increasing – in part through discussions on the Canadian Constitution. In April of 1982 the NWT voted in a plebiscite on the question of division, with about 80 per cent of the Inuit voting in favour. The NWT Legislative Assembly subsequently voted 19–0 in favour of division, and the federal government announced conditional approval, with settlement of land claims and agreement on the boundaries of the new territory as the key conditions. Inuit of the Eastern and Central Arctic have proposed to name their territory "Nunavut",[98] and have formed a Nunavut Constitutional Forum to work with the Western Constitutional Forum in developing a constitution and implementing the concept of Inuit self-government within Canada.

The Inuit Circumpolar Conference (ICC) is a unique international organization which brings together the Inuit of Alaska, Greenland and Canada. ICC was formed in 1977, primarily due to Alaskan Inuit concerns over Beaufort Sea oil and gas developments and Greenlandic-Canadian Inuit concerns over the Arctic Pilot Project.[99] General assemblies are held at periodic intervals to meet and co-operate on matters of cultural, political, economic and environmental importance.[100] ICC has recently been granted non-governmental organization consultative status with the Economic and Social Council of the United Nations. The Siberian Eskimosy, as a possible first step towards joining the circumpolar movement, have authorized the ICC to speak on their behalf regarding seal harvesting issues pursuant to the Convention on International Trade in Endangered Species. The ICC has the potential to play an important future role in regional co-operation for the conservation and protection of renewable resources in Northern waters.

B *Renewable resource harvesting*

The Inuit are a marine-oriented people whose way of life and sustenance are dependent upon the production of renewable resources, especially marine mammals, in Northern waters. Nelson Graburn has estimated that "without the bounty of the seas the Eskimos who could survive on the land would be less than one-tenth of the past and present population."[101] Arctic marine ecosystems are not closed systems and the high productivity, and temporally high standing stocks of large marine mammals are partly explained by migrations carefully timed to utilize peak periods of primary productivity. "The two features that distinguish the environment of the Eskimos and hence their lifeways from that of all other native peoples of North America are residence in the Arctic and dependence on the sea."[102]

Although primarily marine resource harvesters, the Inuit also depend on terrestrial mammals such as caribou, musk-ox and wolf[103] – as well as Arctic char, migratory birds and their eggs. The importance of individual foods naturally varies with time and location, but country food provides by far the greater portion of protein requirements and the essential components for a balanced nutritious diet.[104] Country food also represents a considerable dollar saving in food costs amounting to the equivalent of several thousand dollars of income per family per year.[105] In many Arctic communities the ringed seal is the most common and important resource. With the decline of dog teams, species such as walrus and narwhal have declined in importance as sources of food, but are still valued for their ivory and the prestige of the hunt.[106] Hunting activities are an important part of the cultural heritage of the Inuit, even though hunting methods have changed radically through the introduction of rifles and snowmobiles. Hunting also provides valuable income from the sale of pelts and ivory.

The correlation between Figures 2.14 to 2.18 (distributions of Arctic marine mammals) and Figure 2.19 (extent of the sea ice use by the Inuit) is not coincidental. As opportunistic hunters who prefer a diet of marine mammals, the Inuit have tended to use and occupy the areas where marine mammals are found in abundance. Renewable resource harvesting, therefore, is concentrated in six regions of Canada's North: the Beaufort Sea, Lancaster Sound and vicinity, Eastern Baffin, Hudson Strait, the northwestern rim of Hudson Bay, and the Labrador Coast.

Table 2.1 summarizes data on marine mammal harvests for the Lancaster Sound region. Data should be interpreted with considerable caution, however, due to natural population variability and fluctuation in resource utilization patterns. Ringed seals, harp and other seals are most important to the Inuit economy and diet. Larger mammals such as beluga, narwhal, walrus and polar bear, while valuable to the Inuit, are harvested in relatively small numbers.

Marine mammal harvests are concentrated in the Eastern Arctic. Almost 50 per cent of seals harvested in the NWT are taken by four East Baffin communities, while seal harvests off West Greenland are nearly three times higher. Harvests in Central Arctic waters and the northwestern portion of the High Arctic are low due to biological productivity factors. In the Western Arctic, the community of Holman Island on Amundsen Gulf accounts for a much higher harvest of marine mammals than Beaufort Sea communities.[107] Harvest statistics from the Beaufort Sea region are somewhat surprising, but such data could indicate that the small population of Inuvialuit in the Western Arctic is becoming more reliant on wage employment from Beaufort Sea oil and gas activities.

V Environmental issues

A Introduction

The environmental issues facing Canada's Northern waters today are closely linked to the major non-renewable resource projects being planned for the North. The major projects include: Beaufort Sea oil and gas where extensive reserves have been discovered onshore and offshore and where, depending on world oil prices and market conditions, production may be imminent; Sverdrup Basin oil and gas where substantial reserves have been discovered in the High Arctic and various schemes are being considered for production and marketing; Lancaster Sound where previous applications to drill for hydrocarbons have been opposed; Baffin Bay

Table 2.1. *1979 Marine mammal harvest estimates, Lancaster Sound communities.*

Species	Clyde river (87 hunters)	Grise fiord (22 hunters)	Pond inlet (122 hunters)
Ringed seal	4733	686	2487
Bearded seal	5	25	38
Harp seal	4	166	21
Narwhal	5	15	139
Polar bear	21	24	16
Walrus	—	9	14
Beluga	—	14	9

Source: Kerry J. Finley and Gary W. Miller, 'Wildlife Harvest Statistics from Clyde River, Grise Fiord and Pond Inlet (1979)', LGL Ltd for Petro-Canada Explorations (July 1980), Tables 2–4.

where hydrocarbon exploration is at an early stage and drilling has not yet commenced; South Davis Strait where drilling has commenced and a potentially significant gas discovery has been found; Hudson Bay where seismic exploration for hydrocarbons has begun; and the Labrador Coast where drilling has taken place and both oil and gas have been discovered. Such activities are being planned and developed by Southern Canadians or multinational companies, but the social and environmental impacts of these activities will be felt in the North, and will largely be borne by the Inuit and the Arctic marine environment.

It is no coincidence that numerous projects are being planned just at the time when comprehensive land claims negotiations are taking place and the creation of a Nunavut Territory is being considered. The potential for oil and gas benefits has helped convince Ottawa to enter into comprehensive land claims negotiations to avoid future court challenges. Inuit participation in land claims negotiations, the Berger Commission, National Energy Board Hearings on the Arctic Pilot Project, several Environmental Assessment Review Process (EARP) hearings, the Lancaster Sound Regional Study, negotiations on a proposed land-use planning program, and the Inuit Circumpolar Conference, have greatly increased the political awareness of the Inuit people. In a few short years the Inuit have developed the capability for dealing with the complex institutions and regulatory processes of Canada, while defining their own political and constitutional goals. It remains uncertain, however, how native goals will be integrated into government decision making.

B *Resource development and environmental issues*

Environmental issues in the North, primarily related to hydrocarbon projects but also to hydro and mining projects, may be summarized in eight generic categories:

1. *Environmental impact of hydro projects*

Construction of hydro dams on the northward-flowing rivers of the Soviet Union and Canada could alter the climate of the Arctic Ocean. For example, the proposed Liard River dam could have a significant impact on the Mackenzie Delta and the Beaufort Sea by reducing and delaying the spring flood, thereby decreasing the productivity of the marine/estuarine ecosystems in the Western Arctic.

2. *Deleterious deposits from mining projects*

Lead-zinc mines, such as Nanisivik and Polaris, and uranium mines, such as Baker Lake, may result in deposits of deleterious substances in the

Arctic marine environment. The substances may be toxic to marine organisms and tend to bio-accumulate through the foodchain to marine mammals and man.

3. *Oil under the ice*

Oilspills in ice-covered waters are difficult to clean up with present technology. Blowouts may be impossible to cap until the year after they occur. Oil in the cold Arctic environment will persist for a very long period of time. The impact on marine ecosystems could be catastrophic.

4. *Pollution of polynyas and shore leads*

The productivity of marine ecosystems is greatly enhanced at the ice edges of polynyas and shore leads. Because seabirds and marine mammals concentrate in these areas at certain times of the year, they are highly vulnerable to oil pollution.

5. *Protection of environmentally significant areas*

Critical areas of Arctic marine habitat have never been designated on a systematic basis. No regulations or programmes exist to ensure that hydrocarbon development projects will avoid environmentally significant areas.

6. *Hydrocarbon facilities siting*

The siting of major hydrocarbon facilities such as deepwater ports and supply bases may have significant impacts on the environment, yet no mandatory facilities siting process or regional planning process exists. The debate over Gulf Canada's application for a land-use permit to build a deepwater port at Stokes Point, on the sensitive Northern Yukon coast, exemplifies the need for a formal process.

7. *Artificial islands, dredging, quarrying*

Dredging and quarrying of construction materials, dredging of harbour channels, and the construction of artificial islands, using dredged materials, can affect marine life through direct disturbance or siltation. The presence of numerous artificial islands may alter the ice regime and change the location of the landfast ice edge.

8. *Cumulative impacts on renewable resources*

The wide range of oil and gas activities and facilities associated with exploration and development may result in cumulative impacts which will diminish the supply of renewable resources. This may be of crucial importance to the Inuit who rely on renewable resources, particularly marine mammals, for sustenance and to support their traditional lifestyle.

C *Transportation environmental issues*

Resource development projects such as Beaufort Sea oil and gas and Alaskan North Slope oil, will require product transportation to markets. Although previous Arctic marine transportation has been restricted to a short summer season, the Northwest Passage could be opened for year-round shipping by icebreaking supertankers by the year 2000. The prospect of numerous vessel transits of the Northwest Passage raises a number of environmental issues:

1. *Risk of tanker accident*

The risk analysis of various types of tanker accidents, and the potential impacts of oil and LNG on the environment, are the subjects of considerable debate. A major accident at the wrong place and the wrong time would have catastrophic consequences. Who decides if the risk is acceptable?

2. *Route selection*

Is there sufficient flexibility in route selection and-or the timing of transits to avoid environmentally sensitive areas at particular times of the year?

3. *Impact of noise on marine mammals*

The powerful engines of the tankers, and the smashing of thick sheets of ice, will introduce noise to the marine environment. The Inuit are concerned that such noise will impair the ability of marine mammals to communicate with each other and ultimately drive the mammals away from the area.

4. *Impact of ship tracks on ice and the Inuit*

In the winter months each transit will leave a rough ice rubble which will refreeze quickly into a thicker and rougher formation, so that after several transits the tankers will shift to a new track. The rough tracks will be difficult for Inuit hunters to traverse on snowmobiles. Numerous side-by-side thick tracks may change the location of ice edges and delay break-up. In the early summer, tracks through landfast ice will not refreeze, also causing difficulties for Inuit hunters.

5. *Direct collisions with denning seals*

Ringed seals have their maternity dens in the landfast ice area. Some females and pups inevitably will receive direct hits from tankers. The Inuit are concerned that the numerous ship tracks will result in a significant impact on seal populations along the Passage.

6. *Marine mammals in leads and polynyas*

Tankers would be likely to utilize shore leads and open water areas to increase their efficiency. Marine mammals concentrate in these areas, and they may also be attracted to the open water in ships tracks. This will increase the vulnerability of marine mammals to ship-generated pollution and may lure mammals away from traditional Inuit harvesting locales.

7. *International environmental issues*

The impacts of noise and oil pollution on marine mammals from Northwest Passage shipping traffic have been identified as international issues by the Inuit of Greenland and Alaska.

D The policy muddle

The lack of a comprehensive conservation strategy for the North is part of a policy muddle having direct implications for Canada's Northern waters.[108] The Department of Indian Affairs and Northern Development has prepared a discussion paper as a step towards developing a comprehensive conservation policy and strategy for the Northwest Territories and Yukon, including marine ecosystems.[109] However, the initiative is not likely to become an official policy in the foreseeable future. Draft policy documents serve a useful purpose for government by demonstrating concern for environmental and social issues without actually foreclosing development options. In the meantime, significant marine areas in the North have no protective status.

Many people have identified the policy muddle as the major political problem. For example, during the Lancaster Sound Regional Study the Chairman of the public review phase, Peter Jacobs, in a letter to the Honourable John Munro, stated: "The first issue is the clear need for a national policy across all departmental sectors of government for Canada's High Arctic. The public express an urgent need for integrated national policies with respect to energy, transportation, conservation and development of the High Arctic."[110] In its recent report, *Marching to the Beat of the Same Drum*, the Special Committee of the Senate on the Northern Pipeline chastised the federal government for a lack of policy and planning direction, pointed out the consequences for the environment and the people of the North, and firmly recommended that policy and planning measures be formulated without delay.[111]

Canada's Arctic environment requires special protection. Whether the Canadian government will meet the challenge is likely to depend on a new political will to undertake long-range marine planning rather than incrementally stepping to the tune of industrial initiatives.

Notes

1. Kim Traavik and Willy Ostreng, "The Arctic Ocean and the Law of the Sea", *The Challenge of New Territories* (1974), p. 56.
2. David Sugden, *Arctic and Antarctic: A Modern Geographical Synthesis* (1982), p. 44.
3. Traavik and Ostreng, *supra* note 1, p. 56.
4. Donat Pharand, *The Waters of the Canadian Arctic Archipelago in International Law* (in press).
5. Summary of Canadian Note of April 16 Tabled by the Secretary of State for External Affairs in the House April 17, 9 *Int'l Legal Materials* 607, 612 (hereinafter referred to as Summary of Canadian Note).
6. For a general discussion of the background to Article 234, see D.M. McRae and D.J. Goundrey, "Environmental Jurisdiction in Arctic Waters: The Extent of Article 234", *U.B.C. Law Review* V. 16, No. 2 (1982), pp. 197–228, pp. 210–15.
7. International Union for the Conservation of Nature, *World Conservation Strategy: Living Resource Conservation for Sustainable Development* (1980), section 19.
8. *Ibid.*
9. R.M. M'Gonigle and Mark W. Zacher, *Pollution, Politics and International Law* (1979), p. 109.
10. Howard Hume, "Toward A Canadian Arctic Policy", *Marine Affairs Journal* No. 5 (1978), p. 37.
11. M'Gonigle and Zacher, *supra* note 9, p. 110.
12. Summary of Canadian Note, *supra* note 5, p. 613.
13. John Livingston, *Arctic Oil* (1981), pp. 63–4.
14. Dome Petroleum Ltd, Esso Resources Canada Ltd, Gulf Canada Resources Inc., *Environmental Impact Statement for Hydrocarbon Development in the Beaufort Sea–Mackenzie Delta Region* (1982), V. 3A, p. 1.39 (hereinafter referred to as Beaufort EIS 3A).
15. *Ibid.*, p. 1.40.
16. *Ibid.*
17. Department of Fisheries and Oceans, *Sailing Directions of Arctic Canada*, V. 3 (1981), p. 210.
18. Alberry, Pullerits, Dickson & Associates, "Study of Ice Conditions in Navigational Channels" (on behalf of Dome Petroleum Ltd), (1978), p. 15.
19. Beaufort EIS 3A, *supra* note 14, p. 1.57.
20. Arctic Pilot Project, *Integrated Route Analysis* (1981), V. 1, pp. 2.24, 2.25 (hereinafter referred to as APP).
21. *Ibid.*, p. 2.74.
22. *Ibid.*, p. 2.98.

23. M.J. Dunbar and D.M. Moore, *Marine Life and its Environment in the Canadian Eastern Arctic: A Biogeographic Study* (1978), p. 2.
24. *Ibid.*
25. For a general discussion of current patterns in the Eastern Arctic, see APP, *supra* note 20, pp. 2–148; Beaufort EIS 3A, *supra* note 14, pp. 1.45–1.47; and Department of Fisheries and Oceans, *Sailing Directions of Arctic Canada*, V. 1 (1982), pp. 186–88.
26. Department of Fisheries and Oceans, *Sailing Directions of Arctic Canada*, V. 1 (1982), pp. 124–29.
27. Beaufort EIS 3A, *supra* note 14, p. 1.8.
28. Beaufort EIS 3A, *supra* note 14, p. 1.7.
29. *Ibid.*, p. 1.10.
30. LGL Ltd and ESL Ltd, "Biological Overview of the Beaufort Sea and NE Chukchi Sea" (on behalf of Dome Petroleum Ltd) (1982), p. 2.2.
31. Beaufort EIS 3A, *supra* note 14, p. 1.3.
32. LGL Ltd and ESL Ltd, *supra* note 30, p. 2.2.
33. APP, *supra* note 20, p. 2.27.
34. LGL Ltd, "Biological Overview of the Northwest Passage, Baffin Bay and Davis Strait" (on behalf of Dome Petroleum Ltd) (1982), pp. 2–4.
35. Department of Indian Affairs and Northern Development, *The Lancaster Sound Region*: 1980–2000 (January 1982), p. 47.
36. Dunbar and Moore, *supra* note 23, p. 2.
37. Dome Petroleum Ltd, Esso Resources Canada Ltd, Gulf Canada Resources Inc., *Environmental Impact Statement for Hydrocarbon Development in the Beaufort Sea–Mackenzie Delta Region* (1982), V. 3B, p. 1.14 (hereinafter referred to as Beaufort EIS 3B).
38. Dunbar and Moore, *supra* note 23, p. 16.
39. Beaufort EIS 3B, *supra* note 37, p. 1.15.
40. LGL Ltd, *supra* note 34, pp. 2–4.
41. APP, *supra* note 20, pp. 2–183.
42. Beaufort EIS 3B, *supra* note 37, p. 1.15.
43. Michael Smith and Bruce Rigby, "Distribution of Polynyas in the Canadian Arctic", in Ian Stirling and Holly Cleator (eds), *Polynyas in the Canadian Arctic*, Occasional Paper No. 45, Canadian Wildlife Service (1981), p. 7.
44. M.J. Dunbar, "Physical Causes and Biological Significance of Polynyas and Other Open Water in Sea Ice", in *ibid.*, p. 30.
45. Ian Stirling, Holly Cleator and Thomas G. Smith, "Marine Mammals", in *ibid.*, p. 56.
46. Peter Schledermann, "Polynyas and Prehistoric Settlement Patterns", *Arctic* V. 33, No. 2 (June, 1980), pp. 292–302.
47. Dunbar and Moore, *supra* note 23, p. 17.
48. For a concise description of Arctic climatic conditions see Department of Fisheries and Oceans, *Sailing Directions Arctic Canada*, V.1, Third edition (1982), Chapter VIII.
49. Smith and Rigby, *supra* note 43, pp. 11–12.
50. Lionel Johnson, "Assessment of the Effects of Oil on Arctic Fishes" (Paper prepared for the Arctic Research Directors Committee) (1982), p. 31.
51. Donald A. Blood, *Birds and Marine Mammals: The Beaufort Sea and the Search for Oil* (1977), p. 23.
52. Dunbar and Moore, *supra* note 23, p. 43.
53. Johnson, *supra* note 50, p. 31.
54. *Ibid.*, p. 32.

55. Allan R. Milne and Brian D. Smiley, *Offshore Drilling in Lancaster Sound: Possible Environmental Hazards* (1978), p. 45.
56. Blood, *supra* note 51, p. 87.
57. Beaufort EIS 3A, *supra* note 14, p. 3.3.
58. Johnson, *supra* note 50, p. 38.
59. Milne and Smiley, *supra* note 55, p. 54.
60. *Ibid.*
61. *Ibid.*, p. 55.
62. Johnson, *supra* note 50, p. 48.
63. Blood, *supra* note 51, p. 27.
64. Milne and Smiley, *supra* note 55, p. 55.
65. R.G.B. Brown and David N. Nettleship, "The Biological Significance of Polynyas to Arctic Colonial Seabirds," in Stirling and Cleator (eds), *supra* note 43, p. 59.
66. Arctic Pilot Project, *Gas Supplies and Markets* (1981), V. 2, p. 3–239.
67. R.W. Prach, H. Boyd, and F.G. Cooch, "Polynyas and Sea Ducks", *in* Stirling and Cleator (eds), *supra* note 43, p. 67.
68. Milne and Smiley, *supra* note 55, p. 65.
69. Blood, *supra* note 51, p. 27.
70. Prach, Boyd and Cooch, *supra* note 67, p. 69.
71. Dome Petroleum Ltd, Esso Resources Canada Ltd, Gulf Canada Resources Inc., *Environmental Impact Statement for Hydrocarbon Development in the Beaufort Sea–Mackenzie Delta Region* (1982), V. 1, p. 27.
72. Johnson, *supra* note 50, p. 51.
73. Blood, *supra* note 51, p. 35.
74. Stirling, Cleator and Smith, *supra* note 45, p. 45.
75. Milne and Smiley, *supra* note 55, p. 71.
76. *Ibid.*, pp. 69–71.
77. Rolph A. Davis, Kerwin J. Finley and W. John Richardson, *The Present Status and Future Management of Arctic Marine Mammals in Canada* (January 1980), pp. 25–27.
78. Johnson, *supra* note 50, p. 63.
79. Beaufort EIS 3A, *supra* note 14, pp. 3.8–3.9.
80. Johnson, *supra* note 50, p. 65.
81. Davis, Finley and Richardson, *supra* note 77, p. 32.
82. Environment Canada, *Canada's Special Places in the North: An Environment Canada Perspective for the '80s*; Department of Fisheries and Oceans, Arctic Offshore Development Committee,
"A Classification of Areas in the Canadian Arctic for Use in the Renegotiation of Oil and Gas Exploration Agreements", Working Paper 82–8 (August 1982). Also see J.G. Nelson and Sabine Jessen, *Planning and Managing Environmentally Significant Areas in the Northwest Territories: Issues and Alternatives* (1984). For a general discussion of special areas protection in Canada, see P.M. Taschereau, *The Status of Ecological Reserves in Canada* (1985).
83. Thomas R. Berger, *Northern Frontier, Northern Homeland, The Report of the Mackenzie Valley Pipeline Inquiry: Volume Two* (1978), p. 119.
84. Task Force on Northern Conservation, *Report of the Task Force on Northern Conservation* (December 1984).
85. Department of Indian Affairs and Northern Development, *supra* note 35, p. 13. Parks Canada proposes to develop a marine park in each of eight Arctic Ocean regions. See Parks Canada, *National Marine Parks Draft Policy* (Third Draft, August 1983).

86. Graham Rowley, "Circumpolar Peoples", *Nord* (January–February 1975), p. 6.
87. *Ibid*., p. 13.
88. Gordon Nelson and Sabine Jessen, *Regional Planning in the Beaufort: Scottish and Alaskan Parallels* (1981), p. 89.
89. North Slope Borough, *Coastal Management Program* (Public Hearing Draft) (1983).
90. Bill Hess, "First Whale of the Season", *Inuit*, Inuit Circumpolar Conference, 2/83, p. 4.
91. Ludger Muller-Wille and Pertti J. Pelto, "Political Expressions in the Northern Fourth World: Inuit, Cree, and Saami", *Northern Fourth World*, Inuit Studies, V. 3, No. 2 (1979), pp. 5–16.
92. Rowley, *supra* note 86, p. 10.
93. Lars Toft Rasmussen, "Europe's Reluctant Farewell to Greenland", *Inuit*, Inuit Circumpolar Conference, 2/83, p. 18.
94. Department of Indian Affairs and Northern Development, *In All Fairness: A Native Claims Policy* (1981). In 1985 the federal government created a task force to assist in developing a new policy for the negotiation of comprehensive claims, see Task Force to Review Comprehensive Claims Policy, *Living Treaties: Lasting Agreements* (1985).
95. Government guidelines restrict the Office of Native Claims to seven comprehensive claims negotiations at any one time. Labrador Inuit negotiations cannot commence until some of the existing negotiations are concluded.
96. Milton R. Freeman (ed.), *Report on the Inuit Land Use and Occupancy Project* (1976); and Carol Brice-Bennett (ed.), *Our Footprints are Everywhere: Inuit Land Use and Occupancy in Labrador* (1977).
97. Peter Jull and Nigel Bankes, "Inuit Interests in the Arctic Offshore", *National and Regional Interests in the North*. Canadian Arctic Resources Committee, (Proceedings of the Third National Workshop on People, Resources and the Environment North of 60°) (1984), pp. 557–86. For a discussion of what Inuit use of the sea ice could mean for a Canadian claim under international law, see David VanderZwaag and Donat Pharand, "Inuit and the Ice: Implications for Canadian Arctic Waters", *Canadian Yearbook of International Law* V. 21 (1983), pp. 53–84.
98. Peter Jull, *Nunavut*, Nunavut Constitutional Forum (1983).
99. Jull and Bankes, *supra* note 97.
100. For example, the first Inuit Circumpolar Conference was held at Barrow, Alaska (June 1977), the second at Nuuk, Greenland (July 1980), and the Third General Assembly of the Inuit Circumpolar Conference convened at Frobisher Bay (July 1983).
101. Nelson H.H. Graburn and B. Stephen Strong, *Circumpolar Peoples: An Anthropological Perspective* (1973), p. 137.
102. *Ibid*, p. 138.
103. Heather Myers, *The Use of Biological Resources by Certain Arctic and Subarctic Peoples* (M. Phil thesis, Scott Polar Research Institute, Cambridge) (June 1981), p. 11.
104. Peter Usher, "Fair Game", *Nature Canada* V. 11, No. 1, (1982), p. 10.
105. *Ibid.*
106. Myers, *supra* note 103, p. 13.
107. Beaufort EIS 3A, *supra* note 14, p. 3.66.
108. For a further discussion of the northern policy problem, see Chapter Six, *infra*, David L. VanderZwaag and Cynthia Lamson, "Northern Decision Making: A Drifting Net in a Restless Sea".

109. Department of Indian Affairs and Northern Development, *A Comprehensive Conservation Policy and Strategy for the Northwest Territories and Yukon* (Draft Discussion Paper) (October 1982).
110. Peter Jacobs, *People, Resources and the Environment* (August 1981), p. 3.
111. Special Committee of the Senate on the Northern Pipeline, *Marching to the Beat of the Same Drum* (March 1983), p. 1.

3

The development of Northern ocean industries

CARLYLE L. MITCHELL

I The problem outlined

Northern economic development, whether based on land or sea resources, has been problematic because serious developmental conflicts have not been resolved to the satisfaction of the North's principal interest groups. While corporate pressures promote non-renewable resource development, utilizing high technology and skilled or semi-skilled labour from the South, other groups advocate an entirely different development scenario, preferring development based on the exploitation of renewable resources by the indigenous population utilizing low technology. Government, having the responsibility for developing the North for the benefit of all Canadians, often finds itself in the role of an unwilling arbitrator between industry (large corporations), native peoples, and environmental protection groups. Northern development will, therefore, require making various choices both in terms of alternatives for development, and in resolving and reconciling conflicts between numerous interest groups.

The problems of Northern development are particularly acute in the ocean areas. Canada's Northern marine areas are especially rich in oil and gas resources. Exploration activities are taking place, or are proposed, for the Mackenzie Delta–Beaufort Sea, the High Arctic Islands, Lancaster Sound, and Baffin Bay–Davis Strait; that is, virtually all across the North. The Arctic as a marine region has the largest potential for recoverable oil and gas resources in Canada, but their commercial exploitation requires favourable demand and price conditions, and accelerated development of support technologies such as drill ships and artificial islands.[1] Mineral development is also taking place in the Arctic Archipelago; in the early 1970s a lead-zinc mine began production in Strathcona Sound[2] and, in 1982 Cominco's Polaris Mine began exploitation of the lead-zinc deposits on Little Cornwallis Island.

To date the Northern marine economy has been characterized by exploration for oil in the offshore islands, some mineral production in these islands, and by hunting, trapping and fishing of marine wildlife. Oil exploration has given rise to the concentration of service activities associated with this industry in a few Northern communities. For example, in the Southwestern Beaufort Sea region about 80 per cent of native people live in Inuvik, Aklavik and Tuktoyaktuk — communities in which the people earn more from wage and salary employment than from commercial hunting, trapping and fishing.[3] In contrast, around the periphery of the Beaufort Sea the native economy depends heavily on commercial hunting, trapping and fishing. Economic activities here are rudimentary, and bartering is still a primary mode of exchange. The Northern economy is, therefore, a classic "dual" economy, with a modern technological sector for minerals and oil and gas existing side-by-side with a traditional and non-industrial economy.

Such economic dualism raises a number of challenges. Exploration and exploitation of non-renewable resources, on land or sea, must not severely damage Northern ecosystems, for fish and wildlife provide food and livelihood for many Northerners. A greater integration in the economic structure is required to ensure that Northerners do not become "recognized bystanders to the technology and potential wealth of offshore drilling"[4] and that national economic growth is not thwarted.

Although a number of studies have focused on Northern economic development, few of these have concentrated on the ocean system. Therefore, this chapter offers an analytical framework for Northern ocean industrial development, examines the structure and performance of the Northern economy in the 1970s, and concludes by identifying some of the key factors which will constrain or propel future economic growth in Canada's North.

II An analytical framework for Northern ocean industries development

Despite considerable work on the development problems of the North, an analytical framework or model for the integration of ocean industries in the region has not emerged in the economics literature. It is not possible in a study of this nature to remedy this deficiency, but some of the main theoretical considerations can be outlined.

A The structure of the Northern economy

The North is an area with a series of unique developmental problems. The main reason for this is that Northern development embraces two features which are more pronounced in this region than in other areas of the world. First, the North is associated with an extremely sensitive ecosystem, with the result that environmental mishaps here are expected to have a longer term detrimental impact than elsewhere. Second, the vast size of the region, an inhospitable climate, and small population densities make development a difficult and costly undertaking. These features, particularly the latter, have been responsible for the characteristics and structure of the Northern economy.

The Northern economy is characterized by the economic dualism of an industrial economy and a native economy. The industrial economy consists of three sectors: government, corporate, and resident small business.[5] The corporate sector is based on the exploitation of non-renewable mineral and oil and gas resources, capital intensive and dependent on skilled or semi-skilled labour. Exploitation is conducted by large industrial organizations with their headquarters based in the South, either in Canada or the United States.

The native economy consists of two sectors: a subsistence or domestic sector, and a commodity or exchange sector. The subsistence sector produces goods for direct consumption, while in the exchange sector goods are produced for trade, either for other goods and services, or for cash.[6] The native economy is based on the exploitation of renewable natural resources of the region and, historically, was confined primarily to the subsistence domestic sector. The exchange economy began to develop with the advent of the white man to the North, and this development was fostered by government programmes for welfare and education, as well as by technological changes in hunting activities through the use of mechanized vehicles such as snowmobiles. As a result, the exchange sector has been the more dynamic of the two native sectors, and the differences between them are becoming increasingly blurred. Consequently, Northern communities have been classified as being: traditional (subsistence), transitional (exchange), or modern frontier (industrial).[7]

The structure of the economy, as shown in Figure 3.1, is a narrow one consisting of the following sectors:

(i) non-renewable resource industries;
(ii) government;
(iii) service (small business and limited government-sponsored manufacturing industries); and
(iv) renewable resource industries.

The interactions show the important role that government plays in all major sectors of the Northern economy. However, the interactions and direct linkages between the industrial and native economies are weak.[8]

B Theory and concept of Northern development

Northern development has been triggered by demands for Northern resources, mainly primary staples, from the South of Canada, the United States and other countries. From a theoretical standpoint there are two growth or development theories relevant to the North: first, the staples theory, associated with H. Innis, K. Buckley and M. Watkins, for explaining Canada's historical development,[9] and second, the export-

Figure 3.1. The structure of the Northern economy.

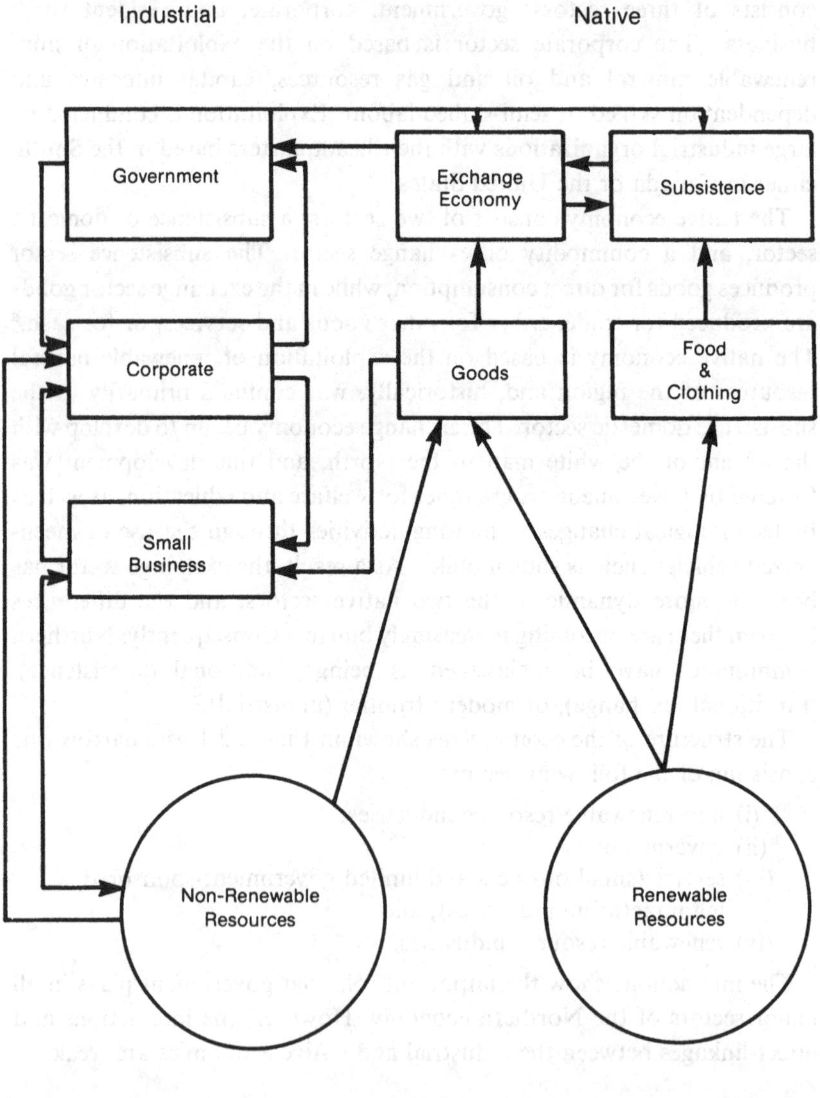

based theory of D.C. North for explaining regional development.[10] Both theories are similar in approach and converge when staples constitute the major exports. Since Canada's North depends on staple exports, the export-based theory will be the model used in this analysis.

The export-based theory revolves around the role that a strong export-based sector plays in a region. The theory refers to: the factors determining the level of absolute and per capita incomes; the factors affecting the cyclical sensitivity of the region and the factors shaping the growth of nodal centres. The export sector exerts direct and indirect effects on the region's economy by means of various linkages. The extent of these effects depends on the income elasticities of the region's exports, their diversity, and the structural diversity of the region. If income elasticities for exports are relatively high, that is elastic, and the structure of the economy is narrow, there is a greater likelihood of violent fluctuations in the export sector, which in turn will affect the cyclical sensitivity of the region. Exports also create nodal centres to cater to, or service, the export sector as a result of locational advantages. The existence of such centres can lower transport or service costs for exports.

The export-based theory is dynamic, helping to explain growth and change in the region. Growth is intertwined with the success of the export sector and with a broadening of the economic structure. However, a decline in staple exports may also cause a region to experience economic decay. Decay or decline can be caused by: changes in foreign demand, exhaustion of resources, increasing costs relative to those of a competing region, and changes in technology.[11] The theory is, therefore, useful in a descriptive sense (describing the growth or decline of a region), as well as in a prescriptive sense (using it as a basis for policy), since it can indicate reasons for the slow rate of growth or stagnation of a region and suggest possible solutions for the attainment of greater growth.

The export-based theory, as pointed out earlier, seems relevant to the North with its dependence on staple exports and its narrow economic structure. However, the use of this model in a Northern context must take into consideration the environmental constraints under which economic activities will have to operate. In fact, the environmental, economic and human development problems of the North indicate the need for a new concept of development for this region. A relevant concept is that of eco-development, which attempts to integrate ecology, economics and development.[12] Applied to the North, it can be defined as development based on the natural, human resource potential and the carrying capacity of the North. Conservation is perceived as an integral part of development in order to ensure the greatest benefit to Northern societies and least damage to the resources, that is, on a sustained yield basis for renewable resources. Given the concept of eco-development, the role that ocean industries can play in this development will be analysed within the context of the staple or export-based sector theory.

C *The role of Northern ocean industries*[13]

The role that ocean industries can play in the North depends on the impact that these industries can, through linkage effects, exert on each other and on other sectors in the Northern and Canadian economies. These linkages are threefold:[14]

- (i) *Backward*, where growth and development in ocean industries induce investment and employment in other sectors, providing inputs to ocean industries (for example, ship and oil rig construction, port facilities, manufacturing, construction and service industries);
- (ii) *Forward*, where ocean industries exert an influence on industries using ocean products as an input (for example, fish for food production and crude oil for energy); and
- (iii) *Demand*, where incomes generated by ocean industries stimulate increased demand for consumer goods and services.

Backward linkages will be felt primarily in the South, but forward and demand linkages will be felt primarily in the regional economy of the North. However, ocean industry development in the North will face a number of serious obstacles.

Ocean industries will be affected by their common property nature, the type of competition that results from this, and by environmental factors. Fisheries and oil and gas activities are based on the exploitation of common property resources. Sea lanes, inasmuch as they can be used by ships of various nations, can be considered international common property. The economic theory of common property resource exploitation, developed by H.S. Gordon[15] and others, indicates that if exploitation of these resources is uncontrolled, there is a tendency towards overexploitation and poor economic returns. Fortunately for the oil and gas industry, governmental arrangements such as leasing, and the high costs of technology, restrict or inhibit unlimited entry.

Apart from their common property nature, ocean industries are highly competitive. There is competition among them for space and resources, as well as for capital and labour inputs. In the North, for example, fisheries, oil and gas, and ocean transportation compete for the use of marine space. Finally, ocean industries tend to carry high environmental risks and exploitation may cause problems ranging from habitat degradation and overexploitation of the living resources, to polluting the marine environment and adversely affecting coastal zones. These industries, therefore, exert strong externalities and diseconomies on one another, and on other industries as well.

In addition to the aforementioned problems associated with ocean industries in general, these industries face particular impediments to their development in the North. Oil and gas exploration and production are costly because of climatic conditions and because of the specialized technology necessary to meet the requirements of Northern conditions. Transportation is also costly. Existing technology only permits shipping operations to be conducted for limited periods during the year. New and expensive technology such as supertankers and more powerful icebreakers will be needed in order to transport oil, gas and minerals produced in the North to markets in the South and elsewhere. Consequently, if oil and gas operations in the North are to become economically feasible, the following conditions are required:

(i) high and increasing world market prices;
(ii) increasing costs and decreased production in traditional and alternative production areas; and
(iii) the existence of large and easily accessible resource pools.

Because of these common property, environmental, and technical difficulties, development of Northern ocean industries will require a more highly integrated management regime than is necessary for most land-based industries.

III The Northern economy: a general perspective

This section provides a general description of the Northern economy and the significant changes that took place in this economy during the 1970s. Because Yukon has a relatively short ocean coastline, emphasis is given to the economy of the Northwest Territories.

A *Economic growth in the North*, 1971–1980

The Northern economy experienced substantial growth during the ten year period between 1971 and 1980. The Gross Domestic Product (GDP) of Yukon and the Northwest Territories increased from $228 million in 1971, to $923 million by 1980 in current dollars; from $469 million to $923 million in constant 1980 dollars; or at an average rate of 7.2 per cent a year (Table 3.1).

The population of Yukon and the Northwest Territories increased from 53,000 in 1971, to 69,000 by 1980. Real per capita incomes increased over this period from $8843 to $14,560, or at an average rate of 5.7 per cent per year. In comparison, GNP per capita for Canada increased from

$8994 in 1971, to $12,132 by 1980, or at an average rate of growth of 3.5 per cent per year. However, because of the narrow structure and dependence on resource industries, per capita incomes in the Northern economy fluctuated extensively during the period in comparison with the national economy.

Even though Yukon possesses a more diversified and developed economic structure, the Northwest Territories made the greatest contribution to the Northern economy, accounting for 72 per cent of the combined GDP for Yukon and the Northwest Territories for the period 1971–1977.[16] Primary resource industries have been the main engines of growth in both economies (Table 3.2).

The primary non-renewable resource industries, consisting mainly of base metal extraction and oil and gas exploration, expanded considerably during the period 1971–1980. In 1971 they accounted for 7 per cent of GDP in the Northern economy but by 1980 their contribution increased to 37 per cent of GDP. Growth in the non-renewable resource sector induced growth in other industries, such as construction, where there was an increase from 24 to 26 per cent of GDP. There was a significant change in the structure of the Northern economy with respect to goods-producing and service industries: goods-producing industries increased their share of

Table 3.1. *Gross domestic product and per capita gross domestic product Yukon and Northwest Territories and Canada, 1971–1980.*

	Yukon and NWT*				Canada**			
Year	Pop. '000	GDP (current $m)	GDP (constant 1980 $m)	GDP Per capita (constant 1980 $)	Pop. '000	GNP (current $m)	GNP (constant 1980 $m)	GNP Per capita (constant)
1971	53	228.2	468.7	8843	21568	94450	194000	8994
1972	57	279.5	552.0	9684	21802	105234	207837	9532
1973	60	378.7	696.8	11613	22043	123560	227350	10314
1974	60	492.3	814.3	13572	22364	147528	244011	10911
1975	63	482.6	720.0	11428	22697	165343	247353	10898
1976	64	482.5	672.1	10501	22993	191031	266106	11573
1977	65	607.1	787.4	12114	23287	208868	270902	11633
1978	67	703.0	849.9	12685	23534	230490	278662	11840
1979	67	848.0	938.7	14010	23769	261576	289564	12182
1980	68	990.0	990.0	14559	24058	291869	291869	12132

* *Source:* Statistics Canada, *Provincial Economic Accounts*, Cat. 13–213 (annual).
** *Source:* Statistics Canada, *National Income and Expenditure Accounts*, Cat. 13–201 (annual).

Table 3.2. *The contribution of selected industries to the Yukon and Northwest Territories gross domestic product 1971–1980 (in millions of dollars).*

	Goods-producing sectors					Services		
	Primary		Secondary			Tertiary		Total GDP
Year	Hunting trapping and fishing	Minerals and oil and gas	Manufacturing and food and beverages	Construction	Utilities electricity gas and water	Accommodation and food	Business government and other services*	
1971	1.6	15.1	2.5	56.9	10.5	10.6	131.0	228.2
1972	1.7	19.0	3.2	88.2	11.7	11.8	143.9	279.5
1973	2.9	84.1	3.0	107.3	11.8	14.2	155.4	378.7
1974	2.5	153.3	3.0	130.6	10.1	17.6	175.2	492.3
1975	2.2	104.8	4.7	121.8	13.9	18.7	216.5	482.6
1976	3.1	68.0	5.1	134.7	17.6	21.2	232.6	482.5
1977	3.8	155.2	5.5	165.5	28.3	24.3	225.5	608.1
1978	4.9	215.2	7.3	146.7	31.8	28.6	268.5	703.0
1979	5.5	262.2	8.1	179.3	34.5	33.7	324.7	848.0
1980	5.6	368.0	8.3	259.5	40.6	36.8	271.2	990.0

* Includes other primary and secondary activities not listed.
Source: Statistics Canada, *Provincial Gross Domestic Product by Industry*, Cat. 61-202 (annual).

GDP from 38 to 68 per cent, while the service industries' share was reduced from 62 to 32 per cent. However, government, banking and business services combined constituted the largest and most stable sub-sector of the Northern economy during the period.

B Employment and incomes in the Northern economy

The population of the North grew at an average rate of 3 per cent a year for the period 1971–1980, with the Northwest Territories population (which accounted for 66 per cent of the overall population) experiencing a growth rate of 3.1 per cent, in comparison with 2.8 per cent for Yukon. These growth rates, which are higher than the national average of 1.2 per cent a year, were brought about by the natural rate of growth of the native Indian, Inuit and white populations, and by the immigration of others (mainly whites) from the South and from other countries. In 1980 others or whites constituted the largest single group in the North — approximately 78 per cent in Yukon, and 47 per cent in the Northwest Territories. The changes in the ethnic distribution of the population in the Northwest Territories between 1971 and 1980 are shown in Table 3.3.

The table shows that the Inuit and Others categories experienced the highest growth rates: 3.6 and 3.3 per cent respectively. Since the white population rate of natural increase is nearly zero for Canada as a nation, the increased proportion of whites has been a consequence of migration to the Northwest Territories during the period. This migration of primarily working-age people, having an age structure highly skewed towards younger people, has led to a disproportionately large work force (see Table 3.4).

Table 3.3. *Ethnic distribution of population, Northwest Territories 1971 and 1980.*

Ethnic group	1971		1980		Growth rate 1971–1980
	No.	%	No.	%	%
Indian	7190	20.6	8488	18.5	1.7
Inuit	11416	32.8	15646	34.1	3.6
Others	16219	46.6	21748	47.4	3.3
Total	34805	100.0	45882	100.0	3.1

Sources: C. M. Lu and D. C. Emerson Mathurin, *Population Projections of the Northwest Territories to 1981* (1973) and Government of the Northwest Territories, *Statistics Quarterly* (1981).

The working force in the North has two components — largely unskilled and relatively immobile native workers who usually reside in small, scattered settlements close to traditional natural resource abundance areas (Figure 3.2), and relatively highly-skilled and mobile non-native workers, many of whom are migrants from Southern Canada, who usually reside close to industrial activities or in the cities and towns. As a consequence of these differences, non-native workers have been dominant in the non-renewable resource industries, in higher government echelons, and in small businesses; that is, in typically higher-income jobs. The native workers, despite corporate attempts to encourage greater participation in industrial activity, predominate in the renewable resource industries and in the lower levels of government services. Since the majority of non-native workers only move to the North if they are assured of a job, this group tends to be fully employed. The native labour force, on the other hand, bears the brunt of the North's unemployment problem.

Government services, such as public administration, defence, education, and health and welfare, provided the largest source of employment in the Northern economy, accounting for 40 per cent of occupational distribution of the labour force in 1971, and 47 per cent in 1981 (Table 3.5). Only 15.3 per cent of the labour force in 1971, and 12.3 per cent in 1981, were employed in primary industries,[17] with mining accounting for the majority of workers. Less than 3 per cent of the labour force were employed in manufacturing, while 6 per cent were employed in the construction sector. In 1981, about 74 per cent of the labour force were classified as service-sector workers.

The Northern economy did not provide full employment for its labour force during the period 1971–1980, although the level of labour force

Table 3.4. *Age distribution of Yukon and Northwest Territories, 1980.*

	Yukon	NWT
	(Per cent of population)	
Infants and school age (0–14 years)	29.4	35.1
Working age 15–64 years)	65.9	61.8
Elderly (65 and over)	4.7	3.1

Sources: Government of the Yukon, *Yukon Economic Review* (1981) and Government of the Northwest Territories, *Statistics Quarterly* (1981).

Figure 3.2. Communities North of 60°.

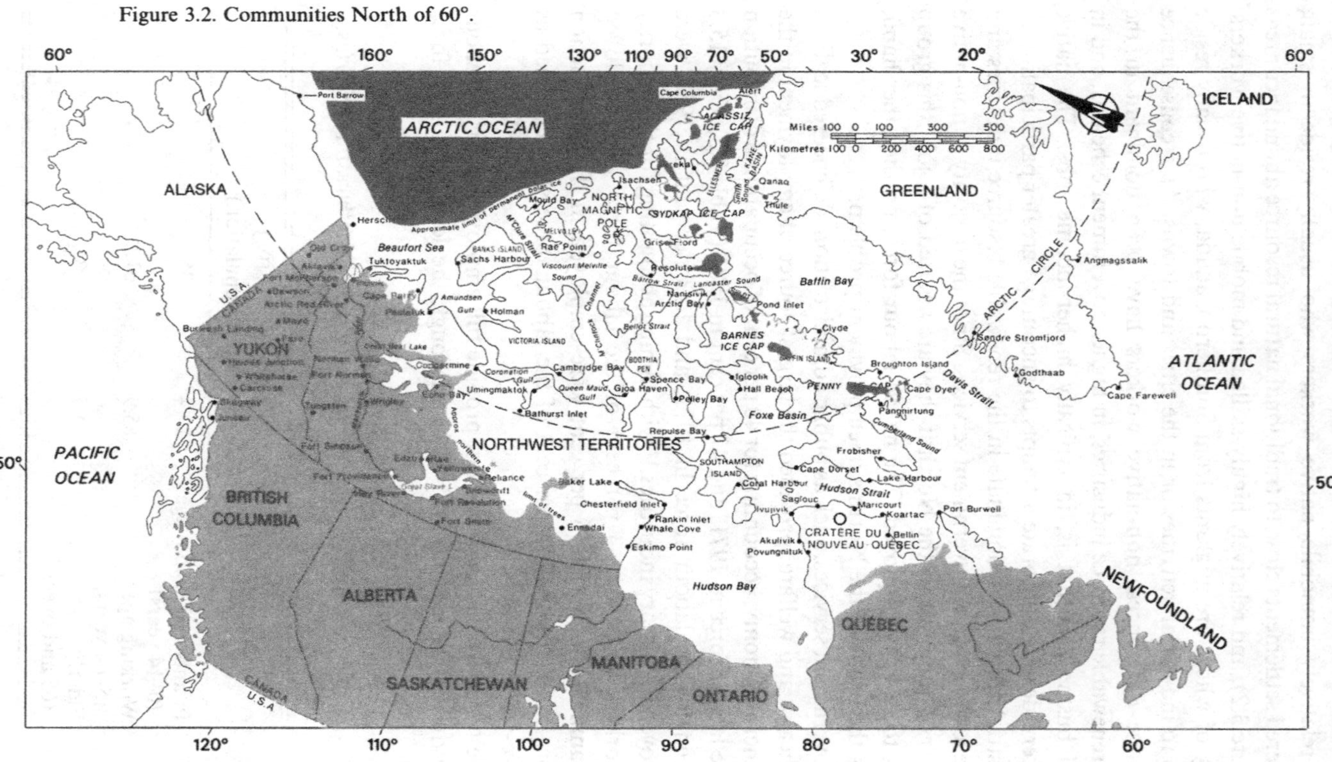

Source: Adapted from Department of Indian Affairs and Northern Development, *North/Nord* (Spring 1983), p. 70.

Table 3.5. *Labour force by industry, Northern economy 1971 and 1981.*

Industry group	Yukon				Northwest Territories				Northern economy			
	1971 No.	% of total	1981 No.	% of total	1971 No.	% of total	1981 No.	% of total	1971 No.	% of total	1981 No.	% of total
Agriculture and forestry	80	1.0	155	1.2	115	1.0	95	0.5	195	1.0	250	0.8
Fishing, hunting and trapping	25	0.4	40	0.4	370	3.4	235	1.2	395	2.1	275	0.8
Mines, quarries and oil wells	1160	14.2	1330	10.1	1180	10.8	2145	11.1	2340	12.2	3475	10.7
Manufacturing	155	1.9	195	2.2	335	3.1	420	2.2	490	2.6	715	2.2
Construction	555	6.8	910	6.9	450	4.1	980	5.1	1005	5.3	1890	5.8
Transportation, communications and utilities	1165	14.3	1590	12.1	1110	10.1	2085	10.8	2275	11.9	3675	11.4
Trade	885	10.8	1710	13.0	905	8.3	2095	10.9	1790	9.4	3805	11.8
Finance, insurance and real estate	185	2.3	495	3.8	110	1.0	690	3.6	295	1.5	1185	3.7
Community, education, health and welfare	1810	22.2	3100	23.6	2380	21.7	4725	24.5	4190	21.9	7825	24.1
Public administration and defence	1035	12.7	2555	19.4	2475	22.6	4970	25.8	3510	18.4	7525	23.2
Unspecified	1095	13.4	955	7.3	1525	13.9	830	4.3	2620	13.7	1785	5.5
Total	8150	100.0	13135	100.0	10955	100.0	19270	100.0	19105	100.0	32405	100.0

Source: Statistics Canada, 'Labour Force by Industry, Demographic and Educational Characteristics', Cat. 92–921 (1971, 1981).

participation was high. The 1976 census showed that the labour force participation rate was 71.1 per cent for Yukon, and 64.8 per cent for the Northwest Territories, in comparison with a 60 per cent rate for Canada on the whole.[18] Unemployment rates, which fluctuated on an annual basis during the period, were higher in Yukon than in the Northwest Territories. In 1976 these rates were 10.2 per cent for Yukon, and 5.1 per cent for the Northwest Territories.[19] Although these rates were not abnormal during the period by Canadian standards, they have serious implications for the native labour force. For example, the 10 and 5 per cent unemployment rates for Yukon and the Northwest Territories respectively mean 45 and 10 per cent unemployment rates for the native labour force.[20]

Incomes are relatively high in the Northern economy and substantial increases were realized during the period 1971–1980 (Table 3.6). Personal disposable incomes increased from $122 million to $537 million, or at an average rate of growth of 17.9 per cent a year, in comparison with a labour force growth rate of 6.1 per cent a year. Thus, per capita disposable incomes increased substantially and this resulted in increased savings: personal savings increased from a dissavings situation of -2.5 per cent of personal incomes in 1971, to about 16 per cent by 1980. However, savings were low in comparison to the investment underlying the growth rates experienced and indicates the dependence of the Northern economy on outside capital from the South and from other countries.

Table 3.6. *Personal income, Yukon and Northwest Territories 1971–1980.*

Year	Personal income	Per cent	Personal disposable income	% of personal income	Personal savings	% of personal income
	(millions of dollars)					
1971	158	100	122	77.2	−4	−2.5
1972	193	100	150	77.7	10	5.2
1973	235	100	183	77.9	15	6.4
1974	293	100	221	77.5	19	6.5
1975	346	100	269	77.7	33	9.5
1976	398	100	313	78.6	51	12.8
1977	460	100	356	77.3	57	12.4
1978	514	100	400	77.8	56	10.9
1979	583	100	455	78.0	77	13.2
1980	680	100	537	78.9	107	15.7

Source: Statistics Canada, *Provincial Economic Accounts 1966–1981* Cat. 13–213 (Annual) Table 10.

IV Northern waters industrial development in the 1970s

Before 1970 economic activities in Arctic waters concentrated primarily on the renewable resources of marine mammals and fish. In the case of the fishery, most development was based on inland or freshwater fisheries rather than on sea fisheries. The Arctic region is rich in mineral and oil and gas resources, but the exploitation of these resources is highly dependent on world demand and prices, as well as on production and transportation costs. High production and transportation costs to major markets from the Arctic have led to the exploitation of only high-grade mineral, large-pool oil and gas resources in this region. This section examines the changes in utilization during the 1970s for both renewable resource industries and non-renewable resource industries.

A The renewable resource industries

The large expanse of Arctic waters and the relative lack of commercial activity in this region explain the dearth of management measures (except licensing for commercial and recreational activities) and biological studies on the extent and status of major stocks. However, it is estimated that most marine mammals are highly exploited, close to or beyond their maximum sustainable yield (MSY), with the exception of some seal species (ringed, bearded and harp seals) and white whales.[21] Most fish species are underexploited and some are not exploited at all.

In the subsistence economy, renewable resources are exploited for food, fuel and clothing. In the exchange economy, they are exploited for recreational purposes and for their commercial value. Their exploitation, along with government transfer payments, are the major means of livelihood in the subsistence-exchange native economy. Virtually the same people fish, hunt and trap. Their level of effort varies with seasonal availability of marine mammals and fish resources, fluctuations in resource abundance caused by environmental conditions, and fluctuations in economic conditions.

There is considerable controversy over the contribution that marine mammals and fish make to the subsistence and native exchange economy.[22] They undoubtedly provide the main source of food and clothing in the scattered native communities, and contribute significantly to meeting food needs in other communities, since they are the cheapest source of animal protein. In the exchange economy, major mammal and fish resources probably contribute between 10 and 40 per cent of total incomes depending on marine community location in relation to resources and industrial activity.[23] The economic importance of renewable resources may be analysed under two categories: commercial activities and recreational activities.

1. *Commercial activities*

Commercial hunting, trapping and fishing for renewable resources made only a small contribution to GDP and employment in the Northern economy. The value of output from these operations increased from $ 1.6 million in 1971 (0.7 per cent of GDP), to $ 5.6 million (0.5 per cent of GDP) in 1980 (Table 3.7).

In terms of employment, many Northerners participate in renewable resource exploitation. For the Northwest Territories the number of trappers (which also include the majority of fishermen) included about 6000 in 1971 and about 4000 in 1981. Gross earnings in 1981 amounted to $ 5 million, of which $ 1 million (or 20 per cent) was derived from the sale of Arctic Ocean marine mammal products such as seal skins, walrus and narwhal tusks, and whale meat.[24]

Earnings from hunting and trapping are low and fluctuate widely from year to year depending on resource availability, market demands and prices. A large proportion of hunters and trappers operate seasonally for short periods during the year. As a result, only a small proportion earn adequate incomes: in 1981 there were only 68 trappers in the Northwest Territories with incomes of over $ 8000 a year.[25] However, some hunters and trappers fish commercially to supplement their incomes.

Table 3.7. *Value of renewable resource industries, Northern economy 1971–1980.*

Year	Commercial fisheries	Hunting and trapping	Total value
		($ million)	
1971	1.3	0.3	1.6
1972	1.3	0.4	1.7
1973	1.5	1.6	3.1
1974	1.5	1.0	2.5
1975	1.5	0.7	2.2
1976	1.6	1.5	3.1
1977	2.0	1.8	3.8
1978	2.3	2.6	4.9
1979	3.2	2.3	5.5
1980	3.2	2.4	5.6

Source: Statistics Canada, *Provincial Gross Domestic Product by Industry*, Cat. 61–202 (Annual).

The commercial fishery in the North is affected adversely by the following:[26]

(i) distance from major markets and consequent high cost of product;
(ii) limited local market since demand is met by subsistence fishing; and
(iii) undercapitalization, unreliable equipment, and unskilled personnel.

To compensate for these factors, commercial fishing operations have concentrated on the more accessible areas and on high-value species. Since there are few areas in the North where these conditions are met, commercial fishing operations have been sporadic and generally unsuccessful. The three main areas in the Arctic Ocean where commercial fishing (mainly for Arctic char and sea-run lake trout) is carried out are the Mackenzie Delta and Beaufort Sea; the Central Arctic and Victoria Island; and the Eastern Arctic and Baffin Island.

There was no growth in output during the 1970s since landings fluctuated between 1000 and 2000 tonnes. The landed and marketed values of fish increased during the period, however, as a result of increased prices. Landed values increased from $ 1.1 million to $ 1.8 million, and market value from $ 1.3 million to $ 3.2 million.[27] The production of Arctic char fluctuated widely during the period 1970–1980 and accounted for only a small proportion (less than 10 per cent) of total volume of landings from Yukon and the Northwest Territories. However, because it is the most highly priced species, it accounted for about 20 per cent of the value of landings and marketed value in 1980.

There is little processing of fish done in the Northwest Territories but two or three plants operated in the Arctic during the 1970s. Processing activities had involved the canning of Arctic char and some marine mammals (for example, whale and seal meat), but because of low returns processing activities now concentrate on the freezing of Arctic char. Processed products are exported and marketed by the Freshwater Fish Marketing Corporation (FFMC), a federal crown corporation which has an exclusive mandate for marketing fish exported from the Northwest Territories.[28]

Fishing effort statistics are not available for the period 1971–1980. However, during this period, between 300 and 400 commercial fishermen utilizing small boats equipped with outboard engines operated on a yearly basis in the North.[29] The majority of fishermen (about 80 per cent) were in the inland fishery (Great Slave Lake) rather than the sea fishery, and received low incomes and returns for their fishing activities. Assuming, for example, 350 commercial fishermen operated in the Northwest Territories in 1980, the gross output per fisherman would average about $5100 and net income would be about $3800. Returns are lowest in Arctic marine fisheries because of higher transportation costs to the FFMC market.[30] Commercial operations for fish and marine mammals in the Canadian Arctic are, therefore, not substantial in monetary terms.

2. *Recreational activities*

Recreational or tourist activities in the North are highly associated with renewable resources and the uniqueness of the Northern environment. Fishing is the main attraction for tourists, and their recreational activities take place mainly in the inland areas (on lakes and streams) of Yukon and the Northwest Territories. The recreational fishing industry in the North is characterized by:

(i) a primary fishing sector where Arctic char, grayling and walleye are the major species sought during the summer months of July and August;
(ii) a direct service sector dependent on the provision of access to the fishery, such as lodges, marinas and outfitters' establishments; and
(iii) an indirect service sector, such as service stations, hotels and restaurants.[31]

The recreational fisheries, through various sectoral linkages, are able to exert a significant contribution to Northern development. In the Northwest Territories, there was substantial growth in recreational fisheries during

Table 3.8. *Licence sales and direct expenditures on recreational fishing, Northwest Territories 1971–1980.*

Year	Resident licences	Non-resident licences	Total licences	Direct expenditures ($ million)
1971–1972	3346	3238	6584	1.3
1972–1973	4772	3352	8124	1.9
1973–1974	5742	3602	9344	2.5
1974–1975	6723	3945	10668	3.3
1975–1976	7716	3659	11375	4.1*
1976–1977	7593	3881	11474	4.6
1977–1978	10198	4264	14462	6.6
1978–1979	9590	3938	13528	6.9
1979–1980	9175	4011	13186	7.6
1980–1981	10656	4468	15124	9.8*

* Actual survey data from DFO, Survey of Sport Fishing in Canada (1975 and 1980).

Source: D. Topoloniski, *Regional Income Analysis of Northwest Territories Fishing Lodges* (unpublished study, Department of Fisheries and Oceans, Western Region) (1982).

the period 1971–1980. The number of anglers increased during this period from 6000 to 15000 or at an average rate of 6.7 per cent per year. Expenditures rose from $ 1.3 million to $ 9.8 million, or at an average rate of 22 per cent per year (Table 3.8).

Direct expenditures were for package tours, food, lodging, and travel costs. Most of these expenditures were retained in the Northwest Territories, although there were some leakages. The impact of these expenditures can be realized from the lodge operations which were studied in 1980. It was found that there were 40 sport-fishing lodges in operation in the Northwest Territories. Three of these, accounting for 5 per cent of capacity, were situated adjacent to Arctic waters. Gross revenues from the lodges amounted to $ 7.3 million, and gross profits of $ 2 million were realized. The economic impact of these lodges to the Northwest Territories economy was found to be as follows:

Value added	$2.3 million
Wages & salaries	$1.5 million
Employment	536 individuals, 264 of whom were residents of the Northwest Territories.[32]

Recreational fisheries, therefore, contribute more to the GDP of the Northwest Territories than the commercial fisheries.

B *The non-renewable resource industries*

1. *The oil and gas industry*

Reconnaissance programmes, both geological and geophysical, for Canadian Arctic oil and gas began in the 1950s in response to a world-wide search for new sources of petroleum. These initially concentrated on land areas, for example Norman Wells and Pointed Mountain in the Northwest Territories, where production was already taking place, but moved offshore during the 1960s. Surveys or exploratory drilling began in the Arctic Islands in 1962, in the Beaufort Sea in 1965, in Hudson Bay and Hudson Strait in 1966, and in Davis Strait and South Baffin Bay in 1969. Discovery of large reserves (12 billion barrels) in Prudhoe Bay in 1968 and increasing world prices for oil and gas from 1972 onwards led to increased drilling activity in the Arctic during the 1970s (Figure 3.3). Exploration effort concentrated on three principal areas: the Beaufort Sea, the Arctic Islands and Davis Strait. The extent and location of offshore activity in Northern waters during 1982 is illustrated by Figure 3.4.

The total production of crude oil and natural gas in the Northern economy during the period 1971–1980 is shown in Table 3.9. The total value of crude petroleum and natural gas increased from $ 1.4 million in 1971 to a peak of $ 39 million in 1977, but declined to $ 11 million in 1980.

The changes in production during the period were brought about by natural gas production which increased in volume and value up to 1977, but declined sharply because of reduced production in 1979 and 1980. Crude petroleum production fluctuated in volume from 141,000 cubic metres to 160,000 cubic metres, but, due to increased oil prices, increased in value from $ 1.2 million in 1971 to $ 10.8 million in 1980.

Crude petroleum is refined into a variety of products (gasoline, stove oil, diesel fuel, heating oil and heavy fuel oil) at the Esso refinery at Norman Wells. This refinery is located on one of the main transport routes in the North, the Mackenzie River. By late spring 1985, a new pipeline from Norman Wells to Zama, Alberta is expected to transport approximately 4400 cubic metres of oil per day.

Figure 3.3. World oil prices, 1960–1983.

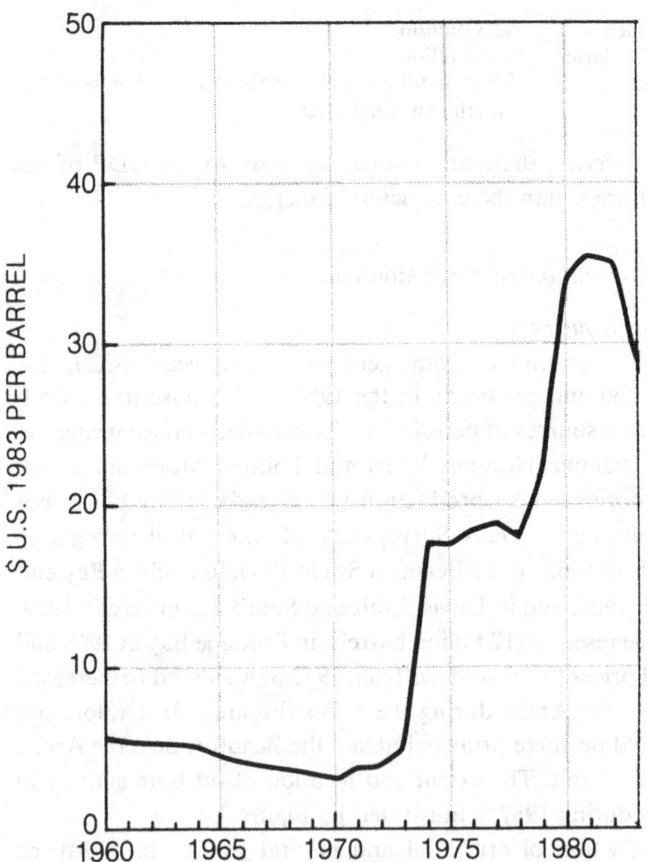

Source: National Energy Board, *Canadian Energy Supply and Demand* 1983–2005, *Summary Report* (September 1984), p. 3.

Figure 3.4. Location of wells drilled for the calendar year 1982.

Legend

- Plugged and abandoned
- Oil well
- Suspended for re-entry
- Gas well
- Drilling
- Suspended for re-entry (oil well)

Moratorium area

Administrative Line of Convenience between the Department of Energy Mines and Resources (below line) and Indian and Northern Affairs (above line)

1 Amoco Pointed Mountain L-68
2 Paramount et al Big Island E-54
3 Esso Norman Wells development
4 Dome Natiak O-44 (glory hole)
5 Dome Superior Orvilruk O-03
6 Gulf et al E. Tarsiut N-44
7 Gulf et al E. Tarsiut N-44A
8 Gulf et al Kiggavik A-43
9 Esso et al Itiyok I-27
10 Dome Hunt Irkaluk B-35
11 Dome Hunt Kenalooak J-94
12 Dome Aiverk 2 I-45
13 Dome Nerlerk M-98
14 Dome Texaco et al Uviluk P-66
15 Esso et al W. Atkinson L-17
16 Esso et al Natagnak O-59
17 Chevron Muskox D-87
18 Panarctic et al Marryatt K-71
19 Panarctic et al Cape Mamen F-24
20 Panarctic et al Whitefish A-26
21 Panarctic et al Cisco C-42
22 Panarctic et al Cisco K-58
23 Panarctic et al Sculpin K-08
24 Canterra et al Ralegh N-18
25 Petro-Canada et al Rut H-11
26 Petro-Canada et al Pothurst P-19
27 Petro-Canada et al Corte Real P-85
28 Mobil et al Linnet E-63
29 Mobil et al Bonanza M-71
30 Mobil et al North Dana I-43
31 Mobil et al West Flying Foam L-23
32 Mobil et al Nautilus C-92
33 Mobil et al Hibernia I-46
34 Mobil et al Hibernia J-34
35 Petro-Canada West Esperanto B-78
36 Petro-Canada et al North Banquereau I-13
37 Petro-Canada et al Banquereau C-21
38 Mobil et al Bluenose G-47A
39 Mobil et al Olympia A-12
40 Mobil et al Venture B-43
41 Mobil et al South Venture O-59
42 Shell et al Shubenacadie H-100

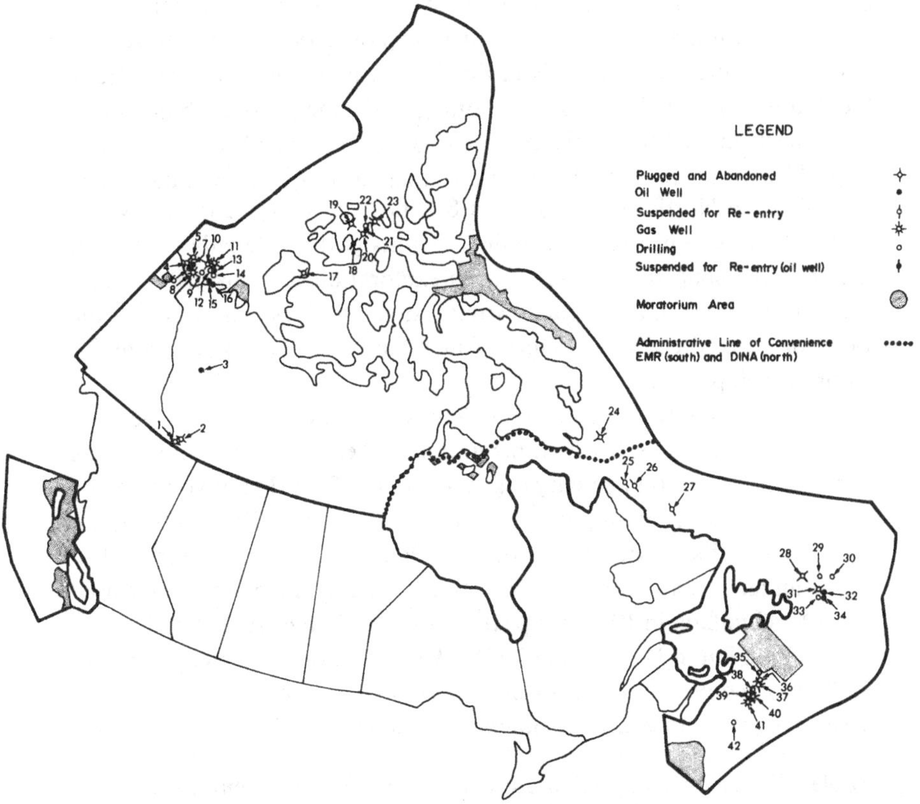

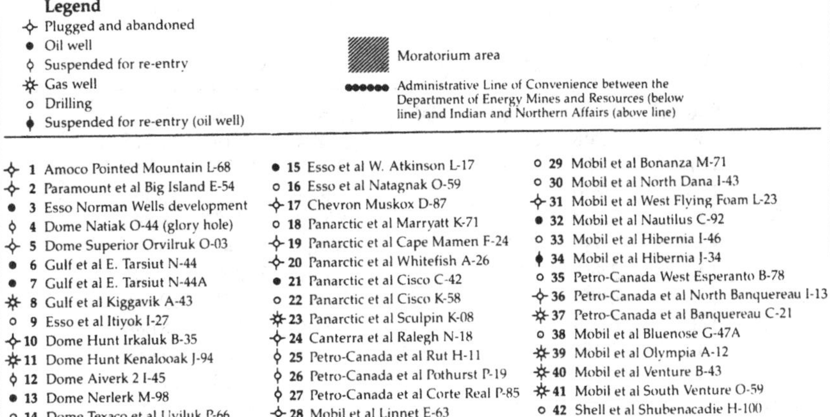

Source: Canada Oil and Gas Lands Administration, *Annual Report* (1982).

The most significant impact of oil and gas on the Northern economy has been made through exploratory activity. Considerable capital investment has been made, not only in direct exploratory activity but also in developing infrastructure. This has induced a movement of labour from the South and provided jobs and training to the indigenous labour force, particularly in the Mackenzie Delta. It has, however, resulted in a number of problems and concerns. For example, there is the question of the extent to which the North and its peoples benefit from these activities. Many believe that the benefits accrue mainly to other parts of Canada or abroad. There is the concern that oil and gas exploration may disturb a rather sensitive ecosystem, as well as disrupt native lifestyles. There is also the question of land claims settlement which has important implications for future development.

It had been known for some time that the Arctic offshore possesses vast reserves of oil and gas, but their magnitude and extent were unknown until substantial exploratory activity began in the late 1960s. This involved many companies, the most important of which are Dome Petroleum, Gulf, Imperial Oil (Esso), Petro-Canada, and Panarctic. The petroleum industry's annual expenditures on exploration activities in the North increased from $149 million in 1971 to a peak of $473 million in 1978, but declined to $237 million in 1980 (Table 3.10). The downturn in expenditures since 1978 was the result of the government's decision, following the Berger inquiry in 1977, to delay the building of the pipeline.[33]

Table 3.9. *Production of oil and gas in the Northern economy 1971–1980.*

	Crude petroleum		Natural gas				
	NWT		Yukon		NWT		Total value
Year	Q	V	Q	V	Q	V	V
	000 m³	$000	m m³	$000	m m³	$000	$000
1971	150	1208	25.3		8.4	117	1415
1972	141	1059	73.6		333.9	1372	2710
1973	153	2240	—		1044.1	—	6564
1974	152	3167	32.2		894.8	5.537	8894
1975	160	4537	53.0		876.2	19742	25589
1976	142	3461	23.6		893.2	32899	37281
1977	138	4295	8.9		759.5	34925	39658
1978	146	6263	—		593.0	32423	38686
1979	142	7455	—		12.0	409	7864
1980	160	10785	—		8.2	384	11169

Source: Statistics Canada, *The Crude Petroleum and Natural Gas Industry*, Cat. No. 26–213 (Annual).

About 90 per cent of petroleum industry expenditures were for exploratory operations, while development accounted for 7 per cent, and operating costs, royalties and returns accounted for 3 per cent.[34] Gross revenues to Yukon and the Northwest Territories from licence, permit, transfer, lease, rental and royalty fees increased from $ 5.3 million in 1971 to $ 11.2 million by 1980, with over 90 per cent of this accruing to the Northwest Territories. The increased expenditures for exploratory operations reflect higher costs per well drilled, since the number of wells drilled has been declining since 1973 (Figure 3.5). The higher costs per well drilled have been the result of operations moving offshore from more accessible drilling sites on land (see Table 3.11).

Table 3.11 shows that the Arctic offshore became the predominant area for exploratory drilling during the 1970s. Drilling commenced in this area in 1973 and, since then, expenditures increased from $ 10 million to a peak of $ 330 million in 1978, but declined to $ 53 million by 1980. Drilling costs are much higher in the offshore than onshore: for example, the costs per metre of well drilled were over six times as high as those on land during the last four years of the period.

Oil and gas exploration and development activities are capital, rather than labour, intensive. Since virtually all the technology and most of the labour skills required are not available in the North, only the expenditures for infrastructural works, wages and salaries made an impact on the

Table 3.10. *Oil and gas expenditures and revenue, Yukon and Northwest Territories, 1971–1980* ($000's)

Year	Oil and Gas exploratory expenditures	Well drilling expenditures	Gross revenues		
			N.W.T.	Yukon	Total
1971	149039	76807	5310	475	5785
1972	221842	127575	4925	534	5459
1973	266426	190512	6156	440	6596
1974	253791	174521	5175	434	5609
1975	234556	173950	7305	385	7690
1976	292825	222916	7610	249	7859
1977	276567	237174	8176	221	8397
1978	473411	437196	8642	346	8988
1979*	413446	388056	10762	344	11106
1980*	236870	193457	10723	456	11176

* Estimated
Source: Department of Indian Affairs and Northern Development, *Oil and Gas Activities* (1976; 1980).

Figure 3.5. Wells drilled, Yukon and Northwest Territories, 1969–1980.

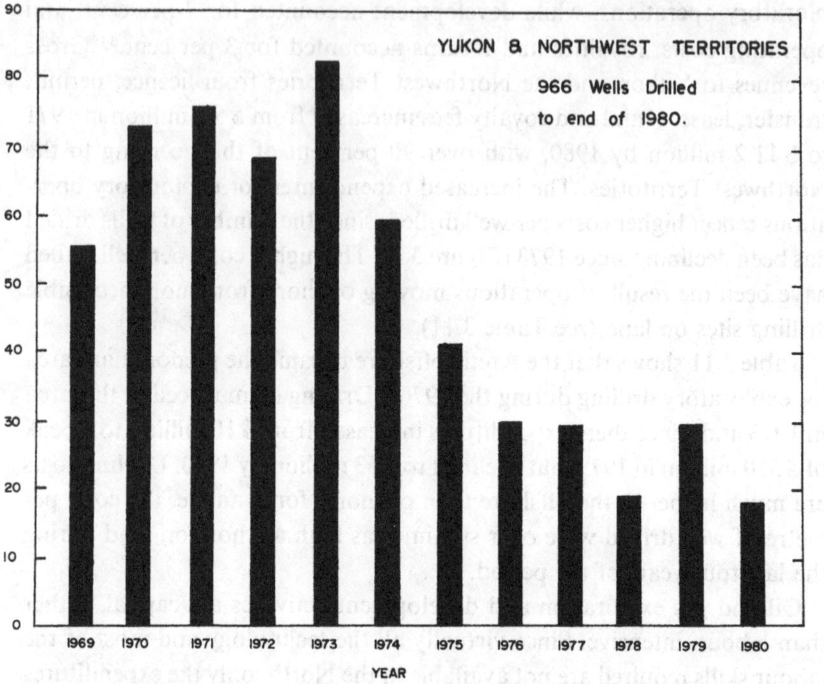

YUKON & NORTHWEST TERRITORIES
966 Wells Drilled
to end of 1980.

Source: Department of Indian Affairs and Northern Development, *Oil and Gas Activities*, 1980 (1981).

Table 3.11. *Expenditures and cost per metre of wells drilled, onshore and offshore, Northwest Territories 1971–1980.*

	Onshore				Offshore			
Year	No. of wells	Total cost ($000)	Metres drilled	Cost per metre ($)	No. of wells	Total cost ($000)	Metres drilled	Cost per metre ($)
1971	10	15379	27032	569	0	—	—	—
1972	15	47090	49990	942	0	—	—	—
1973	28	67374	79324	849	1	10503	2708	3879
1974	19	50800	50614	1005	2	13181	7094	1858
1975	18	57840	52253	1107	5	59074	16400	3602
1976	12	57272	42364	1352	6	115032	17861	6440
1977	13	60672	42292	1435	5	139791	16036	8717
1978	3	21239	8313	2555	4	330946	13974	23683
1979	1	3233	1960	1650	5	276571	18019	15349
1980	1	6440	3093	2082	2	53599	4232	12665

Source: Department of Indian Affairs and Northern Development, Northern Programs Branch (unpublished data).

Northern economy. The geographic distribution of expenditures for salaries, purchases and services in the Northwest Territories for the period 1975–1979 indicated that only 11 per cent were retained in the local economy; 76 per cent went to other parts of Canada, and 13 per cent went to other countries.[35] There is a 100 per cent leakage, as far as the North is concerned, for capital expenditures for heavy machinery and equipment, and, since most of the Arctic drill ships, platforms and icebreakers were imported from abroad, a considerable proportion of these expenditures also constituted a leakage from the Canadian economy.

Direct employment in the oil and gas industry is not high, but average incomes are. In 1979 there were just over 1000 full-time employees in the industry, but their incomes averaged $ 23000 (Table 3.12). However, the industry benefited the non-native labour force more than the native labour force due to the fact that the native community was largely unskilled, and wage-earning opportunities were concentrated in initial clearing and construction activities. Employment opportunities for natives were, therefore, generally high during the early construction phase, but tapered off as more technical expertise was required. Although oil company efforts since the early 1970s to encourage greater participation of natives have been responsible for increased employment and stability, only 235 native workers were employed by the industry on a full-time basis in 1979.[36]

2. *The hard mineral industry*

The large and diversified reserves of hard minerals in the North have been exploited for many years. As a result, this is the North's largest and most

Table 3.12. *Employment in oil and gas industry 1975–1979.*

Year	No. of native employees	Total direct employees	Wages and salaries ($000)	Average wage per year ($)
1975	23	347	3680	10605
1976	136	885	11929	13479
1977	201	943	15711	16661
1978	197	1030	19080	18524
1979	235	1113	26305	23634

Source: The Oil and Gas Industry in the Northwest Territories 1975–1979. Prepared for the Department of Indian Affairs and Northern Development by Price Waterhouse Associates (1980).

important resource industry and has been responsible for many of the infrastructural developments that have taken place in this region. During the 1970s, the industry increased its output and exploratory activity considerably. In Yukon, mineral exploration expenditures rose from about $4 million in 1971 to about $18 million in 1972, and to an estimated $40 million in 1981. In the Northwest Territories, these expenditures were higher and estimated to be around $50 million in 1981.[37] Some of these exploration expenditures have been made in the Arctic Islands. Exploratory activities in this area have concentrated on zinc, lead, copper, lithium, diamond, iron, barite and soapstone, whereas on land uranium exploration is the largest activity.

The value of mineral production increased from $207 million in 1971 to $799 million in 1980, or at an average rate of 16.2 per cent a year (Table 3.13). This high rate of growth was the result of the diversity of, rather than increased prices for, base metals which made it possible for the industry to respond to changing world demands and prices.

The Arctic contributed about 10 per cent of the value of Canadian hard minerals produced in 1980,[38] but the mineral industry is mainly land-based. Of the nine mineral plants operating in the Northwest Territories in 1982, only two plants were situated adjacent to the Arctic coast (Figure 3.6). The Nanisivik mine on Baffin Island employed about 200 workers and produced zinc, lead, silver and cadmium. This plant had the highest

Table. 3.13. *Value of mineral production, Northern economy.*

	Minerals ($000's)		
Year	Yukon	Northwest Territories	Total
1971	93020	114228	207248
1972	106502	117905	224407
1973	150667	158925	309592
1974	171348	214346	385694
1975	228840	182070	410910
1976	123813	188254	312067
1977	209460	216439	425899
1978	218804	270953	489757
1979	299244	436975	736219
1980	364104	435458	799562

Source: Department of Indian Affairs and Northern Development, *Mines and Mineral Activities* (Annual).

mill-rate capacity and produced the largest quantity of tonnes milled in 1980. Cominco's Polaris mine on Little Cornwallis Island commenced production in 1982.

The mining industry in the North does not provide considerable employment since it is highly capital intensive. The number of employees ranged between 2000 and 3000 workers during the period. However, mining activities do provide year-round and high-paying jobs: the average income per worker in the industry was $ 25000 in 1980.[39]

C *Theoretical evaluation*

The export-based model of development seems adequate for explaining Northern development during the 1970s. This development has been based on export demands for Northern resources, particularly the non-renewable mineral and oil and gas resources. World demand and prices for base metals and oil and gas led to the expansion in exploration and output in the Northern mineral industry, and increased exploratory activity for oil and gas. These activities were confined mainly to land areas prior to the 1970s, but moved offshore during the 1970s. Although renewable resource activities in the Arctic were not significant in terms of the industrial economy, they provided a source of livelihood and income to the native people which conformed to their traditional way of life. Even in the renewable resource sector, the growth segments, such as recreational fisheries, were dependent on export demands.

Export demands for the resources of the North were responsible for the heavy inflow of capital from the South and other countries, and a concomitant increase in government expenditures on infrastructural works to encourage development. This inflow of capital (and also of labour) into the North was responsible for the high levels of growth experienced by the Northern economy during the 1970s. The major sources of capital investment for Northern development are, in order of importance: government (federal and territorial); the mineral industry; and the oil and gas industry. The expenditures made by these sources on Northern activities are given in Table 3.14. These expenditures, not all of which were made in the North, give some idea of the level of capital investment made, their impact on the growth of GDP in the Northern economy, and the heavy dependence of this economy on capital from outside sources.[40]

The expenditures made on the resource industries brought about growth in output in the mineral industry, rather than in the oil and gas industry. However, both industries were able to induce growth in other sectors of the Northern economy, particularly those with strong linkages such as construction, transportation, and services. In the past, the development of

Figure 3.6. Mineral exploration and mining, Northwest Territories, 1981.

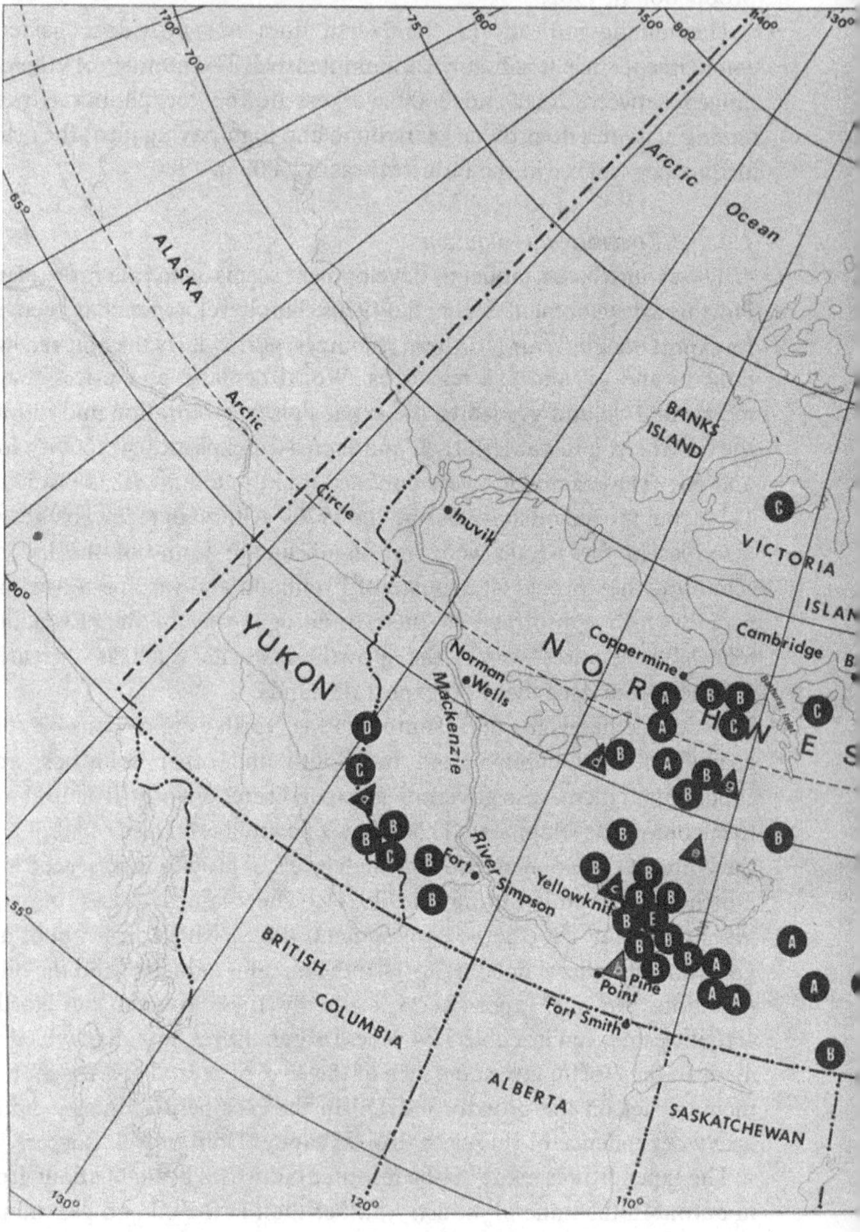

Source: Department of Indian Affairs and Northern Development, *Mines and Mineral Activities*, 1981 (1982).

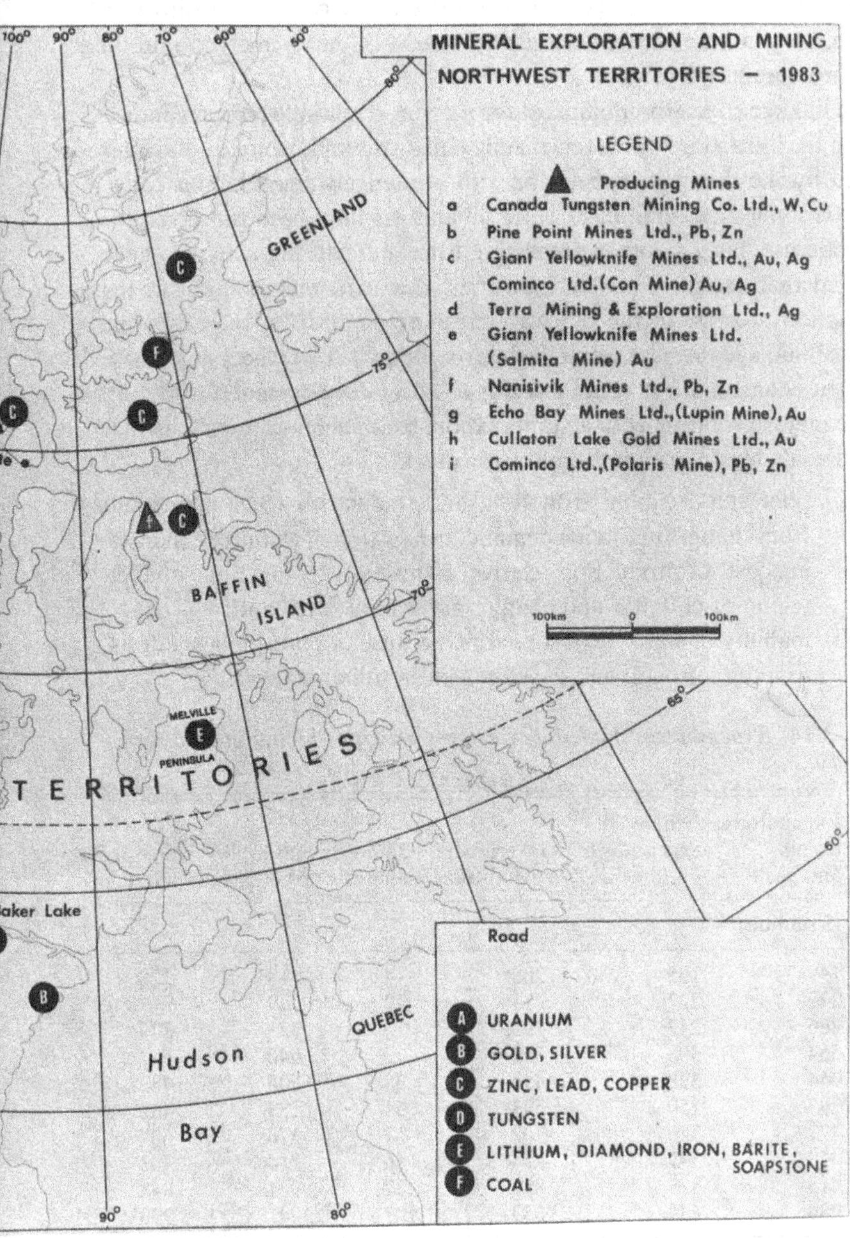
MINERAL EXPLORATION AND MINING
NORTHWEST TERRITORIES – 1983
LEGEND
Producing Mines
a Canada Tungsten Mining Co. Ltd., W, Cu
b Pine Point Mines Ltd., Pb, Zn
c Giant Yellowknife Mines Ltd., Au, Ag
Cominco Ltd.(Con Mine) Au, Ag
d Terra Mining & Exploration Ltd., Ag
e Giant Yellowknife Mines Ltd.
(Salmita Mine) Au
f Nanisivik Mines Ltd., Pb, Zn
g Echo Bay Mines Ltd.,(Lupin Mine), Au
h Cullaton Lake Gold Mines Ltd., Au
i Cominco Ltd.,(Polaris Mine), Pb, Zn
100km 0 100km
Road
A URANIUM
B GOLD, SILVER
C ZINC, LEAD, COPPER
D TUNGSTEN
E LITHIUM, DIAMOND, IRON, BARITE, SOAPSTONE
F COAL
GREENLAND
BAFFIN ISLAND
MELVILLE PENINSULA
TERRITORIES
Baker Lake
QUEBEC
Hudson Bay

the base metal industry was responsible for considerable infrastructural development, such as roads and power, and this continued during the 1970s. Although exploratory activities for oil and gas are transitory in nature, certain base areas, such as Tuktoyaktuk, have emerged and an infrastructure has developed.

The linkage effects, predominant for the non-renewable resource industries in the North, are, however, mainly backward and demand; the latter caused by the the high incomes paid to a relatively small labour force. There is some forward linkage, since oil and gas have been processed for domestic use, but few other processing or manufacturing activities have emerged that use the resource industries' raw materials as input in the production process. In the renewable resource industries, there is some forward linkage in that some fish processing takes place, but these industries constitute too small a sector to affect development.

The major problems with a staple export-based economy, such as that of Canada's North, emanate from two causes:

(i) poor and declining growth in staple industries which may result from changes in foreign demand, exhaustion of resources, increasing costs of production relative to those of a competing product, region or country, and changes in technology;[41] and,

(ii) inability of staple export sectors, because of poor linkage effects, to induce growth and development in other sectors.

Table 3.14. *Expenditures made on Northern activities by major sources 1971–1981.*

Year	Expenditures on oil and gas	Estimated expenditures on minerals*	Government expenditures	Domestic savings	Sub total expenditures	GDP
	($ million)					
1971	149	165	208	−4	518	228
1972	222	227	281	10	740	279
1973	266	247	293	15	821	379
1974	354	308	359	19	1040	492
1975	344	328	494	38	1204	483
1976	293	250	593	51	1187	482
1977	277	340	702	57	1376	607
1978	473	391	794	56	1714	703
1979	413	589	882	77	1961	848
1980	236	639	1059	107	2041	990

* Based on 80 per cent of the value of production.

Sources: Department of Indian Affairs and Northern Development, *Oil and Gas Activities* (Annual) and *Mines and Mineral Activities* (Annual); and Statistics Canada, *Provincial Gross Domestic Product by Industry*, Cat. 61–202 (Annual).

An analysis of Northern development during the 1970s indicates that the second causal factor applied rather than the first. World demand and prices for oil and gas and minerals led to growth in Northern staple industries but, due to poor linkage effects, staple industries did not induce significant growth in other goods producing sectors.

V The future of Northern ocean industries

In the 1980s the attainment of growth and development in Arctic marine resource industries will be difficult, with the possible exception of the hard mineral industry. With respect to the development potential of the renewable resource sector, the prospects for commercial fisheries development are limited because of resource availability, location and transportation costs. Recreational fisheries, however, should continue to experience growth. The major problem industry in the North will be the oil and gas industry.

Although hydrocarbon resources in the Canadian Arctic have been estimated at between 7 and 13 per cent of recoverable oil, and 59 per cent of recoverable gas reserves in Canada (Table 3.15), the development problems are staggering. Industry must innovate with respect to technology. In the Beaufort Sea, drilling is done mainly by drill ships, while in the Sverdrup Basin in the Arctic Islands, drilling is done from artificially thickened platforms. In both cases, the technology is costly and can be applied only for limited periods during the year. This applies also to transport facilities. Only prolific wells which can sustain a high level of

Table 3.15. *Oil and gas potential, Arctic and East Coast.*

	Oil		Gas	
		Per cent of Canada's total		Per cent of Canada's total
Arctic	10^9 m^3	recoverable	10^{12} m^3	recoverable
Beaufort Sea–Mackenzie Delta	1.5	3.9–7.4	3.2	27.8
High Arctic	0.8	2.1–3.9	2.9	25.2
Other	0.3	0.8–1.5	0.7	6.1
Sub total	2.6	6.8–12.8	6.8	59.1
East Coast	2.0	5.2–9.9	1.7	14.8
Total	4.6	12.0–22.7	8.5	73.9

Source: Adapted from the presentation of the Department of Energy, Mines and Resources to the Special Committee of the Senate on the Northern Pipeline. *Proceedings of the Special Committee of the Senate on the Northern Pipeline*, Issue No. 32 (June 29, 1982), pp. 49–51.

production – in the order of 5000–6000 barrels a day – can justify the high initial capital cost. Production and transport costs are higher in the Arctic than in other accessible areas on land and sea. For example, cost comparisons between production in the Beaufort Sea and production in the Hibernia field, located approximately 170 nautical miles southwest of St John's, Newfoundland, in 1980 revealed the following:[42] production and transport costs, that is total costs, in the Beaufort Sea were close to the world oil price in 1980, but were about 70 per cent higher than in the Hibernia field on the East Coast. Oil prices, however, which rose in 1981 and 1982, began to decline in 1983.

In view of the uncertain world market picture and the high costs of developing adequate transport technology for the North, it is unlikely that significant development will take place in the oil and gas industry in the Arctic during the 1980s. The major oil and gas companies, which planned to commence commercial production of oil in 1987 and gas in the early 1990s, are now in a "holding pattern" with these plans. They have indicated that, without significant government assistance, planned production will be pushed further back, probably to the mid- and late 1990s depending on market conditions. However, the most significant change is the switch in activities from the Arctic to the East Coast, where commercial production from the Hibernia field is expected to commence in 1988.

A slowdown in Arctic oil and gas activity need not be a great disaster. It will enable the North to come to grips with certain issues which have to be resolved such as land claims, environmental protection and integration of Northerners into all aspects of industrial development. Balanced, long-term economic growth will require emphasis on both renewable and non-renewable resources, a strengthening of industrial linkages through improved infrastructure (mainly transportation networks) and, where possible, increasing raw material processing in the region.

Table 3.16. *Estimated production costs of Hibernia and Beaufort Sea Fields.*

	Hibernia	Beaufort Sea
	(one million barrels recoverable) 1980$ per barrel	
Investment	5–7	6–7
Operating costs	6	6–8
Transport (ship)	2	10–11
Total costs	13–15	22–26
World price per barrel	34	34

Because the Arctic is a storehouse of multiple resources, a highly integrated marine management regime will be required to balance user conflicts in the future. Unfortunately, ocean management as a science is still in its infancy, and ocean management measures which treat the ocean and its activities within a systems framework have not been practised to any great extent throughout the world. Canadian Northern waters could, however, provide a valuable testing ground during the 1990s.

Notes

1. Dome Petroleum Ltd, Esso Resources Canada Ltd, and Gulf Canada Resources Inc., *Environmental Impact Statement for Hydrocarbon Development in the Beaufort Sea–Mackenzie Delta Region* (1982), V. 2 (hereinafter referred to as Beaufort EIS).
2. Canadian Arctic Resources Committee, *Northern Perspectives* V. 9, No. 2 (1981).
3. Department of Fisheries and the Environment, *Birds and Marine Mammals: The Beaufort Sea and the Search for Oil* (1977), p. 89.
4. W.D. Brakel, *Socio-Economic Importance of Marine Wildlife Utilization*, Beaufort Sea Project Technical Report No. 32 (1977).
5. P.J. Usher, "A Northern Perspective on the Informal Economy", *Perspectives*, The Vanier Institute of the Family (1980), pp. 1–23.
6. *Ibid.*
7. Beaufort EIS, *supra* note 1, V. 5, pp. xiii–ix.
8. P.J. Usher, *supra* note 5.
9. For a description of the origin and application of this theory, see K. Buckley, "The Role of Staple Industries in Canada's Economical Development", *Journal of Economic History* V. 18 (1958), pp. 439–50.
10. D.C. North, "Location Theory and Regional Economic Growth", *Journal of Political Economy* V. 63 (1955).
11. *Ibid.*, p. 254.
12. K.E. Coulianos, *Concepts of Eco-developmental Tourism For Small Scale Caribbean Islands* (M.A. thesis, School of Natural Resources, University of Michigan) (1980).
13. This section is based on C.L. Mitchell, "National Legislation and the National Development Plan" (a paper presented at the International Ocean Institute "Pacem in Maribus XI", Mexico City, October 26, 1982).
14. A.O. Hirschman, *The Strategy of Economic Development* (1965).
15. H. Scott Gordon, "The Economic Theory of a Common Property Resource: The Fishery", *Journal of Political Economy* V. 62 (1954), pp. 124–42.
16. Amazingly, there is no official statistical breakdown between Yukon and the Northwest Territories, even though they are separate political entities and display few economic linkages. For a general discussion of the economy of the Northwest Territories, see D.C. Emerson Mathurin, "Northwest Territories Economic Circumstances and Opportunities" (unpublished paper, Department of Indian Affairs and Northern Development) (1976).
17. The number of workers in the hunting, fishing and trapping category is probably understated in the census, because many who engage in these activities consider them to be recreational activities rather than work or occupation-related tasks. Thus, there is a large discrepancy between the census data and data provided by the Territorial governments. It is thought, however, that most of the unspecified workers should be in the hunting, fishing and trapping category.

18. *The Canada Year Book* (1980–81), p. 256.
19. *Ibid.*, p. 256.
20. Mathurin pointed out that in 1970, "there were only 530 gainfully employed native workers even though there were 1,375 natives of labour force age in the Yukon in that year". See D.C. Emerson Mathurin, "Yukon Territories Economic Circumstances and Opportunities" (unpublished paper, Department of Indian Affairs and Northern Development) (1979).
21. F. Friesen and J.G. Nelson, "An Overview of the Economic Potential of Wildlife and Fish Resources in the Canadian Arctic", in R.F. Keith and J.B. Wright (eds), *Northern Transitions* V. II (1978), pp. 163–180.
22. See J.C. Stabler, "The Report of the Mackenzie Valley Pipeline Inquiry, Volume 1: A Socio-Economic Critique", in Keith and Wright (eds), *ibid.*, p. 190.
23. *Ibid.*, and Heather Myers, *The Use of Biological Resources by Certain Arctic and Sub-Arctic Peoples* (M. Phil. thesis, Scott Polar Research Institute, Cambridge) (June 1981).
24. D. Devine (ed.), *NWT Data Book* 1982–83 (1982), p. 62.
25. *Ibid.*, p. 62.
26. L.D. Corkum and P.J. McCart, *A Review of the Fisheries of the Mackenzie Delta and Nearshore Beaufort Sea*, Canadian Manuscript Report of Fisheries & Aquatic Sciences, No. 1613 (1981).
27. See Department of Fisheries and Oceans, *Annual Statistical Review*, Cat. Fs 1–9 (Annual). Such figures represent landings and values for the Northwest Territories. Annual Statistical Reviews do not provide statistics for Yukon fisheries except for the contribution of fisheries to commodity-producing industries.
28. In addition to the Northwest Territories, other participants in the marketing programme are the provinces of Manitoba, Saskatchewan, Alberta and Ontario. Freshwater Fish Marketing Corporation, *Annual Report* (1982). The *Freshwater Fish Marketing Act* limits the corporation to marketing certain freshwater and anadromous species listed in a Schedule to the Act, which includes Arctic char, but not marine species. J. Fielding Sherwood, *Canada-Denmark Fisheries Relations in Davis Strait: Domestic and International Management of a Subarctic Fishery* (Masters of Laws Thesis, Dalhousie University) (September 1984), pp. 115–117.
29. This is based on the number of commercial licences issued a year by the Department of Fisheries and Oceans.
30. It has been pointed out that "the price set for Arctic fish products has often been less than the operational and transportation costs of delivering fish to the FFMC market", see L.D. Corkum and P.J. McCart, *supra* note 26, p. 17.
31. D. Topolinski, *Regional Income Analysis of Northwest Territories Fishing Lodges* (unpublished study, Department of Fisheries and Oceans, Western Region) (1982).
32. *Ibid.*
33. Thomas R. Berger, *Northern Frontier, Northern Homeland, The Report of the Mackenzie Valley Pipeline Inquiry: Volume Two* (1977).
34. Department of Indian Affairs and Northern Development, *Oil and Gas Activities* (1976).
35. Even taxes on wages to migrant labour from the South accrued to their province of origin, rather than to the NWT or Yukon Territorial governments. See *The Oil and Gas Industry in The Northwest Territories* 1975–79 (prepared for the Department of Indian Affairs and Northern Development by Price Waterhouse Associates) (1980).

36. Serious attempts to involve native people directly in oil and gas exploratory work began in 1971 when Panarctic initiated a Native Employment Program. "{S}ince that year, the company has transported willing residents of Pond Inlet and Arctic Bay to and from the job site. By 1981, a total of 92 Inuit from seven Arctic communities were employed by Panarctic. Many of these workers have upgraded their skills through practical experience and on-the-job training." See Special Committee of the Senate on the Northern Pipeline, *Marching to the Beat of the Same Drum* (1983), p. 61.
37. Department of Indian Affairs and Northern Development, *Mines and Mineral Activities* 1981 (1982), p. 26.
38. Statistics Canada, *General Review of the Mineral Industries*, Cat. 26–201 (Annual).
39. *Ibid.*
40. For example, most of the capital expenditures for heavy machinery and equipment for oil and gas exploratory work are made in the South or in other countries, see *The Oil and Gas Industry in the Northwest Territories, supra* note 35.
41. D.C. North, *supra* note 10, p. 254.
42. Adapted from *Proceedings* of The Special Committee of The Senate on The Northern Pipeline, 1st Sess., 32nd Parliament, Issue No. 20 (March 23, 1982), pp. 40–41.

4

Arctic marine transport and ancillary technologies

ERNST G. FRANKEL

I The challenge of Arctic marine transportation

A Introduction

The Beaufort Sea is expected to be one of the world's largest reservoirs of petroleum and gas, and large deposits of hard minerals may be found on surrounding islands. However, from the perspective of Arctic resource exploration, production, and transportation, the region poses a number of unique technical problems. In part because of the array of obstacles, and in part because of the novelty and scale of the task, many of the world's most competent technologists have been drawn to the challenge of designing and testing of Arctic drilling structures, artificial islands, terminal equipment, and transport systems. For example, artificial islands (constructed in up to 18 m of water) have become the most popular drilling platforms for use in the shallow waters of the Mackenzie River Delta. As exploration proceeds into deeper waters other types of rigs and platforms are planned, including the "Conical Rig" currently under construction for Gulf Canada and designed to operate in 55 m water depths (see Figure 4.1), and a type of movable or relocatable drilling platform, shown in Figure 4.2.

This paper reviews existing Arctic marine transportation technologies and identifies the most critical operational and interface problems which must be resolved before routine Arctic shipping is possible. For example, traffic routing, close-in manoeuvring, cargo transfer and crew replacement place special demands on Arctic operators as compared to transport operations conducted in more moderate climates. "Arctic transportation" is here defined as a total logistics system from resource extraction to delivery, including:

(i) resource collection;
(ii) resource storage;

(iii) physical form change of resource or resource processing;
(iv) terminals and other interfaces;
(v) resource transfer and cargo handling;
(vi) resource transport;
(vii) navigation and communication; and
(viii) sensing and control.

B *Environmental problems*

Arctic environmental and atmospheric conditions pose a complex array of technical problems for transportation system designers:

> One may ask what is the difference between arctic and non-arctic conditions. Apparently this must be related to physical conditions

Figure 4.1. Conical drilling platform.

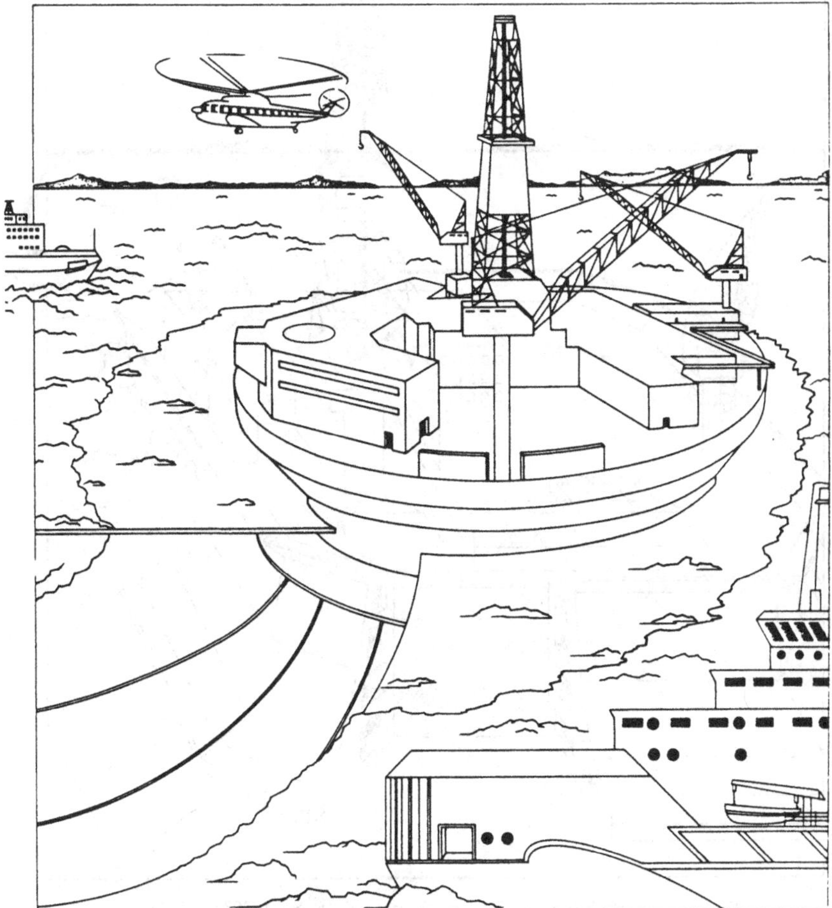

Source: American Bureau of Shipping, *Surveyor*, V. 16, No. 1 (1982) p. 14.

including climatic factors. That means temperatures, winds, waves, currents and their mutual interference causing deviations from non-arctic conditions. This includes changes in transfer of action and reaction of forces. Low temperatures transform water to ice. Ice is solid water and behaves very different from water, influencing flow pattern circulations and force interaction. When waves and ice join, wave action changes and so do wave forces which are now a combination of forces by water and by solid mass. Material properties also change character. Most materials become more brittle. Also tear and wear of materials change. Effluents

Figure 4.2. Movable drilling platform.

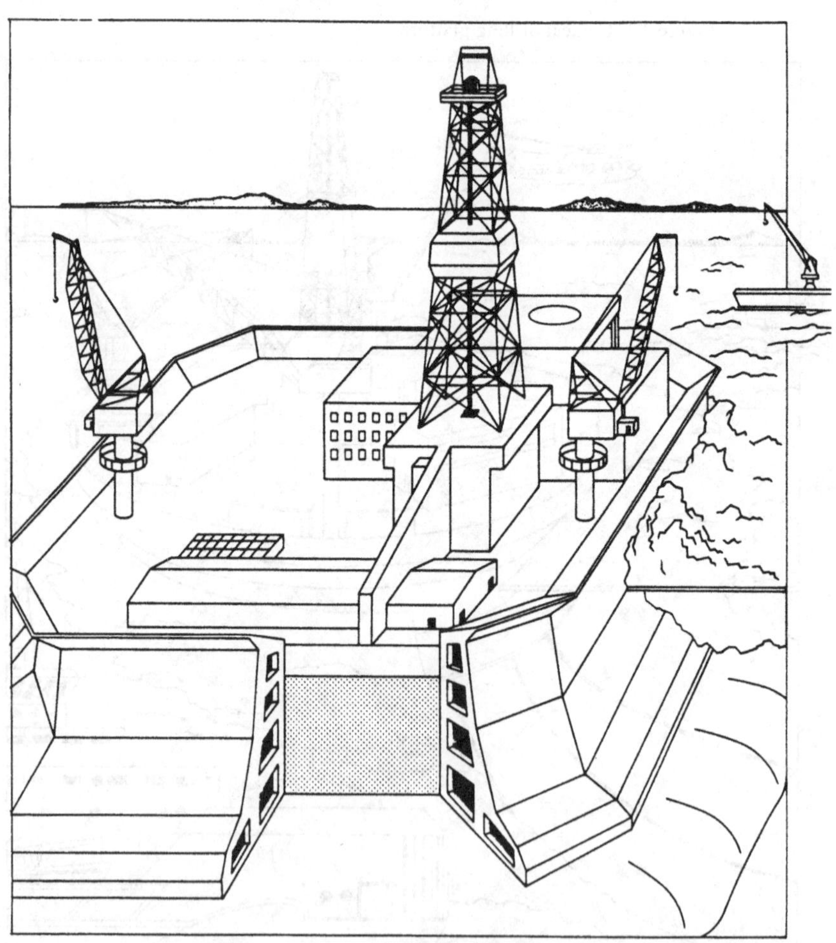

Source: American Bureau of Shipping, *Surveyor*, V. 16, No. 1 (1982)

behave differently and pollutants will usually break down much slower. Ice always puts extra loads on the structure and the entire load distribution for design changes compared to a no-ice condition.[1]

Ice is perhaps the central problem of Arctic resource exploitation and transportation. Ice cover determines the way in which resource extraction-exploitation can occur, and what primary processing may be required to effect delivery to a transport facility. Since the Canadian Arctic consists largely of islands, air and sea are the dominant transport modes. Both require special adaptation to ice conditions and associated meteorological problems. For example, pack ice, from 0.5 to 2.0 m thick, can occur in comparatively open forms, posing few problems to navigation; alternatively, it can compact into virtually impenetrable pressure ridges up to 10 m in height. Pack ice prevents the year-round use of conventional aids to navigation. The presence of landfast ice must also be taken into account, and the probabilities of ice covers of given magnitudes must be calculated and incorporated into the design of Arctic terminal and offshore facilities. In addition, icebergs are navigational hazards and pose a threat to offshore facilities. To mitigate potential damage from collisions, ice surveillance on heavily used navigation routes is routine; radio, radar, aerial stereophotography, laser, scanning radiometers, and satellite imagery are the primary surveillance technologies currently available, but observation capabilities are continuously being refined.

The inimical climate of the Arctic has two major effects on the applicability of technology: low temperatures alter the physical properties of many materials, and the overall environment severely degrades human performance unless elaborate precautions are taken.

Arctic temperatures slow the speed of chemical reactions, change the viscosity of fuels, and cause brittle-fatigue in metals. Low temperatures promote icing of structures, which poses risks in terms of stability, visibility, and operability of moving parts. Remote operating devices may be required to perform physical tasks in open environments.

Human performance is affected by confinement and sensory deprivation. Long winter nights and long summer days disrupt circadian rhythms. Visibility is often reduced by fog or whiteout. Arctic workers, monitoring duplicated or remotely operated equipment, may be subject to stress and inefficiency, depending on the frequency of problems and the ergonomic design of the equipment-system.

The last major category of environmental problems confronting Arctic technology arises from the nature of Arctic ecosystems and social systems. The ability of the Arctic to support life is so low that vast tracts are required to maintain comparatively small animal populations and the indigenous hunting cultures. Species interdependency is very marked. The technical challenge of permafrost has been met, but Arctic flora and fauna require further study if they are to be preserved and protected from man-made environmental changes. Social problems, including the management of cultural conflict and provision of social services for transient and fluctuating work forces, can be acute in populated areas, and may hinder or prohibit the development of isolated areas.

C *Navigation problems*

Arctic navigation is complicated by a number of factors, including lack of a reliable compass; ice; maintenance of navaids and radio stations; and communication and control problems.

1. *Lack of a reliable compass*

This is the greatest single problem of Arctic navigation. The horizontal component of the earth's magnetic field is weak. Over much of the Arctic, magnetic compasses may not operate or may require frequent adjustment. Gyro-compasses may be similarly affected. Alternatively, convergence of the magnetic meridians may cause significant problems resulting in inaccuracy and wander. Possible technological answers to this problem are: cryogenic nuclear magnetic resonance gyroscopes; single-degree-of-freedom gyros (used in ships' inertial navigation systems (SINS)); and electrostatically supported, superconducting, or dynamically tuned gyros. Spinning masses, however, are not the only potential sensors which could provide the basis for a reliable heading indicator. Tuning forks, fibre-optic interferometers, superfluid masses, and ring lasers have all been, or are being, investigated.

2. *Ice*

Any understanding of navigation by high-powered vessels within the pack requires a detailed knowledge of Arctic ice.[2] Multi-year ice provides the most severe obstruction to navigation and, therefore, routes covered with first-year ice are usually selected as navigational passages. As recently discovered, broken ice passages do not provide a natural and easy path for subsequent ship passages because broken ice usually re-freezes into a thicker mass and often leads to the formation of ice ridges.

Ship resistance in ice, even in uniformly thick ice, is still largely unpredictable. However, experts generally agree on the following:

(i) overall resistance is a linear function of the ships beam;
(ii) icebreaking itself consumes only a small part of total ship resistance; and

(iii) the strength of the ice, including its age, does not materially affect ship resistance in ice.

Marine transportation in the Arctic is technically and operationally feasible. The main problem is to predict effectively the performance of ships navigating in Arctic waters. The importance of viewing Arctic transportation as a system is highlighted by the Soviet experience in using landfast ice to access areas closed off by shoal water during the open season.

3. *Maintenance of navaids and radio stations*

Equipping each station is logistically costly, and poses a continuing problem of balancing cost-effectiveness and system reliability. The most desirable system is probably long-range and flexible with respect to transmitter siting, but more research on Arctic propagation is required.

Table 4.1 summarizes navigation systems available in the Arctic. The oldest and most basic form of navigation is manual dead reckoning, using the speed and heading references available aboard ship or other vehicle. This is being supplemented, if not supplanted, by electronic navigation systems.

4. *Communications and control problems*

Marine transportation is the most demanding Arctic transport mode and requires special communication and control facilities. Extension of shipping seasons to facilitate Arctic resource development, as well as the prevention of polluting incidents, places heavy demands on communications and control systems.

Table 4.1. *Arctic navigation systems.*

Dead reckoning systems
Doppler:
radar
sonar
satellite
Inertial navigation systems (SINS)
Radio aids
Multiple bearings – DF
Bearing and range – Radar
Hyperbolic:
Decca
Loran-C (characteristic mode of operation)
Omega
Multiple ranges:
Omega (characteristic mode of operation)
Loran-C

Arctic radio communications are conventional single-side band (SSB) in the MF and HF bands. Although groundwave propagation is impaired by ice and permafrost, it is used extensively to avoid problems caused by the variability of ionospheric conditions. Radio telegraph A-1 Morse is an alternative radio communication mode. At present, however, there is no single communications system available in the Arctic to provide reliable, uninterrupted 24-hour service. In addition to the interruptions caused by ionospheric instability and HF absorption (polar cap absorption), intermittent congestion also hinders system reliability. While satellite technologies hold great promise in terms of mitigating operational problems, additional research and development activity is required to enhance the reliability of communications systems in the Arctic.

There is also a need to design and implement a traffic management control system in the Arctic to support precision inshore navigation and to safeguard against loss or damage from environmental hazards. A vessel traffic system (VTS), while expensive to operate – costing several millions annually – can provide reliable, real-time information to ensure that: vessels are in compliance with regulations; vessel movements are co-ordinated; and emergencies are promptly contained. Existing VTSs display a range of functions, but none are adequate for projected Arctic developments. For example, ship-to-ship systems, in which users periodically broadcast their positions, courses, and observations, are voluntary, unenforceable and *ad hoc* – possibly even counterproductive in medium traffic densities or confined waters.

Shore participation is essential to an effective VTS. In the simplest type of ship-to-shore VTS the shore station may act in an advisory role, giving clearance to enter the regulated area, monitoring communications and movements, and broadcasting navigation information. More detailed VTSs control vessel movements within their range of coverage, and require position-status reports from vessels in the system. The most advanced VTSs in use actually survey and record vessel movements on port radars, and manage close-quarters situations. This level of VTS, with high concomitant expenses to install and maintain elaborate surveillance equipment and to support a team of expert traffic controllers, is envisioned for a limited number of circumstances, such as in ports with high traffic densities or where high transit risks are common.

II Arctic transportation as a systems problem

Arctic transportation forms an integral part of all the activities in the Arctic, and the Arctic transport problem must therefore be considered as a systems problem.

A Operational requirements

A fundamental operational requirement is appropriate ship design for service in expected ice conditions. Important design features are: a hull with a circular or oval midship section; a shoulder type wedge-shaped waterplane section, and a curved stem with an entrance angle and flare at the bow adapted to the expected ice conditions (age, type, thickness). Reliable and self-sufficient transport systems are essential since variable Arctic conditions do not permit a high degree of dependence on operational assistance or support. To prevent breakdowns, Arctic ships are designed with built-in redundancies such as twin screws, twin rudders, and excess fuel capacity. Equipment such as capstans, anchor and mooring winches, are generally stored in heated spaces below deck.

A number of other critical adjustments are necesssary for Arctic operations. For instance, living spaces must be well insulated. Life saving equipment must be designed for temperatures as low as -60°C. Lifeboats must be able to maintain ambient temperatures of 15°C, so that occupants may survive extended exposure to Arctic conditions. Stability of lifeboats must be enhanced. Propulsion plants must be instantaneously available, able to sustain shock loadings and full power reversal, and be fully protected against icing. Fuel systems must be able to handle cold, viscous fuel. All ship fittings and structures must be Arctic-proof. Crews, capable of responding to the special challenges of navigation, operation, and working in the Arctic, must be trained.

B Modelling and systems analysis

Arctic transportation modelling is required to evaluate the number, size, capacity, form, method of operation, characteristics, and design details of ships — the lifeline of the Arctic development system. However, such modelling is particularly difficult because ship performance and interfaces are hard to predict and, therefore, to simulate. For given conditions, ship resistance in ice has been successfully simulated, but forecasting the range of conditions under which a ship will operate is extremely difficult. Interface modelling of Arctic transportation systems is complicated by great variations in local conditions. Although a number of Arctic transportation simulation models have been built, their primary purpose has been to test the sensitivity of system performance to changes in ship and interface design and operation, rather than to improve design parameters. Most existing models use mathematical simulation, but several queuing network and stochastic scheduling and routing models have also been developed.

A systems analysis of Arctic transportation must incorporate the uncertainties of environmental conditions and ship performance. Systems analysis must extend to the coupling of interfacing components because excess inventories and-or insufficient cargoes will contribute inefficiencies and higher costs to the transportation system.

III Arctic shipping technology

To date, most development work performed by industry and government agencies on Arctic transportation has not advanced to levels which permit adequate technical and economic comparisons. A number of alternative marine transportation options have been proposed, but no single option is yet more convincingly attractive than any other. The list of options includes icebreaking tankers; cargo submarines; tug-barge systems; icebreaking LNG carriers; and icebreakers and icebreaking service vessels.

Arctic ship technology is about 200 years old. Since World War II, large numbers of welded steel icebreaking and ice-strengthened vessels have been built. Major research efforts have been made to effect technological improvements in ship design and construction, and advances have been made in at least five key areas, including:

(i) icebreaking bows;
(ii) ship hulls which are self-freeing;
(iii) ship hulls with low resistance when traversing ice-covered waters;
(iv) methods of propulsion and thrust in ice; and
(v) methods of steering and manoeuvring in ice.

The goal is to develop ships capable of year-round ice navigation. The *MV Arctic*, a 28000 dwt oil bulk ore (OBO) carrier operated by Canarctic Shipping Co. Ltd, is a preliminary model already in service, which transports lead and zinc concentrate from the northwestern shore of Baffin Island to southern and European ports. The journey includes a passage of 200 miles through ice which reaches thicknesses of 2 m during July. In 1985, the ship assisted in transporting the first cargo of Northern hydrocarbons to Southern markets. The experiences of the *Arctic* provide excellent data for the design of much larger vessels capable of traversing even thicker and often older ice.

Icebreaking bows have undergone a gradual evolution. Bows were designed originally to cut and split the ice and thereby open a passage. Later bows, such as the bow of the *Manhattan*, were designed to ride up on the ice and break it by the ships' weight. More recent research, based largely on the experiences of the *Canmar Kigoriak*, has led to the development of spoon-type bows which bend and snap, rather than crush, the ice.[3] Designed to withstand pressures up to 2900 lb per inch, the spoon-shaped bow has proven to be superior to the broad wedge-type bow design of most US and Canadian icebreakers. To facilitate self-freeing vessels, three structural adaptations are usually adopted: a bulge-type projection near the bow, a backwards tapered hull form, or a rounded hull underwater body form.

Ship resistance to ice may be reduced by using air-bubbling systems, special hull coatings, and hull lubrication. Propulsion innovations include high thrust nozzles, controllable-pitch propellers, and other types of thrusters. Steering mechanisms include multiple rudders, bow and stern thrusters and other manoeuvring devices.

A *Icebreaking ships*

While the Soviets rely primarily on ice-strengthened ships, led by very powerful icebreakers in Arctic voyages, the United States and Canada have concentrated on the development of icebreaking ships such as the *Manhattan*, a converted icebreaking 120000 dwt twin-screw tanker put into Arctic service in 1969 for the purpose of testing the feasibility of year-round transit of the Northwest Passage by large ships. The most recent example of a proposed icebreaking ship is the 140,000 m^3 LNG carrier designed for the Arctic Pilot Project (Figure 4.3). Capable of year-round operations through ice of up to 3 m thick, the ship is equipped with a spoon-shaped bow and "reamers" behind the bow. An underwater "ice island" is incorporated in the hull towards the stern to spread broken ice and move it away from the twin nozzle propellers.

Figure 4.3. Icebreaking LNG carrier.

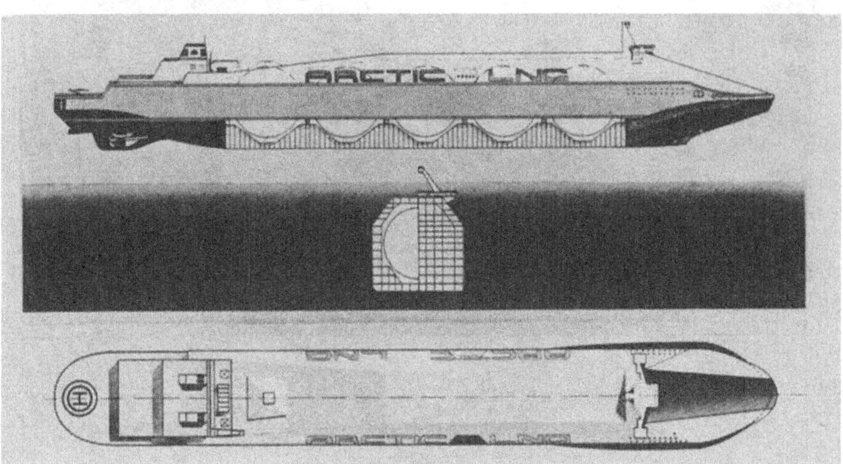

Source: American Bureau of Shipping, *Surveyor*, V. 16, No. 1 (1982)

B Submarine tankers

Several proposals for the use of submarine tankers to transport LNG or petroleum from the Arctic have been developed in recent years. General Dynamics Corporation first proposed a nuclear cargo submarine in 1964. This involved a conventional pressure vessel containing the nuclear reactor, propulsion plant, life support systems, accommodations and ships auxiliaries, while cargo tanks were placed in pressure-balanced compartments outside the vessel. A later, improved version of the General Dynamics' cargo submarine is designed to sail at depths of up to 183 m and at speeds of about 14 knots. Design features include six pressured tanks with a cargo capacity of about 1.3 million barrels of oil and about

Figure 4.4. Submarine tanker.

Source: American Bureau of Shipping, *Surveyor*, V. 16, No. 1 (1982)

580000 barrels of LNG. The submarine is also designed to sail below the ice and will surface only upon reaching its destination, that is an ice-free loading terminal. The submarine will utilize three conventional steam boilers in which distillated hydrocarbon fuel is burned with the assistance of liquid oxygen. Exhaust gases will be ejected from the ship after processing, recycling, and compression.

Other designs for submarine tankers have been proposed, including one put forward by Arctic Enterprises Inc. of Annapolis, Maryland, which would utilize alternate propulsion systems. Specifically, propulsion could be achieved by utilizing electric motors which derive their power from large methanol-phosphoric acid-fuel cells using liquid oxygen as the catalyst to create the electric current. The designs of underwater loading arrangements are still in the conceptual stage.

An E.G. Frankel-Seatrain concept allows for an extension interface of a seabottom-mounted terminal and a near-bottom hovering submarine. Panarctic Oils Ltd has plans to land two submerged submarines at an underwater terminal containing concrete storage tanks. Similar systems have been developed for use in the North Sea to eliminate construction of expensive pipeline manifold and pumping systems. An artist's conception of a submarine tanker is shown in Figure 4.4.

C Icebreakers

The USSR is the major user of icebreakers and there are at least eighteen icebreakers of 10,000 SHP or more currently in use on the Northern Sea Route (see Table 4.2). Among these are the two most powerful icebreakers in the world, the *Leonid Brezhnev* (formerly *Arktika*) and the

Figure 4.5. *MV Robert LeMeur.*

Source: Institute of Marine Engineers, *Marine Engineers Review* (December 1982), p. 40.

Table 4.2. *Icebreakers of 10000 shp or more known to operate on the Northern sea route.*

Name	Where built	Power shp	Displacement tonnes	Entered service	Remarks
Kapitan Belousov	Finland	10500	5360	1954	
Kapitan Voronin	Finland	10500	5360	1955	
Kapitan Melekhov	Finland	10500	5360	1956	
Lenin	USSR	44000	16000	1959	nuclear
Moskva	Finland	22000	13290	1960	
Leningrad	Finland	22000	13290	1961	
Kiyev	Finland	22000	13290	1965	
Murmansk	Finland	22000	13290	1968	
Vladivostok	Finland	22000	13290	1969	
Yermak	Finland	36000	20240	1974	
Admiral Makarov	Finland	36000	20240	1975	
Arktika	USSR	75000	23400	1975	nuclear
Krasin	Finland	36000	20240	1976	
Sibir'	USSR	75000	23400	1977	nuclear
Kapitan Sorokin	Finland	22000	14900	1977	shallow-draught
Kapitan Nikolayev	Finland	22000	14900	1978	shallow-draught
Kapitan Dranitsyn	Finland	22000	14900	1980	shallow-draught
Kapitan Khlebnikov	Finland	22000	14900	1981	shallow-draught
Under construction					
Dikson	Finland	10000		1982–83	not primarily for the Northern sea route
Magadan	Finland	10000		1982–83	not primarily for the Northern sea route
Mud'yuga	Finland	10000		1982–83	not primarily for the Northern sea route
Rossiya	USSR	75000		1985–86?	nuclear

Source: Terence Armstrong, The Northern Sea Routes Today, in *Arctic Energy Resources*, Louis Rey (ed.). Proceedings of the Comité Arctique International Conference on Arctic Energy Resources (Oslo, September 22–24, 1982), p. 253.

Sibir. These nuclear-powered 75,000 SHP vessels follow the successful experience with the smaller 44,000 SHP icebreaker *Lenin* built in 1959. Even many fossil-fuelled vessels are equipped with 22,000 or 36,000 SHP power plants. In fact, a third of the USSR's Northern Sea Route icebreakers have SHPs in excess of 22,000.

While the USSR makes extensive use of icebreakers to guide ships through Northern waters, the United States and Canada more often rely on icebreaking ships. A new type of icebreaking supply and service vessel is the *MV Robert LeMeur* (Figure 4.5). This Arctic Class 3 vessel has a cargo capacity of 1200 tons and, through the use of special hull coatings, hull lubrications and air-bubble systems, is capable of continuous passage through ice up to 1 m thick. This type of vessel is expected to be exceedingly versatile and to serve as an icebreaker, tug, suppply vessel, service platform, and cargo ship.

D Arctic tug-barge operations

Tug-barge operations have become a popular and economic method for use in coastal shipping of bulk commodities. Barges are either towed by a tug, using a towing hawser, or are pushed, in which case the tug is usually coupled mechanically to the barge to form an integrated tug-barge system. Couplings between tug and barge can be rigid, semi-rigid, flexible, or loose depending on method used and the type of service. In recent years tug-barge operations have been introduced into Arctic transportation and now provide the major means of freight transport to the Northern shores of Alaska, such as Prudhoe Bay.

Initial operations used icebreaking tugs towing ice-strengthened cargo barges. Consideration is now being given to the use of icebreaking barges pushed by an integrated tug. Because the tug and barge are flexibly coupled, their relative trim can change. The icebreaking barge can ride up on the ice while the tug stays horizontal. This not only improves thrust efficiency, but also results in major savings in structural weight. Experiments are under way to study the effect of the tug-barge relation, in terms of size, form and power, on icebreaking and ice operations performance.

The advantages of tug-barge operations are: lower acquisition costs, lower operating costs, and drop and swap operations. Experience shows that an integrated tug-barge has capital costs which are usually only about 70 per cent of those of a self-propelled ship designed for the same speed and capacity. Similar savings are attainable in operating costs. Perhaps the system's major advantage is the ability to drop off the cargo compartment barge and proceed with the tug alone or in conjunction with another barge.

E *Technological evaluation*

There is little consensus about Arctic shipping technology. The form of ships and method of operations are under continuous re-evaluation. For example, the controversy between using ice-strengthened cargo ships led by icebreakers or using independent icebreaking ships is not expected to be resolved for some time. Additional research in icebreaking methods and ship operations in ice must be performed to obtain more reliable data on which to base system design decisions.

Integrated tug-barge technology or hinged decouplable ship technology may offer major advantages. Large ships, such as the 140000 m³ LNG carriers under development for Dome Petroleum, will continue to be developed with spoon-shaped bows but inverted wedge-shaped hull forms may become more popular. Cargo submarines are probably too expensive an alternative transport mode to build and to operate, but there are indications that a semi-submerged ship – a hybrid between a large submerged ship and a small waterplane spoon-bow type of surface vessel – will offer new opportunities for Arctic ship technology development.

IV **Arctic interface technology**

Interface and terminal technology are important components of an Arctic transportation system. Terminals accommodate ships, barges, tugs, and other transport, and provide capability for the effective accumulation, form change, and transfer of cargo from shore to ship, or other types of vessels. To date, little experience has been acquired in the use of alternative terminals (for example, sand-filled and gravity caisson-retained islands and sea bottom-mounted telescoping submerged terminals). The problems confronting Arctic marine terminal design are numerous, including: ice and climatic conditions such as temperature and wind, waves, ice flow, water current, seabed instability, and water depth. Interface problems are also posed by ships and other terminal users. Certain operational procedures such as navigation and manoeuvring, for example, pose different types of problems and challenges in the Arctic than they do under more temperate conditions and must be accounted for in any terminal design that proceeds to the construction phase.

The most important problem of interface technology is the provision of adequate resistance to ice forces. Ice forces can be exerted by:

(i) floating or moving ice sheets of 0.5–2 m thickness;
(ii) rubble formation of built-up grounded ice;
(iii) thick ice sheets consisting of interlocking sheets of multi-year ice with thicknesses of 2–16 m; and
(iv) ridges of pack ice.

The type of ice providing dominant forces depends on water depth. For example, in shallow waters, rubble formation of built-up ice exerts the greatest ice force, while in deep water, ice ridges or thick ice sheet-induced forces dominate.

A *Terminal requirements*

Arctic terminals are distinct from conventional terminals because of severe environmental conditions, physical constraints, and differences in required functions. Terminals must be located and designed to provide a safe harbour for ships, particularly during cargo transfer. Facilities must maintain ship position and protect the ship from wave, floating, or built-up ice. The terminal must also provide sufficient depth for safe navigation and cargo handling operations. It must also prevent freezing of the vessel to docking facilities, which may occur through icing-in of the stationary vessel (with ice built-up against the vessel), and-or by actual freezing against the terminal face.

Terminals must provide for adequate ship handling, storage facilities for cargo and supplies, material and cargo handling as well as transfer, communications, and various support and service systems.

B *Submarine pipelines*

Submarine pipelines are the most effective means for liquid cargo transfer and short-distance transport in the Arctic marine environment. Submarine pipelines usually start and end on land, and have to traverse land-based and shallow water permafrost at the land-sea interface. Laying pipe over permafrost is extremely difficult on land and even harder in the sea. In deeper water where a permafrost-free seabottom exists, trenching and laying of submarine pipe is a real challenge because of the cold temperatures, as well as the problems imposed by ice or ice ridges, over the pipe location. To prevent damage from ice and to maintain reasonable temperatures, submarine pipelines are often deeply trenched and buried. Trenching is now often accomplished by using remotely controlled bottom-crawling trenching machines.

Even greater difficulties are experienced in pipe-laying. Conventional pipe-laying surface ships and barges cannot be effectively used except during a few short summer months, or in locations experiencing little or no ice formation. Novel approaches to pipe-laying include submerged or submergible barges using bottom-crawling pipe-layers, and neutrally buoyant vessels capable of hovering over the bottom but well below surface ice. Arctic submarine pipelines have been laid in numerous locations and now handle large quantities of oil per year.

C *Island terminals*

Island terminals, used in the Arctic, can be constructed in a number of different ways, including:

(i) reclaimed islands made of sand or soil excavated from the sea bottom with concrete, stone or ice armor;
(ii) prefabricated gravity caissons floated in place, sunk and locked together;
(iii) steel caisson retained sand-filled islands;
(iv) floating caisson terminals with tension legs;
(v) fixed terminal structures (steel or reinforced precast concrete) with large submerged storage tanks;
(vi) ice gravity island terminals; and
(vii) semi-submerged gravity platform terminals.

While reclaimed island terminals are attractive, they are usually prohibitively expensive (particularly if installed in deeper water), take a long time to build and are not relocatable, a characteristic of most of the prefabricated and floated-in-place types of terminals.

Subsequent to the actual design and construction of the terminal, a difficult problem is assurance of approach to the terminal. On the one hand, sloping terminal sides, which help to break up the ice and allow a pile-up of the rubble ice, is advantageous from a structural and position-keeping point of view. On the other hand, this design interferes seriously with vessel approach. In most cases island terminals are designed to provide shelter for marine vessels without a requirement for close approach or contact between ship and terminal. The ship is maintained at some safe distance, and both mooring-retaining equipment and cargo handling-transfer equipment are designed to bridge the gap between terminal and ship.

Cargo transfer can occur by submarine pipeline or telescoping steel bridge girder-supported pipeline. Dry bulk transfer is usually achieved by pipe (slurry form), or via a bridge girder supporting a fully enclosed telescoping conveyor.

D *Submerged terminals*

A number of concepts, based to some extent on technological developments in the design and engineering of submerged oil production platforms, have been developed for fully and permanently submerged loading terminals. All of these concepts assume a terminal structure placed on the bottom of the sea and tied to a submarine pipeline delivery system. Most concepts would also incorporate submerged storage tanks.

Connection with the ship may be made via an underwater delivery-receiving station at the bottom of the ship. The terminal is equipped with a telescoping or linkage arm extension structure which encloses the delivery piping and which fits into a receiving coupling at the bottom of the ship. To locate itself correctly above the submerged terminal, a ship would use sonar transducers and targets as well as bottom anchors. Some of these designs use technology developed for matching submarine rescue vessels to submarines and have been tested to depths of up to 183 m.

E *Sensors and controls for interfacing*

Remote sensing – using sonar, radar, and satellite imagery – will become increasingly important for position fixing, berthing, and ice surveillance. Formerly a largely subjective exercise, interpretation is now gaining speed and accuracy from computer assistance and, eventually, system users will be able to obtain numerical estimates of specific parameters of interest.

All-weather radar systems, such as side-looking airborne radar (SLAR) and synthetic aperture radar (SAR), have been used extensively for ice surveillance since the early 1970s. NASA's 1979 Sea Ice Radar Equipment project (SIRE), generated extensive data on their comparative performance and provided an analytical frame for interpretation of SLAR data. However, SLAR possesses a number of shortcomings for ice analysis; for example, it has a narrow geographic range, a 12-mile swath, compared to the 115-mile swath available through satellite imagery. It also lacks the inherent positioning capabilities of satellite, and is unable to distinguish reliably between first- and multi-year ice, and open water.

Because no single radar sensor, active or passive, is capable of identifying all ice types, a combination of techniques is often utilized. For example, SLAR is often used in conjunction with satellite images. Combinations of like- and cross-polarized scatterometers and microwave radiometers have been used in ice analysis to create a 3-D "feature space" with ice types displaying separately. Wideband impulse radar is theoretically capable of determining ice thickness, although the technique requires practical refinement. Wideband seabed profiles provide a better example of the potential of this branch of remote sensing.

F *Cargo storage*

Storage in the Arctic is required for oil, gas, and minerals. In addition, storage facilities are required for construction materials, drilling mud, drilling pipe, drilling and construction equipment, fuel, spare parts, transport equipment, as well as various human support and service supplies.

The most important storage problem is posed by oil, gas, and minerals. Oil and gas are usually contained in surface-mounted cylindrical or spherical tanks. Because of permafrost on land and filled surfaces, special "floating" foundations or deep pile foundations are often required to support such tanks. Because oil and gas would usually be transported by submarine pipelines, submerged tank storage has become an increasingly popular concept. Floating storage, using ice-strengthened tankers, barges, or specially built storage vessels, is another attractive approach. Such floating storage vessels may also serve as floating terminals and could be designed to permit the approach and docking of vessels.

Inflatable storage tanks and storage shelters capable of standing up to temperature, snow, ice, and wind conditions in the Arctic, are also in the process of development. The most difficult problem is expected to be provision of reliable pumping to and from storage tanks. Coverage of dry bulk cargo will generally be required to prevent freezing of the cargo. Various types of flexible and permanent stockpile storage covers are available.

V Transfer and cargo handling

Cargo transfer and handling are among the most difficult problems for designers of Arctic transportation systems, largely because of the environmental and physical characteristics of the region. The principal types of cargo expected to be handled are: petroleum; natural gas; coal; iron ore; bauxite; and copper, zinc, and nickel ores.

Petroleum, under Arctic conditions, has a high viscosity and, depending on its particular characteristics, in many cases will require pre- or continuous heating to allow pumping and transfer. Natural gas handling is not affected by Arctic conditions. Minerals, on the other hand, can become very difficult to handle when ice-up occurs. Most have some water content and an affinity for water which accelerates icing.

Cargo handling in the Arctic can be difficult for a variety of reasons. Environmental conditions may affect handling procedures and operations, and may cause alterations to the physical condition of the cargo itself. To alleviate problems associated with the uncertain Arctic environment, novel cargo handling equipment, methods and storage facilities are required.

Arctic cargo handling equipment, such as pumps, cranes, conveyors, stackers, reclaimers, and various types of mobile equipment, consist largely of standard handling equipment which has been modified or converted for use under extreme conditions. Typical modifications are the addition of protective coverage and heating elements for the purpose of reducing the potential for freezing, and to maintain the required viscosity of fuel, lubricating oil, and other liquids. Handling methods for mineral transfer in the Arctic consist of the same grab, or bucket, type of reclaimers and loaders used elsewhere. To prevent cargo freezing during transfer or handling operations, as well as during storage, some type of vibrating or continuous impact equipment may be necessary to agitate the cargo throughout the transfer process. Because mechanical belts and chain conveyors are prone to breakdowns, vibrators or vibrating conveyors may mitigate against cargo freezing, and they may facilitate subsequent stacking and reclaiming operations. Impact separation techniques also function to loosen cargo and to improve handling characteristics during the reclaiming process.

VI Arctic navigation, communications and control technology

Arctic navigation and communications capabilities have been dramatically extended in recent years. The two functions have become more closely allied with advances in multipurpose satellite coverage, and the costs of maintaining remote land stations subsequently have been reduced. The clear priorities for Arctic navigation and communications are the provision of a single, full-coverage, inexpensive air-marine navigation system, and extension of ANIK (satellite) ground terminals.

A Dead reckoning navigation systems

Two types of electronic navigation systems are employed in the Arctic, namely, dead reckoning and radio-based systems.

Dead reckoning systems – whether manual, Doppler or inertial – track heading and speed from a known departure point and integrate the velocity vector over the journey time to arrive at an updated position. The overall error of a dead reckoning system ranges from about 10 per cent (manual system) to 2 per cent (Doppler system). Inaccuracy arises principally from poor estimates of the speed-motion of the sea or air travelling medium.

Airborne Doppler radar integrates and averages many readings of Doppler shift to produce a speed estimate. The need to average multiple headings of the Doppler shift leads to "spectrum error", that is, inability to track rapid accelerations. Doppler radar is also subject to "overwater error", the cumulative result of calibration shift and motion of the sea surface. The usefulness of airborne Doppler is recognized in helicopter operations, where hover control is facilitated by Doppler's ability to measure low-zero velocity precisely.

Doppler sonar, the marine analogue, has additional limitations. High sea states affect sonar, as they do radar and, in addition, the beam is attenuated in water over 183 m in depth. Backscattering can occur from thermal layers, fish shoals, or other misleading causes. Doppler errors are, however, well-documented, thus allowing manual or mental correction. The system error arises largely from the heading reference error; the Doppler sensor itself can be upgraded to virtually any required degree of accuracy. Table 4.3 summarizes the accuracy of various heading references.

Doppler satellite technology has proved accurate (within 3 ft) in comparative field studies of methods of position fixing, and has been used

Table 4.3. *Accuracy of heading references.*

		Limitations	Accuracy
	Compass	Subject to large static and dynamic errors	2° static 60° dynamic
	Directional gyro indicator	Subject to attitude errors and drifts	10°/h
Air	Gyro magnetic compass	End of development	2°/h free gyro +0.75° magnetic
	Stable platform	Needs azimuth reference for initial alignment	0.2°/h to 0.4°/h Azimuth drift rates
	Inertial	Requires independent velocity to inflight gyro-compass	0.01°/h
	Compass	Subject to large static and dynamic errors	2° static 60° dynamic
Sea	Gyro-compass		1°
	Inertial	Used only on military ships	Classified
	Compass	Must be calibrated	5° for hand held 0.1° for magnetic survey
Land	Gyro-compass	Must be gyro-compassed at zero velocity and used as free gyro *en route*	0.1° gyro-compass 0.1°/h free gyro
	Gyrotheodolite	Fixed survey only	0.001°
	PADS	Many zero-velocity updates required	0.01°

Source: Captain M. Walker, 'A Review of Navigation Systems and Their Applicability to the Arctic', in P. J. Amaria, A. A. Bruneau and P. A. Lapp (eds), *Arctic Systems* (1977) (Proceedings of a Conference on Arctic Systems, St. John's, August 18–22, 1975), p. 584.

by industry to generate data on ice movements near Arctic oil and gas anomalies, and to provide reliable fixes for drilling operations conducted through the ice.

The principal application of Doppler technology to the problems of Arctic navigation is in the various military and civilian satellite navigation systems. These allow fixes every 30 minutes in high latitudes, although the precision obtainable needs to be improved for navigation in the most confined waters. The NAVSTAR satellite system, when fully operational, will provide additional accuracy.

B *Radio-based navigation systems*

There are over 200 different types of radio positioning devices.[4] All generate lines of positions (LOPs), and some generate ranges. The intersection of any two constitutes a fix, with a third LOP-range usually employed for a check or to allow statistical averaging of error. Coverage of a radio navigation system is dependent on three factors: the quality and power of the receiver and signal transmission, and the extent of propagation losses over the signal path. Within a coverage area, accuracy depends on the precision with which the LOP can be measured, the angle of cut between LOPs, and the type of LOP.

Accuracy contours can be drawn up by using the error budget of the signal. This is related to the propagation of the radio wave, a function of the location on the spectrum of radio frequencies (as shown in Table 4.4). The modes of propagation – direct, ground-reflected, ionospherically reflected, and surface – are affected differently by Arctic conditions. Groundwave propagation is most adversely affected, which in turn affects otherwise ideal systems such as Loran-C.

The Loran-C system is a sophisticated extension of the fundamental concepts of the now-obsolete Loran-A system, but the two differ in frequency and time measurement techniques. The Loran-C operates in the band centred around the carrier frequency of 100 kHz, with a spectrum contained within a band of 90 kHz to 110 kHz. This low-frequency transmission gives a considerably extended groundwave range over both land and sea, and enables the stations to be located with longer baselines. Due to the type of transmission, skywave contamination is reduced, which allows both course and fix measurement of the time differences.

Loran-C groundwave ranges of 700 to 1000 nautical miles over seawater are typical, depending on transmitter power, receiver sensitivity, propagation losses over the signal path, and distance from the transmitters. Variation in propagation loss and in velocity increase with the distance from the transmitter. Taking into account the above factors, accuracy is

Table 4.4. *Radio waves: spectrum of frequencies.*

Acronym	Meaning	Wave size	Frequency	Mode of propagation	Range	Error/limit	Remarks
EHF	Extremely high frequency	0.1–1.0 cm	3C–300 GHz				
SHF	Super high frequency	1.0–10 cm	3–30 GHz			Self-noise	High antennae gains possible if directional 'beams' are employed
UHF	Ultra high frequency	0.1–1.0 m	300–3000 MHz	Groundwave, from earth stations, skywave from satellite	1.2 H where *H* = height of transmitter		
VHF	Very high frequency	1–10 m	30–300 MHz				
HF	High frequency	10–100 m	3–30 MHz				
MF	Medium frequency	0.1–1 km	300–3000 kHz				
LF	Low frequency	1–10 k	30–300 kHz	Groundwave	1200 miles, at 100 kHz	Attenuation-related	Attenuation is maximal over Arctic tundra
VLF	Very low frequency	10–100 km	3–30 kHz	Duct: spherical wave guide formed by earth and D layer of ionosphere	Round the world at 16 kHz	Diurnal – due to movement of D layer	

quoted in the region of 50 to 200 feet at 200–500 nautical miles, changing to approximately 500 feet at 1000 nautical miles. Accuracy of 50 to 200 feet is sufficient for ice navigation. Ship wave "splitting", which is common and useful in Loran-A, does not occur from a practical standpoint at the Loran-C frequency of 100 kHz. However, useful one-hop E skywaves have been observed during both daylight and darkness, at ranges of 2300 nautical miles; while two-hop waves have been observed as far as 3400 nautical miles. Multihop skywaves have been monitored at even longer ranges, but an almost complete darkness path between users and transmitter is required for stable nighttime multihop operation. In general, multihop skywaves do not permit high accuracy navigation.

C *Satellite communications*

The development of satellite technology has greatly assisted exploration and operations in remote regions. At present, satellite systems provide support for Arctic communications, navigation and atmospheric monitoring and surveillance activities. Technology has tended toward higher-powered satellites and smaller land stations which have been proven under Arctic conditions.

1. *ANIK*

ANIK (or "Brother" as the name signifies in Inuktitut) is the synchronous communications satellite network covering the Arctic to 78°N, and has brought reliable communications to the Arctic for the first time. ANIK is able to bypass polar ionospheric turbulence to provide telephone, television, telex, facsimile, and other services. ANIK is also able to assist navigation by transmitting ice imagery on facsimile channels. ANIK's reliability is enhanced by its independence of any need for repeater stations, or for manning of earth stations.

One significant advantage of being a synchronous (or geostationary) communication system in the Arctic is the ability to use a fixed antenna instead of a tracking antenna with its exposed moving panels. On the other hand, synchrony limits satellite coverage to 80°N, even when maximal allowance is made for the fading due to ionospheric and tropospheric propagation losses.

2. *United States Navy Navigation Satellite System*

The United States Navy Navigation Satellite System (NNSS), known as the TRANSIT System, became operational for military use in January 1964. Its main function was to provide an all-weather precision guidance system for the Polaris submarine fleet. In 1967 operational details were released, paving the way for the development of a commercial system for both US and foreign users. TRANSIT is now classified as an international navigation system. Compared to the highly sophisticated and costly military equipment, the commercial system has accepted lower accuracy in order to secure lower cost, reduced size, and lower maintenance requirements without losing dependability and essential accuracy.

Due to rapid technological advances in recent years, the user's equipment has altered considerably. The commercial satellite equipment was originally the size of a large radar display unit, but the introduction of micro-processors has reduced the size (and price) of the units so that it is now feasible to install such equipment on fishing and service vessels and small boats. The smallest units, still providing the full range of functions, can now be obtained in packages about the size of a shoe box, weighing only a few kilos. The larger units, which may incorporate Omega, have outputs to integrated navigation systems and spare computer capacity which may be used to solve a variety of navigational problems.

The navigation message stored in the satellite's memory system is transmitted to the ground as a phase modulation on the transmitted carrier frequency. This message is transmitted continuously and is timed to last exactly two minutes, starting and finishing at an exact even minute. Because of this precise time control, the beginning and end of each message provides an accurate even-minute time-mark. From this navigation message, the user can obtain the Doppler shift of the received signal, the satellite's navigation message, and an accurate time control. This information enables the computer to derive a position fix.

If the user is moving during the satellite pass, the motion during each counting interval must be measured and accounted for in the computation. Any error in representing the user's motion to the computer will result in a position-fix error. Each knot of unknown velocity will typically cause less than 0.2 nautical miles position-fix error. Thus, although the position fix is expressed as a latitude and a longitude at a specific time, the navigation computation is in fact one of finding where on the earth's surface the user's total path best matches the Doppler-measured range differences.

System error in satellite navigation consists of inherent error and unknown user error. However, user error swamps the inherent error, and makes this system less than perfect for precise ice navigation unless the vessel is equipped with accurate speed-input and interpolation tables. The system is, however, extremely reliable by comparison with other navigation systems.

D *Navaids*

Radar reflectors, radar transponder beacons (RACONs) and aero-marine omni-directional radio beacons are among the more useful aids to navigation in the Arctic. These aids are mounted ashore and located to minimize refraction and interference. RACONs provide an especially precise and reliable fix. Many conventional navaids such as buoys and lights are not reliable under Arctic conditions because they may be physically displaced, or their mechanisms rendered inoperable (particularly if they have moving parts). For example, icing is a constant threat to aerials and support structures, and frequent fog and low cloud cover restricts visual identification of conventional aids. In addition, there are intrinsic limitations where ice cover may alter shoreline and channel configuration, or where chart details, needed for land-referenced radar and coastal navigation, are lacking. These factors, and low traffic density, make it unlikely that conventional navigational aids will be extended further than the existing routes (primarily in the Eastern Arctic).

Future Arctic navigation, with its requirement for precision, will have to be based on a combination of:

(i) inexpensive, preferably unmanned and proximity-triggered signalling devices which can be buried or submerged in a fixed position;

(ii) circumpolar satellites; and

(iii) range-range radio-based position fixing systems.

E *Sensors and submerged markers*

Submarine storage and submarine terminal operations are becoming increasingly attractive options and will require the development and installation of submerged markers and sensing equipment. Since, in most locations, the installation of fixed surface types of navigational aids is impractical, hazardous, or outright impossible, submerged sensors and markers may be required to guide ships navigating in close waters or manoeuvring towards or away from a terminal.

In general, submerged acoustic transponders and optical or RP positioning systems on the ice surface which tie back to some geographical reference point on land provide feasible systems, but the Arctic environment poses some unique problems for acoustic signal propagation. For example, signals are affected by water depth, ice layer, and ceiling topography drift. The upward refraction of sound rays, usually limits the range of the direct horizontal acoustic path through the effects of "bottom grazing". In addition, the ice-ridge keel or ceiling topography structure is typically complex and continuously changing. However, because light transmission in Arctic water is excellent, and as photosynthetic processes are largely dormant in ice-covered seas, light sensors or markers may be more effective than acoustic sensing devices.

F *Traffic routing*

The sensitivities of Arctic ecosystems, the problems of manoeuvring in confined waters and fog, and the difficulty and expense of synthetic aperture radar (SAR) would seem to require implementation of a multimodal Arctic traffic management system. In addition, the dynamic character and vast extent of ice cover requires that Arctic shipping be ice-routed for optimum economic performance. The characteristics of an appropriate traffic management system depend, however, on the type and levels of future traffic. It should be noted that virtually every route for transportation of Arctic non-renewable resources involves a sea leg (see Figure 4.6).

The most futuristic visions see submarine tankers, air cushion vehicles (ACVs), and ice-strengthened ships playing increasingly significant roles in Arctic transport. Traffic levels would be appreciably higher than at present, but still not approach the levels in waters south of 60°.

A majority of analysts are inclined to see Arctic development proceeding along present lines, in an evolutionary fashion. This implies a continuation of the existing pattern of local traffic management systems of varying intensities. More systems will probably develop in the Western Arctic, and the well-established systems of the Eastern Arctic may move toward more detailed supervision. There is a strong desire to use presently available satellite technology to integrate management of marine traffic with management of the subsidiary land and air transport modes.

Marine traffic routing would localize the need for hydrographic survey work in the Arctic, and could help to minimize SAR infrastructure by concentrating traffic in lanes which could be intensively monitored and maintained. Routing would be essential if the hybrid, transshipment schemes for shipment of Arctic oil, gas, and ores are implemented. Such plans generally involve scheduled "shuttle" services which would require routing based on real-time ice information relayed from satellites and other sources. Indeed, the economics of most proposals for Arctic transportation are sensitive to average speed through ice, which implies at least the availability of an ice-routing service.

VII Future requirements

Arctic transportation is a systems problem involving infant technologies and major technological gaps in five areas — icebreaking, transport vessel design, terminal design, cargo handling and transfer, and navigational aids and communications. Specifically, further research is needed in a number of component areas, including:

(i) hinged icebreakers;
(ii) laser, vibratory, or shock methods to fracture ice and facilitate ice breakage;
(iii) ship hull coatings;
(iv) integrated icebreaking tug-barge systems;

Figure 4.6. The Arctic Ocean, sea routes.

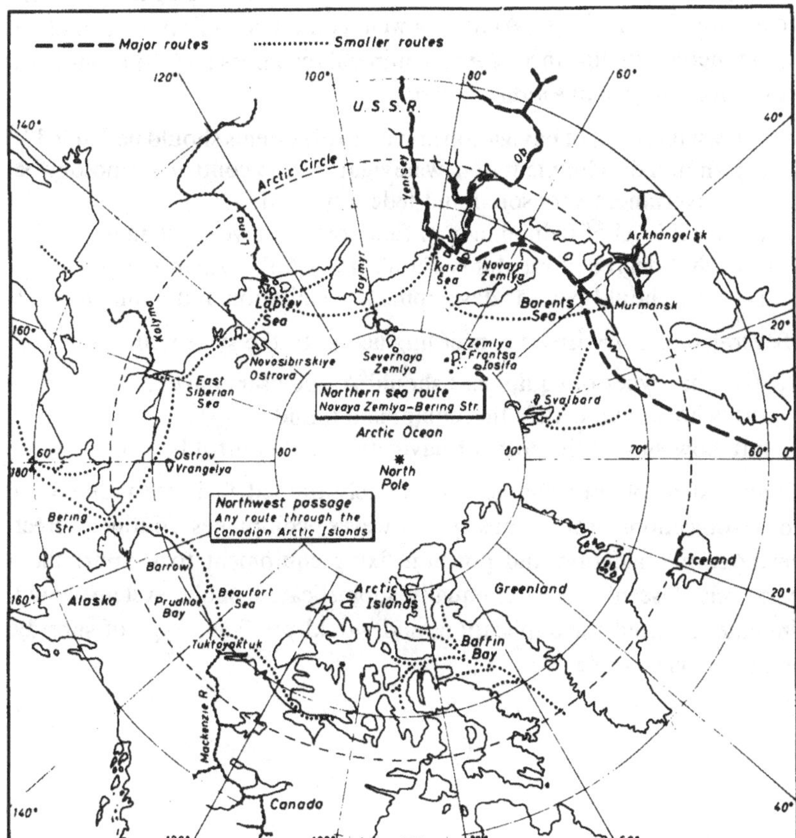

Source: R. Maybourn, "Arctic Navigation Past, Present and Future", *Journal of Navigation* V. 34, No. 1 (1981), p. 10.

(v) semi-submerged icebreaking cargo vessels;
(vi) ice islands;
(vii) submerged terminals;
(viii) tower terminals;
(ix) specialized vibration-ship cargo handling equipment for mineral handling;
(x) submersible equipment designed for underwater cargo handling;
(xi) protective coverings for dry cargoes;
(xii) underwater storage for bulk liquids and gases; and
(xiii) underwater light sources and markers.

Subsequent Arctic transport systems research must be conducted in an integrative fashion to achieve an optimal blend of operational factors such as cost, reliability, safety, capability and flexibility. However, many questions about the design and development of Arctic marine transportation systems are basically policy issues. Given the facts of low resident human populations in the Arctic, as well as the uncertainty surrounding the commercial significance of Arctic mineral resources and the economics of exploitation, key questions include:

(i) what type(s) of navigation and control systems should be funded?;
(ii) should development of new navigation and control technology be government-sponsored and federally-funded?;
(iii) what level of redundancy of function in a given system is 'safe'?;
(iv) what degree of reliability should the whole system possess?;
(v) who should pay for development-installation and maintenance?

Several subsidiary issues concerning access to the system are:

(i) what types of equipment should Arctic users have?;
(ii) who should certify this equipment?; and,
(iii) how should the costs of navigation and control be allocated?

Any system of user charges on publicly funded facilities is likely to arouse objections from industry. However, companies routinely erect private communication and position fixing equipment to support their operations. Use of a secure, equally sophisticated public system would eliminate the need for duplication, and would provide a type of subsidy for commercial traffic.

Notes

1. P. Bruun and G. Moe, "Design Criteria for Nearshore and Offshore Structures Under Arctic Conditions", *POAC* 81: *Sixth International Conference on Port and Ocean Engineering Under Arctic Conditons* V. 1 (Quebec City, 27–31 July 1981), p. 2.
2. R. Maybourn, "Arctic Navigation Past, Present and Future", *Journal of Arctic Navigation* V. 34, No. 1 (1981), p. 4.
3. Daniel F. Kelly, "The Chilling Challenge of Canada's Last Frontiers", *Surveyor* (February 1982), pp. 6–15.
4. M. Walker, "A Review of Navigation Systems and Their Applicability to the Arctic", in P.J. Amaria, A.A. Bruneau and P.A. Lapp (eds), *Arctic Systems* (1977) (Proceedings of a Conference on Arctic Systems, St John's, 18–22 August 1975), pp. 573–650.

Additional Sources

1. Amaria, P.J., A.A. Bruneau, and P.A. Lapp (eds). *Arctic Systems* (1977) (Proceedings of a Conference on Arctic Systems, St John's, 18–22 August 1975).
2. Armstrong, Terence. "The Northern Sea Route Today", in Louis Rey (ed), *Arctic Energy Resources* (1983) (Proceedings of the Comité Arctique International Conference on Arctic Energy Resources, Oslo, 22–24 September 1982), pp. 251–57.
3. Bailey, J.S. "ONR's Remote Sensing Program", *Naval Research Reviews* V. XXIV, No. 10 (1973), pp. 1–10.
4. Clark, H.L. "Some Problems Associated with Airborne Radiometry of the Sea", *Journal Applied Optics* V. 6, No. 12 (1967), pp. 2151–57.
5. Cracknell, A.P. (ed). *Remote Sensing in Meteorology, Oceanography and Hydrology* (1981).
6. Kelly, Daniel F. "The Chilling Challenge of Canada's Last Frontiers", *Surveyor* (February 1982), pp. 6–15.
7. Markham, W.E. *Ice Atlas*: *Canadian Arctic Waterways* (1981).
8. Moll-Christensen, E. "Oceanographic Satellite Data User's Workshop", *Bulletin American Meteorological Society* V. 62, No. 2 (1981), pp. 262–63.
9. Noble, V.E. and Felt, R.Y. "US Navy Planning for Satellite Oceanographic Data Exploitation" (Proceedings International Space Technical Applications, 19th Goddard Memorial Symposium, V. 52, Science and Technology) (1981), pp. 117–30.
10. Skarborn, S. "Marine Transportation in Arctic Waters", in *POAC* 81: *Sixth International Conference on Port and Ocean Engineering Under Arctic Conditions* V. 1 (Quebec City, 27–31 July 1981), pp. 117–35.
11. Transport Canada. *Ships Navigating in Ice*: *A Selected Bibliography* (1982).
12. Twitchell, P.F. "Meeting on Physical Oceanography and Satellites", *Bulletin American Meteorological Society* V. 60, No. 3 (1979), pp. 225–31.

5

Canadian arctic marine transportation: present status and future requirements

WILLIAM J. H. STUART
CYNTHIA LAMSON

I Introduction

The future of Canadian marine transportation development in Arctic waters remains uncertain. Although proponents of the Arctic Pilot Project (APP) – a proposal to ship LNG through the Northwest Passage – have withdrawn their application from the National Energy Board, a smaller-scale project involving production and transportation of oil from the Bent Horn field on Cameron Island by the marine mode is being promoted by Panarctic Oils Ltd. Beaufort Sea project proponents (Dome, Gulf and Esso) are still considering the feasibility of both tanker and overland pipelines and have not made final decisions on the preferred mode of hydrocarbon transport. The Beaufort Sea Environmental Assessment Panel, in policy recommendations to the federal government in July 1984, has cast further uncertainty over marine transportation of Arctic hydrocarbons by advising the government of Canada to withhold approval of the tanker option until the completion of two evaluation stages – general government research and preparation, and experimental testing of two Arctic Class 10 oil-carrying tankers.

If Canada chooses the marine option, will government and industry be prepared to meet the challenge? This paper examines the question of present Canadian shipping capacities and future shipping requirements from six perspectives: administration, icebreaking vessels, hydrographic knowledge, operational support services (including navigation and communications, vessel traffic management, search and rescue), crew training, and shipbuilding.

II Administration

Transport Canada is organized along functional lines with operating administrations for air, surface, and marine modes. The Marine Group includes the Canadian Coast Guard, the Canada Ports Corporation, the St Lawrence Seaway Authority, four Pilotage Authorities, nine Harbour Commissions, and the Canarctic Shipping Company (the federal government owns a controlling interest in the company, with the remaining shares owned by North Waters Navigation Ltd, a consortium of Federal Commerce and Navigation Ltd, Canada Steamship Lines and Upper Lakes Shipping).[1]

The Canadian Coast Guard (CCG) is the largest component of the Marine Group. CCG Headquarters are located in Ottawa, with day-to-day operations handled through five regional offices in St John's, Dartmouth, Quebec City, Toronto, and Vancouver (see Figure 5.1). To fulfil its mandate to "ensure safe and efficient navigation in Canadian waters", the CCG administers a number of current and planned programmes including:[2]

(i) Icebreaking and Escorting Services;
(ii) Ship and Port Safety;
(iii) Northern Resupply;
(iv) The Provision of Navigational Aids;
(v) Maritime Mobile Communications;
(vi) Vessel Traffic Management (VTM);
(vii) Search and Rescue;
(viii) Pollution Countermeasures;
(ix) Pilotage;
(x) Hydrography;
(xi) Ice Reconnaissance;
(xii) Arctic Shipping Control Authority;
(xiii) Proposed Coast Guard Northern Region;
(xiv) Training; and
(xv) Satellites.

Although the Coast Guard is the Canadian government's principal presence in Northern waters, at least five other federal departments share responsibility for maintaining and operating Northern marine services. The Canadian Coast Guard provides a support role to the Canada Oil and Gas Lands Administration (COGLA) for inspections of installations, structures and vessels not otherwise subject to inspection under the Canada Shipping Act; to the Department of Fisheries and Oceans (DFO) for hydrographic services; to the Department of Environment (DOE) for ice reconnaissance and meteorological services; to the Department of

National Defence (DND) for search and rescue; and to the Government of the Northwest Territories (GNWT) for marine pollution emergencies.[3]

Due in part to the seasonal nature and limited volume of marine traffic in Northern waters, responsibility for handling Northern operations has been, until recently, shared by the regional Coast Guard offices. The Director General of Newfoundland was responsible for the Labrador Coast. The Director General of the Western Region had authority over Mackenzie River and Western Arctic marine traffic. Coast Guard headquarters in Ottawa coordinated Eastern Arctic icebreaking and Northern resupply activities.[4]

Recent proposals for year-round shipping, however, have been catalysts for a number of administrative changes. In 1981, a Director, Coast Guard Northern (DCGN) was appointed to organize, develop and implement a Northern Region capable of providing year-round facilities and services. The creation of a CCG Northern Region was a response to the perceived requirements of industry and government. Industry, facing a confusing array of governmental committees, requested a one-window approach to the Coast Guard for addressing Northern development problems. In addition, existing regional divisions within the Coast Guard precluded the

Figure 5.1. Canadian Coast Guard administrative regions.

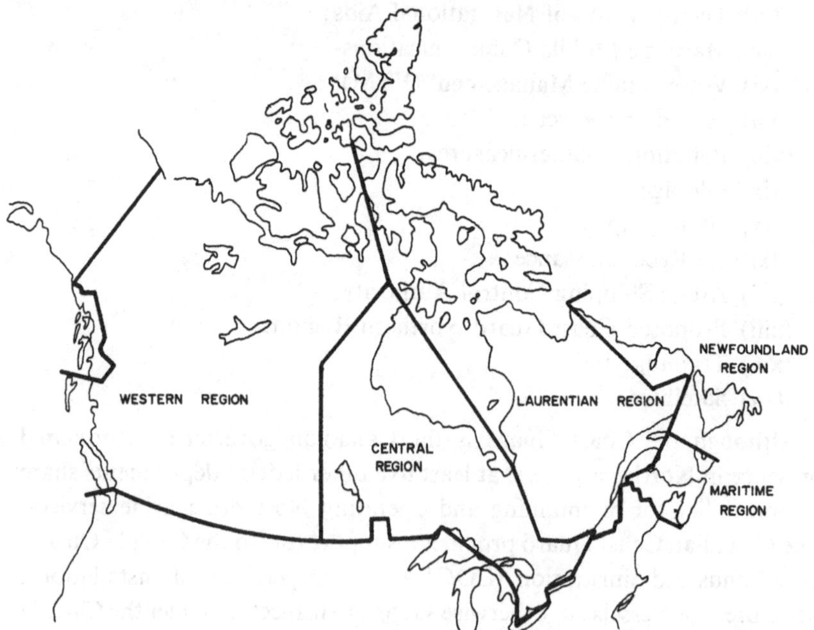

Source: Transport Canada, Coast Guard, Newfoundland Region, *Regional Profile of the Region* (1981).

administration of Northern services with the required central focus. Further, the creation of a separate Northern Region was a political demonstration of commitment to maintain authority for Northern waters decision making on behalf of the Ministry of Transport.

Numerous problems face the Northern Directorate. Perhaps the greatest difficulty is the lack of a clear Northern development policy, which leaves officials in an uncomfortable dilemma. The Northern Directorate must be ready to manage year-round shipping, if and when government and industry desire. But the lack of a clear policy, combined with the present economic downswing and the potential development of East Coast oil and gas fields (for example, the Venture gas field off Nova Scotia and the Hibernia oil field off Newfoundland) makes it difficult to justify the financing of Northern research and planning. As a result, the Northern Directorate's staff is rather skeletal, consisting of a director, staff from the Ship Safety Branch and Special Projects, and persons developing plans for a Polar 8 icebreaker. Moreover, industry has tended to overtake government in technical expertise, causing many shipping regulations to be outdated or based on best guesses.

If year-round shipping does occur in the North, the Northern Directorate will face an additional difficulty of choosing an appropriate organizational response. A full-scale regional office could be established in the North, but such a response would carry three negative ramifications. First would be the problem of finding a politically acceptable location. Yellowknife would probably be considered too far west and Frobisher Bay too far east, but perhaps a central locale such as Resolute would be an acceptable compromise. Second, in order to attract senior management to the North, government would likely have to make major expenditures developing recreational and social amenities. Third, since industrial managers have largely operated out of major Southern cities such as Vancouver, Calgary and Toronto, unless industry also chose to establish offices in the North, government managers would suffer a major travel disadvantage. As another option the regional office might operate from a Polar 8 icebreaker. While offering the attraction of a mobile office travelling all across the North, the option also raises the difficulty of keeping senior managers in the North. Perhaps two managers would be required for each position in order to retain a "fresh" management team aboard ship at all times. A third option, a compromise between the above two, would be to retain administrative functions in Ottawa but implement programmes through a commander and a small team of managers aboard a Polar 8 icebreaker. But this leaves unanswered a crucial, but politically sensitive, question: why is the Northern Region not located in the North?[5]

In response to recommendations made by the Environmental Assessment Review Panel for the Arctic Pilot Project (Northern component), an Arctic Shipping Control Authority and an Environmental Advisory Committee on Arctic Marine Transportation were also formed in 1981. The Control Authority, chaired by the Commissioner of the Canadian Coast Guard, is to monitor, assist and regulate ship movements in the Arctic on behalf of the government of Canada. The Environmental Advisory Committee, headed by the Director, Coast Guard Northern, is to advise the Control Authority concerning environmental and social effects of ship movements.[6] Such administrative innovations are laudable, but may, in actuality, be a case of placing the cart before the horse. No clear legislative or regulatory reins are present to guide the Control Authority on relationships with other departments and review processes. As emphasized by the Special Committee of the Senate on the Northern Pipeline, implementation of an adequate management regime for Canadian Arctic waters will require major expansion of financial resources and personnel:

> The Committee is sympathetic to the difficulties with which the Canadian Coast Guard is faced in spreading its meagre financial and personnel resources across the whole gamut of year-round marine services in arctic waters ... The Committee recommends: That in order to upgrade the Federal Government's year-round arctic response capability, the Canadian Coast Guard be provided with adequate financial and personnel resources to conduct R and D, to supply marine support services and to meet emergencies.[7]

III Icebreaking vessels

A substantial icebreaking fleet presently operates in Canada's Northern waters in order to exert a sovereign presence, provide Northern resupply services, and support industrial activities.

The Canadian Coast Guard icebreaking fleet consists of seven heavy icebreakers (see Table 5.1), one heavy icebreaker-cable ship, seven medium icebreakers-navaid tenders, three light icebreakers-navaid tenders, nine ice-strengthened navaid tenders and 16 non-ice-strengthened navaid tenders.

Canadian industry contributes greatly to icebreaking capability. The $36.8 million *MV Arctic* is a Canadian designed and built ice-strengthened cargo ship, owned and operated by Canarctic Shipping Company. The federal government owns a controlling interest in the ship, which transports lead zinc ore from Nanisivik Mine on North Baffin Island. Built in 1978 as the world's first heavy icebreaking bulk cargo ship, the *Arctic*

Table 5.1. *General particulars of CCG heavy icebreakers (metric measurement).*

Name	Region	Length	Breadth	Draft	Full-load displacement	Power kW	Range miles	Max. cruising speed	Fuel capacity	Crew	Completed	Shipyard
Louis S. St. Laurent	M	111.7	24.38	9.45	14509	17896	16000	17.7 Kts	3632	81	1969	Canadian Vickers, Montreal, PQ
John A. Macdonald	M	96.01	21.34	8.58	9307	11185	20000	15.5 Kts	2245	80	1960	Davie Shipbuilding Lauzon, PQ
Pierre Radisson	L	98.15	19.50	7.16	7721	10142	15000	16.2 Kts	2215	55	1978	Burrard Drydock Vancouver, BC
Sir John Franklin	N	98.15	19.50	7.16	7721	10142	15000	16.2 Kts	2215	55	1979	Burrard Drydock Vancouver, BC
des Groseilliers	L	98.15	19.50	7.16	7721	10142	15000	16.2 Kts	2215	55	1982	Port Weller Drydock Port Weller, Ontario
Norman McLeod Rogers	L	89.92	19.05	6.10	6506	8948	12000	15.0 Kts	1095	55	1968	Canadian Vickers Montreal, PQ
Labrador	M	81.99	19.44	9.30	7051	7457	23000	16.0 Kts	2641	85	1953	Marine Industries Sorel, PQ

Note: M – Maritimes
N – Newfoundland
L – Laurentian.

is a 28000 dwt bulk carrier having novel features including an icebreaking hull form (a compromise between icebreaking and efficient use of cargo space), an air bubbler system (which reduces friction between the ship and ice), a controllable-pitch propellor and a fixed propellor nozzle (which enhances manoeuvrability).[8] Dome Petroleum's experimental Arctic Class 3 icebreaker, the *Kigoriak*, has been operating in the Beaufort since its completion in September 1979. After transiting the Northwest Passage from the Atlantic to the Beaufort, it has been assisting Beaufort drilling operations and has been undergoing continuous performance testing. Experiments with ships like the *Arctic* and *Kigoriak* have provided engineers with new design features for the next generation of industry-built ships including Gulf Oil's Class 4 icebreakers, the *Kalvik* and *Terry Fox*, as well as two Class 4 supply vessels, the *Miscaroo* and *Ikaluk* (see Figures 5.2 and 5.3).

Serious questions arise, however, about the adequacy of Canadian icebreaking capacity to meet future shipping demands. The CCGS *Louis S. St Laurent*, currently the most powerful icebreaker in the fleet, is restricted to operating in the summer and autumn months. The Coast Guard fleet is rapidly approching maximum capacity, and close to one-third of the heavy- and medium-class vessels are nearing the end of their 30-year operational life.[9] In 1983 the Subcommittee on National Defence of the Standing Senate Committee on Foreign Affairs issued a report, *Manpower in Canada's Armed Forces*, which noted that Canada should maintain a more significant Northern presence to support its sovereignty claims:

Figure 5.2. Class 4 icebreaker/ice management vessel.

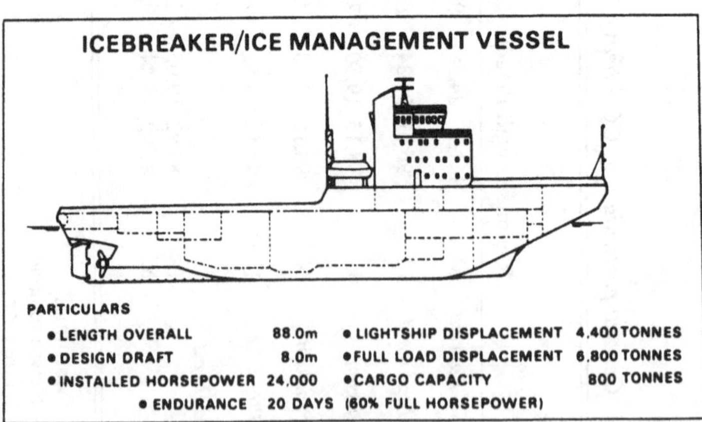

Source: Dome Petroleum Ltd, Esso Resources Canada Ltd, Gulf Canada Resources Inc., *Environmental Impact Statement for Hydrocarbon Development in the Beaufort Sea–Mackenzie Delta Region* (1982), V. 2, p. 5.4.

> The country must be able to control access to and enforce its jurisdictional claims over Arctic Waters. In peacetime, this presence should be maintained by icebreakers with capabilities equal or superior to those of commercial vessels now being designed for use in the North (and likely to grow in numbers as commercial exploitation of Arctic resources increases). Polar-8 icebreakers capable of operating nine to ten months of the year will undoubtedly be required in the next five to seven years.[10]

In order to provide essential icebreaking services, the Canadian government has been evaluating designs and costs of adding more powerful icebreakers (that is, Polar Classes 7, 8 and 10) to the CCG fleet. In 1981 after a five-year feasibility study, the nuclear design option was abandoned but approval was given to proceed with a design update for a conventionally-powered Polar 8 icebreaker. However, even a Polar 8 is foreseen to have limited operational utility:

> The Polar 8 will be able to take care of the foreseeable ice-breaking needs in the Arctic archipelago and Northwest Passage commencing in the late 1980s. In the 1990s, it is expected the development of high Arctic resources will eventually require the services of a Class 10 Polar icebreaker. Such an icebreaker would have year-round capability, without restriction, in all areas of the Arctic.[11]

As noted by Captain Tom Pullen, Canada has fallen far behind the Soviet Union in Arctic icebreaking capacity. The USSR, the world leader

Figure 5.3. Class 4 supply vessel.

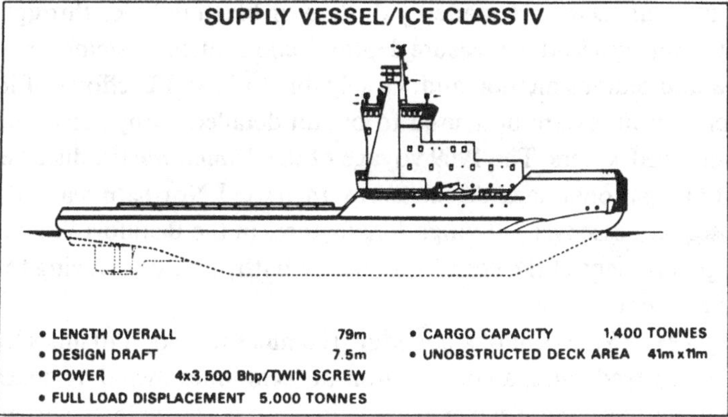

Source: Dome Petroleum Ltd, Esso Resources Canada Ltd, Gulf Canada Resources Inc., *Environmental Impact Statement for Hydrocarbon Development in the Beaufort Sea–Mackenzie Delta Region* (1982), V. 2, p. 5.4.

in icebreaking capability, has a fleet of 43 icebreakers. The three largest Soviet icebreakers have nuclear propulsion for unlimited endurance, displace more than 20000 tonnes, and develop 75000 horsepower. Canada, on the other hand, seems content to meet her Arctic responsibilities, for the next ten years at least, with ships which displace 7600 tonnes and develop only 13600 horsepower.[12]

The Canadian Coast Guard, however, faces difficulties in calculating future fleet requirements because no clear policy directives have been forthcoming to clarify the Coast Guard's expected role in assisting expanded marine transport in the Northwest Passage. A number of serious questions are still outstanding. For example, will industry build ships requiring government icebreaker assistance? Or, will industry construct ships able to manoeuvre independently through the ice? The "need to know" by the Coast Guard is critical, given the lengthy time necessary, seven to ten years, for ship design and construction.[13]

IV Hydrography

The Canadian Hydrographic Service (CHS) operates from DFO headquarters in Ottawa and four regional offices in Dartmouth, Nova Scotia (Bedford Institute of Oceanography), Burlington, Ontario (Bayfield Laboratory), Quebec City (Champlain Centre for Marine Science and Surveys) and Patricia Bay, British Columbia (Institute of Ocean Sciences).

Hydrographic data acquisition has been a slow process, although technological advances have made surveying projects easier and charts more reliable, and international political developments have stimulated more intensified hydrographic work. In 1959 the first surveys of the polar continental shelf were made by drilling or blasting holes through the ice and lowering a lead to measure depths. Echo sounder systems replaced this slow and tedious method and, largely due to R and D efforts of industry, tracked vehicles are now used to obtain detailed bathymetric surveys of ice-covered waters. The 1969 voyage of the *Manhattan* produced evidence that navigational hazards existed in uncharted Northern waters, and the subsequent discovery of pingo-like features in the Beaufort Sea convinced the government of the need to accelerate bathymetric surveying to achieve 100 per cent coverage.

By 1982 85 per cent of a corridor, 100 miles wide by 170 miles long, had been surveyed using a combination of techniques including spaced echo sounding lines and electronic sweeping. However, with the exception of Lancaster Sound, Amundsen Gulf, sections of Hudson Strait and the Labrador Sea, substantial areas of Northern waters remain either partly surveyed or are not surveyed to modern standards[14] (see Figure 5.4). To meet anticipated demands of increased marine traffic, and to strengthen Canadian claims to sovereignty, comprehensive surveying and charting will be essential.

The Canadian Hydrographic Service, however, faces four major problems: shortage of trained experts in hydrography, shortage of survey vessels (most icebreakers must be borrowed from the Coast Guard), lack of financial resources to hire adequate staff and to purchase sufficient equipment, and lack of a systematic, long-term plan for hydrographic research. As a result, major gaps remain with respect to the hydrographic knowledge of Canada's Northern waters.

Solution to the knowledge problem may depend on three factors. First, the Hydrographic Service is considering the use of new technologies to increase the accuracy and coverage of surveys. For instance, remote-controlled submersibles, dropped through holes in the ice, may cruise over large areas and collect data on tides, currents and sea floor characteristics. Data would be retrieved manually or perhaps be transmitted via satellite

Figure 5.4. Status of surveys (1981).

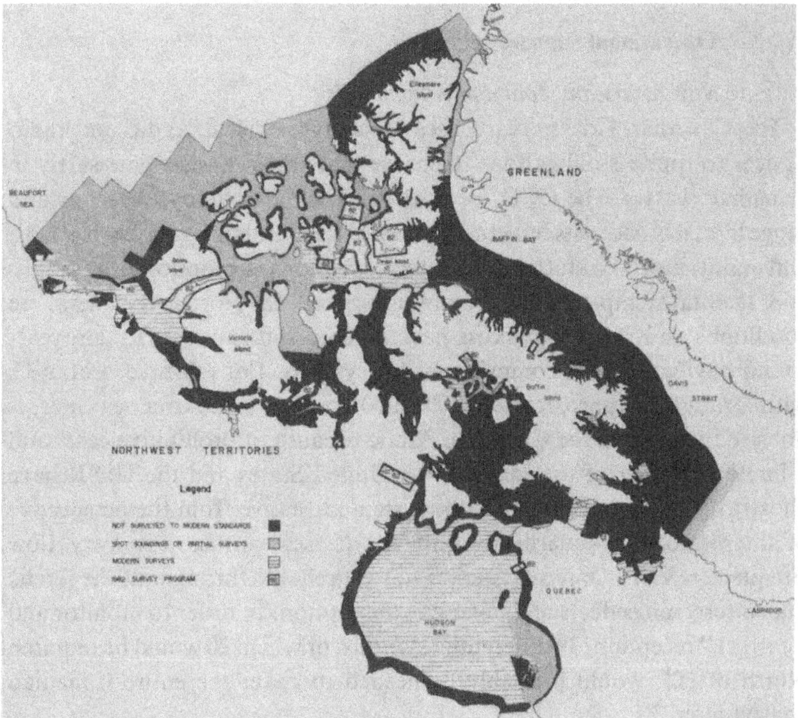

Source: Canadian Hydrographic Service, M-270.

to a ship or shore base. Second, the Canadian Hydrographic Service must be given a high governmental profile to assure adequate funding from the Treasury Board. Four options for reorganization have been suggested: combining the Hydrographic Service and the Coast Guard, perhaps with a separate Minister of Marine Transportation, under the Department of Transport; joining the Hydrographic Service and the Coast Guard under the Department of Fisheries and Oceans; uniting the Hydrographic Service and the Coast Guard under a new government agency called the Department of Oceans; or making the Hydrographic Service an arm of the Canadian Navy. On the negative side, such reorganization could lead to greater fragmentation (for example, by splitting transportation management into marine and air-surface sectors) and further administrative confusion such as that produced by the combination of the Department of Regional Economic Expansion and the Department of Industry, Trade and Commerce into the Department of Regional Industrial Expansion. Third, government must formulate a long-range plan for Northern development and, within such a plan, establish priorities for hydrographic research.[15]

V Operational support services

A Navigation and communications aids

The Canadian Coast Guard uses an array of visual, audio and radio devices to fulfil its obligations to ensure that ships can navigate safely in Canadian waters. The CCG provides and maintains buoys, beacons and range lights, broadcasts bulletins on weather, ice and navigational hazards, and maintains radio stations at strategic locations to monitor distress calls and facilitate ship-to-shore communications. In Northern waters, ice conditions and restricted access have made it difficult to maintain year-round navigation and communications systems. For example, Loran-C, with a range of approximately 700–1000 nautical miles over seawater, is not yet available in the Canadian Arctic because of prohibitive costs and a limited number of users. (Both the United States and the USSR have chosen Loran-C, which makes the system attractive from the perspective of international standardization.)[16] At the sea surface, the Very Low Frequency (VLF) Omega system is not yet reliable throughout the Arctic due to screening effects and polar cap absorption. In order to monitor and "correct" reception, 39 differential stations, of which 20 would be required North of 60°, would probably be needed to cover the entire Canadian coastal area.[17]

An additional problem to navigators is that the operational capability of many components of shipboard navigational equipment, such as echo sounders, compasses, and radars, is reduced in Northern waters, particularly when ice is present. As noted by the *Pilot of Arctic Canada*, such conditions require personal skill and experience on the part of operators, who are forced to navigate without the benefit of conventional aids:

> Navigation in the waters of the Arctic regions is beset with many problems not encountered in lower latitudes. These problems are created through various factors such as weather conditions, ice charts, and lack of navigation aids {...} . The difficulties so far mentioned are not insurmountable if the navigator accepts the fact that unconventional methods are required to meet unconventional situations. It may be impossible to fix a ship's position geographically, but fixing relative to land masses, is usually possible and the fact that the land mass itself is inaccurately fixed is immaterial.[18]

Environmental and atmospheric obstacles to accurate positioning and navigating may be remedied by satellite technologies. The CCG expects to become increasingly involved with other federal departments, Energy, Mines and Resources, Communications, and National Defence, who have pioneered satellite R and D in Canada. Several satellite systems for communications, navigation, surveillance and search and rescue are either in use, or are in final stages of design. The CCG currently leases three route circuits, provided by the ANIK A satellite, for remote operation of high-frequency radio sites. Additional use of the ANIK A system is expected as demand increases for mobile communications systems in Northern waters.

Two US designed and operated navigation satellite systems promise to be of use to Northern mariners. The six-satellite TRANSIT system assists position fixing, but its utility is restricted because fixes are not available on a continuous basis. NAVSTAR, a satellite system operated by the US Air Force, could be fully operational by 1989, and based on preliminary experiments could provide greater navigational accuracy than Loran-C, as well as global coverage. Civilian access to a somewhat lesser degree of accuracy, without compromising military security, is intended. A user's fee has been proposed.[19]

Other satellite systems to aid Arctic communications and navigation exist. RADARSAT, a satellite with synthetic aperture radar, as well as visual and microwave sensors, is being developed to provide 24-hour information on ice conditions and vessel movements. SARSAT, an experimental satellite programme sponsored by Canada, France and the

United States, is providing a search and rescue alerting capability to Arctic mariners. INMARSAT (International Maritime Communication System), while not giving communication coverage to the Northwest Passage proper, does cover the Beaufort Sea in the west and the waters up to Baffin Bay in the east. MSAT (Mobile Satellite), a joint US–Canadian effort, is presently in the detailed design stage and could fill the satellite communication gap in the Central Arctic by 1988.[20]

In sum, limited demand, the high costs of installation and maintenance of sophisticated navigation and communications systems, as well as the rapidity of technological change, have combined to delay comprehensive support services to Northern mariners. Still outstanding are questions of standardization, and a pressing need exists to establish some type of forum whereby all potential users can discuss and debate their requirements, as well as their willingness to bear some of the costs associated with R and D, installation, and maintenance.

B *Vessel traffic management*

Vessel traffic management (VTM), for purposes of environmental protection through "safe, expeditous and orderly traffic flow", was inaugurated in Canada along the St Lawrence Seaway in 1966. Since the tanker *Arrow* ran aground in 1970 and spewed her cargo of heavy oil along the Cape Breton and Nova Scotia coasts, vessel traffic management systems have expanded in all marine regions and may consist of four levels of ship control:

Level 1. *Ship-to-Ship Information System*: *At designated calling-in points, vessels broadcast on specific frequencies information on their position and intentions.*

Level 2. *Shore-to-Ship Advisory System*: *Vessels are required to obtain clearance from the VTM shore station prior to entering the area designated as a traffic management area. This clearance is related to regulations currently in effect concerning the capability of a vessel to navigate safely without pollution risk while in Canadian waters.*

Level 3. *Shore-to-Ship Regulating System*: *Traffic management is enforced to the extent of scheduling vessel movements through the VTM area by instructing vessels to anchor, to hold their present position, to leave an area or to proceed.*

Level 4. *Shore-to-Ship Control System*: *Radar surveillance is maintained over the entire VTM area to allow positive identification of targets and the actual plotting is based on real-time observed data. With radar surveillance of clearly identified targets, it will be possible eventually to introduce a system of traffic regulations which will include direction from a shore station to a vessel requiring that vessel to take positive and direct action in circumstances of an especially hazardous nature. This capability will be assigned only to those VTM control centres having fully trained and qualified traffic regulators.*[21]

In the Arctic, vessel movements are currently managed under the NORDREG system, in effect since August 1977. NORDREG is voluntary in nature and operates seasonally from July to November. Vessels transmit information to the Operations Centre located in Frobisher Bay via Coast Guard radio stations; and in turn, the information is processed by experienced marine traffic regulators to ensure detection of deficient and defective vessels and to ensure compliance with applicable Canadian regulations.

In 1981 Transport Canada, in an Arctic Marine Services Policy, called for a strengthening of vessel traffic management in Canada's North. Besides urging application of the NORDREG system to all Arctic waters and the designation of Arctic shipping routes, taking account of navigational safety, system efficiency and environmental concerns, the Policy suggested provision for:

- (i) monitoring and control of Arctic vessel movements;
- (ii) direction and control of Canadian Coast Guard assistance and support to shipping;
- (iii) pre-clearance of vessels for Arctic entry based on cargo, destination, etc.;
- (iv) enforcement of operational procedures;
- (v) initiation of Search and Rescue and other emergency response;
- (vi) provision of information concerning navigation, weather and ice; and
- (vii) preparation, coordination and monitoring of the "NOTSHIPS" broadcast.[22]

For purposes of controlling shipping pollution in waters adjacent to the Arctic mainland and islands, the Arctic Waters Pollution Prevention Act established 16 shipping safety zones. Restrictions to movement are based on ice thickness as well as on ship design and capability. For example, only a Polar Class 10 icebreaker would have the capability to move in all 16 zones on a year-round basis; a vessel in category 14 (little or no ice-strengthened capability) is allowed to operate in zone 13 (Lancaster Sound) between 15 August and 20 September, and at no time of the year is it permitted to enter zones 1–6 (Gulf of Boothia and areas to the north and west of Resolute Bay).[23]

Constraints on implementation of a fully operational, year-round and mandatory VTM system in the Arctic include: enforcement capability, costs of installation, operation and maintenance, as well as resistance on the part of some mariners to relinquish their traditional freedom and authority to be encroached upon by an external control authority. In addition, the uncertainties of ice movement make the establishment of designated traffic lanes problematic. Icebreakers always seek the line of least resistance and to achieve this, they must steer a sinuous course. The Arctic, unlike more temperate areas, does not adapt kindly to such things as VTM.[24]

C *Search and rescue/salvage*

The Canadian Armed Forces are responsible for coordinating all search and rescue (SAR) activities in Canada. Operations are directed from four Rescue Coordination Centres, and the Coast Guard operates two Search and Rescue Emergency Centres (see Figure 5.5). However, none of the existing Centres are located to deal effectively with Arctic operations. Indeed, for part of the year, SAR capability in Northern waters is non-existent. Fortunately, the number of incidents requiring assistance

Figure 5.5. Rescue Coordination Centres (RCC) and Search and Rescue Emergency Centres (SAREC).

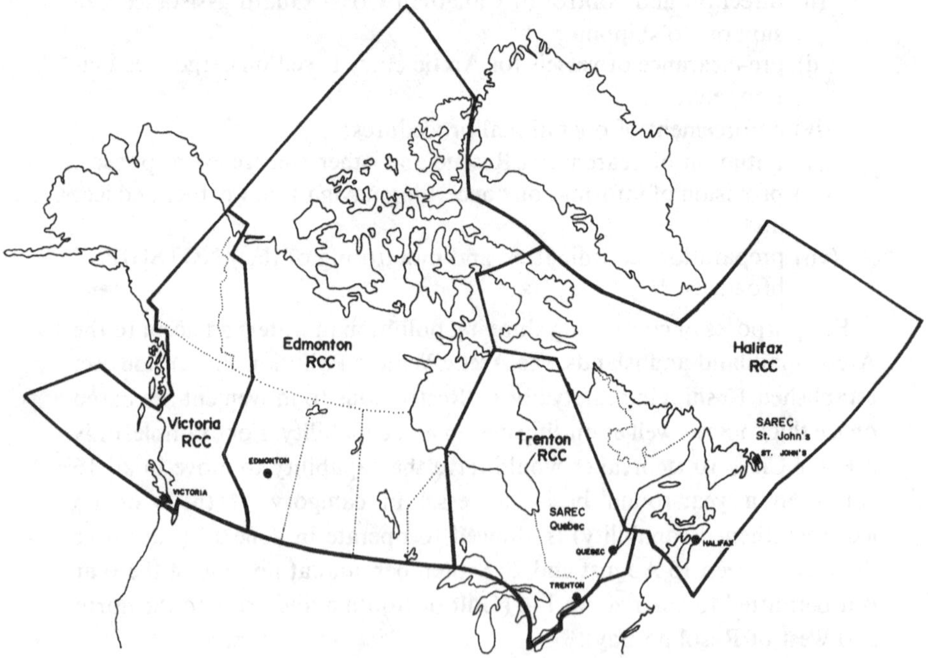

Source: Government of Canada, *Search and Rescue in Canada*.

have been relatively low since use of Northern waters has been limited. For example, in 1978, approximately 18 distress incidents were reported in relation to commercial shipping.[25]

As noted by the Brief of the Department of Transport to the Senate Committee on the Northern Pipeline, the implementation of adequate search and rescue services will depend upon an increased volume of shipping traffic:

> Over the years, many commercial ships and even Coast Guard ice-breakers have suffered damage to hull platings in the ice, and others have been severely damaged. There have been some potential disasters, where for instance a tanker suffered hull damage. There is great difficulty in planning marine rescue in the North. Even as the potential for a major incident grows, the lives-at-risk per thousand square miles are so few, and the operating conditions so variable, as to preclude primary SAR (Air/Marine) resources on standby for some years yet.[26]

Salvage is a variation on search and rescue which pertains to property. The Department of Transport formerly subsidized salvage companies to maintain a national capability, but support was withdrawn fifteen years ago, and services have diminished significantly. In the event of a late season casualty, helicopters may be utilized to evacuate crews and some cargoes when conditions permit. Temporary repairs are possible if the hull is damaged, but one must wait until the following year to refloat a vessel. In Canada no lightering barges are capable of operating in ice should a tanker sustain damage requiring discharge of cargo, and should discharge to another tanker be impracticable. For offshore drilling a certain towing facility is likely to be available, but it will probably be fully dedicated to its prime task of exploration. For reasons ranging from the law of the sea to anti-pollution measures, it is essential that Canada be able to meet any and all such eventualities.[27]

For search, rescue and salvage operations, satellite technology again offers a remedy to problems associated with delayed response time in the detection and location of vessels and craft in distress. Emergency Positioning Indicating Radio Beacons (EPIRBs), carried aboard certain classes of marine vessels, will be capable of detection by the planned SARSAT system which can relay positioning information back to ground stations which, in turn, can notify rescue operations. The SARSAT system is connected with meteorological and environmental observation spacecraft launched by the US National Oceanic and Atmospheric Administration (NOAA).[28]

The federal government has established an Interdepartmental Committee on Search and Rescue (ICSAR) to assess the need for SAR services throughout Canada and to help ensure efficient and smooth operations. Members of the committee include representatives from the Departments of Defence, Transport, Indian Affairs and Northern Development, and Energy, Mines and Resources, and the Royal Canadian Mounted Police. Each year the federal government formulates a national SAR plan which recommends improvements based on regional assessments. However, a Cabinet Committee on Foreign Defence Policy, tabling a report on search and rescue, criticized Canadian SAR response capability and management and cited four areas of concern: lack of priorities and national SAR standards and goals, the lack of major maritime disaster contingency plans, insufficient investigation, analysis and follow-up, and overlapping managerial jurisdictions. The Report emphasized three priority issues: the desirability of designating overall responsibility for SAR to a single Minister, other than the Minister responsible for National Defence, the need to improve weather services, and the need to improve SAR response-training programmes.[29]

VI Manpower and training

Recognizing that even the most superb ship will be of little value if she is operated by an ill-trained crew, the federal Cabinet in 1975 established the National Advisory Council on Marine Training (NACMT) to review and assess present and future human resource requirements of the shipping and fishing industries. Regional Committees of the NACMT were established from coast to coast to provide valuable assistance to federal and provincial governments on ways and means of carrying forward NACMT's mandate. On the advice of all parties concerned, the first line of attack on the shortage of qualified marine personnel has been to commence the upgrading of existing navigation training facilities and the provision of specialized marine emergency training facilities and equipment.[30] Although some upgrading of facilities and extension of training has been achieved, responsibility for training qualified marine personnel remains divided among several government departments and industry. Even within a single department decentralization of training and educational facilities may be an obstacle to comprehensive preparation for Northern operations. For example, the Canadian Coast Guard College is located in Sydney, Nova Scotia, while the Transport Canada Training Institute is situated in Cornwall, Ontario.

Canada still faces some manpower shortages in various marine industries, and a recent assessment of future officer and crewing personnel requirements for the shipping and fishing industries, as well as the federal fleets, indicates that shortages are expected, due to intensified offshore activity, fleet replacement and expansion.[31]

The introduction of year-round shipping and related marine activity in Northern waters will place new demands upon Canadian mariners. A new breed of sea-going professionals, who have been trained to cope with the technical and social demands placed upon them, is urgently needed. These individuals will be working in the most hostile environment ever faced by mariners with the most complex merchant ships ever built. Stable, intelligent, and technically competent individuals will be required, even though a two-crew rotational system will be adopted to relieve the strain of winter and isolated operations.

Because real-time training is an exceedingly slow and costly process, feasibility studies are being conducted to examine the benefits of acquiring a Canadian shiphandling simulator. The need for a simulator appears to be great, since trainees must presently be sent to the United States for navigational training on a somewhat outmoded computer-based simulator, and costs of extensive simulator training would likely be recouped in avoidance of future ship groundings and associated environmental and property damage. To justify the $ 20-million price tag, a Canadian simulator would be likely to have to perform double duty – as a training tool and as a research and development aid (for example, a simulator could be an ideal vehicle for designing marine terminals). Practical problems which will have to be resolved are: where should the simulator be located? who should run it? and, can a simulator adequately model the movements and characteristics of ice?[32]

VII Shipbuilding

The Canadian shipbuilding industry consists of approximately 12 yards, which in 1979 employed some 12,000 workers. Production comprised 163,200 gross tons of new building.[33] Although accelerated activity in Northern waters has the potential of stimulating the shipbuilding industry through construction, repair and maintenance of exploration and supply vessels, it is beyond existing Canadian capacity to construct large ice-breaking tankers and LNG carriers. Capital investment in excess of $ 50 million would be needed to expand an existing yard to enable construction of every transportation vessel required.[34]

Three major problems have plagued the Canadian shipbuilding industry. First, Canadian shipyards lack the physical capacity to handle approximately 20 per cent of predicted vessel requirements, particularly large vessels with the newest technologies, such as LNG carriers and supertankers. Second, the National Energy Program through PIP (Petroleum Incentives Program) grants made it more advantageous for oil companies to lease rather than purchase rigs and ships. Instead of purchasing vessels from Canadian shipbuilders, the oil companies leased foreign-owned rigs and ships. Third, an economic downswing has caused a surplus of shipping capacity resulting in a corresponding decline in ship-construction orders. In 1982 some 6000 persons in the shipbuilding industry were laid off.[35]

The oil and gas industry, particularly Petro-Canada in support of the Arctic Pilot Project and Dome Petroleum in support of Beaufort Sea development, have taken major initiatives in Arctic shipbuilding design and engineering. For example, from 1976–1985, Dome allocated $200 million for transportation research and development.[36]

Due to the lack of domestic capability, the Canadian industry has been forced to acquire foreign expertise and collaborate with foreign yards fairly extensively. As part of an effort to acquire and foster domestic capacity, the National Research Council is building a $56-million Arctic Vessel Marine Research Centre on the campus of Memorial University in St John's, Newfoundland. The facility is expected to be the finest of its kind anywhere in the world, and will be the principal Canadian research centre to tackle problems associated with ice-vessel interactions.

Beleaguered by lack of domestic expertise, capacity, and high labour costs, the Canadian shipbuilding industry is also affected by industrial policies which constrain the acquisition of essential materials such as steel. Problems of quality, supply, and ability to deliver within a reasonable time-frame are often cited by industry spokesmen as constraints on competitive and efficient production.[37] According to the Transportation Subcommittee Report of the Major Projects Task Force, the percentage of Canadian material, machinery and equipment which could be provided under existing supply conditions is only 38 per cent.[38]

The Special Committee of the Senate on the Northern Pipeline responded to industry petitions for expanded shipbuilding capacity by recommending that, "immediate consideration be given to developing a Canadian large-vessel shipyard capability to supply not only all vessel requirements for Arctic development but also compete for similar undertakings".[39] Total shipbuilding expenditures required by Canadian projects from 1980–1991 are listed in Table 5.2.

VIII Summary

In 1981 the Canadian Marine Transportation Administration issued an Arctic Marine Services Policy which identified four development priorities

Table 4.4. *Total projected shipbuilding expenditure profile, by project* (*$ million, 1980*).

Project	1980	1981	1982	1983	1984	1985	1986	1987	1988	1989	1990	1991	Total	Rating	Percentage
Beaufort Sea Oil and Gas	—	461	857	1 544	1 598	1 647	1 620	1 189	1 320	604	—	—	10 840	H	60
Arctic Pilot Project	—	46	117	480	344	241	72	—	—	—	—	—	1 300	H	7
Canadian Flag Deep Sea Fleet	—	—	57	151	213	542	421	426	356	265	169	—	2 600	L	14
Great Lakes Fleet Expansion	—	10	27	111	81	55	16	—	—	—	—	—	300	H	2
Commercial and Government Fisheries	45	184	199	213	228	217	23	23	24	25	—	—	1 181	M	6
Coast Guard – Fleet Investment	30	40	49	55	57	58	57	58	57	58	57	58	634	M	4
C.P.F. Programme	—	—	—	26	69	98	249	193	195	164	122	78	1 194	H	7
Total annual	75	741	1 306	2 580	2 590	2 858	2 458	1 889	1 952	1 116	348	136	18 049		100

Source: Ship Transportation Sub committee Report to the Major Projects Task Force on Major Projects to the Year 2000 (October 1980), p. 32.

to meet anticipated expansion of Northern shipping. Research and development respecting Arctic vessel design and performance, navigation and communications, pollution effects and control, and search and rescue, must be expanded. Standards, regulations and codes governing the design, construction and operation of Arctic vessels must be updated. Coast Guard and industry personnel must receive specialized crew training for Arctic service. An appropriate administrative structure must be established to ensure the timely implementation of Arctic marine policies and services and to facilitate liaison with industry and other government departments.[40]

Cabinet and Treasury Board have partly responded to such pleas. A five-year R and D plan, submitted by Transport Canada, Fisheries and Oceans and Environment Canada, requested approximately $51 million for development of Arctic marine transportation[41] (see Table 5.3).

The Departments argued that unless an enhanced and integrated Arctic marine R and D programme was initiated, Canada would incur unacceptable technological and economic risks.[42] Government responded by allocating approximately $21.9 million for Arctic marine R and D from 1981–1982 to 1985–1986.[43] Whether such a budget is adequate is likely to depend on the timing of industrial development.

Clearly, the feasibility of year-round Arctic marine transportation is rapidly becoming a reality; however, the capability of the Canadian government to support, regulate and manage year-round commercial marine traffic effectively remains uncertain. As numerous government committee and task force reports have stressed, Canada cannot afford to delay in making critical decisions about its willingness to support extension

Table 5.3

Programme area	Total costs (1981–1982 to 1985–1986)
	($000)
Arctic class vessel research	$16,720
Icebreaking	2695
Arctic ways information	12,050
Navigation support systems	11,145
Ports, terminals, and platforms	550
Oil spill detection, clean-up and containment	2500
SAR, emergencies and crew training	5700
	$51360

and expansion of services in support of Arctic marine transportation. The era of assessment and reassessment of Canadian capability is rapidly passing. It is now time for thoughtful but tough decision-making.

Notes

1. Transport Canada, *Canadian Marine Transportation Administration* (1981).
2. Proceedings of the Special Committee of the Senate on the Northern Pipeline, 1st Sess., 32nd Parliament, Issue 30 (15 June 1982), p. 30A:14. (hereinafter referred to as *Proceedings*, SCSNP).
3. Transport Canada, Canadian Marine Transportation Administration, *Position Statement to the Beaufort Sea Environmental Assessment Review Panel* (16 August 1982), pp. 3–6 (hereinafter referred to as *Position Statement* TC/CMTA).
4. Personal communication to Professor Douglas M. Johnston from Carol Stephenson, Director, Coast Guard Northern (30 May 1983).
5. Commentary from participants at a Workshop on Arctic Marine Transportation convened by the Dalhousie Ocean Studies Programme on 10 March 1983 at the Chateau Laurier, Ottawa.
6. *Proceedings*, SCSNP, *supra* note 2, pp. 30A: 46–54. See also *Memorandum of Understanding Between the Department of Transport and, Jointly, the Department of Environment and the Department of Fisheries* (April 1982).
7. Special Committee of the Senate on the Northern Pipeline, *Marching to the Beat of the Same Drum*: *Transportation of Petroleum and Natural Gas North of* 60 (March 1983), p. 40. (hereinafter referred to as *Report*, SCSNP).
8. Transport Canada, *Canadian Marine Transportation Administration* (1981); *APOA Review*, V. 6, No. 1 (Spring/Summer 1983), p. 21; Dome Petroleum Ltd, Esso Resources Canada Ltd, Gulf Canada Resources Inc., *Environmental Impact Statement for Hydrocarbon Development in the Beaufort Sea–Mackenzie Delta Region*, V. 2 (1982), pp. 5.4–5.5.
9. *Position Statement* TC/CMTA, *supra* note 3, p. 16.
10. Sub-committee on National Defence of the Standing Senate Committee on Foreign Affairs, *Manpower in Canada's Armed Forces* (1983) (hereinafter referred to as *Manpower in Canada's Armed Forces*).
11. *Proceedings*, SCSNP, *supra* note 2, p. 30A:54.
12. See Captain T.C. Pullen, "Arctic Marine Transportation Issues", A Report to the Canadian Arctic Resources Committee (1983), pp. 7–8; see also *Manpower in Canada's Armed Forces*, *supra* note 10; and Terence Armstrong, "Northern Sea Route", in Louis Rey (ed.), *Arctic Energy Resources* (1983) (Proceedings of the Comité Arctique International Conference on Arctic Energy Resources, Oslo, 22–24 September 1982), pp. 251–57.
13. *Supra* note 5.
14. See Canadian Hydrographic Service, 1983 *Operational Plan*, *Hydrography*, and *Status of Surveys* (1981) (Map 270).
15. *Supra* note 5.
16. M.A. Turner, "Arctic Traffic Management, Navigation and Communications Systems", in P.J. Amaria, A.A. Bruneau, and P.A. Lapp (eds), *Arctic Systems* (1977) (Proceedings of a Conference on Arctic Systems, St John's, 18–22 August 1975).
17. *Ibid.*
18. *Ibid.*, pp. 31–33; see also, Captain M. Walker, "A Review of Navigation

Systems and their Applicability to the Arctic", in *Arctic Systems*, *supra* note 16, pp. 573–649.

19. *Proceedings*, SCSNP, *supra* note 2, pp. 30A:57–58.
20. *Supra* note 5.
21. Turner, *supra* note 16, pp. 26–31.
22. *Proceedings*, SCSNP, *supra* note 2, p. 30A:58; see also, *Position Statement* TC/CMTA, *supra* note 3, Appendix 1, p. 33.
23. Turner, *supra* note 16, p. 43.
24. Captain T.C. Pullen, *supra* note 11, p. 14.
25. George G. Leask, "Overview of Arctic Activities of the Canadian Marine Transportation Administration", in *Marine Transportation and High Arctic Development: Policy Framework and Priorities* (Symposium Proceedings, Montebello, Quebec, 21–23 March 1979) p. 35.
26. *Position Statement* TC/CMTA, *supra* note 3, p. 5.
27. Leask, *supra* note 25, p. 35.
28. *Proceedings*, SCSNP, *supra* note 2, pp. 30A:58–59.
29. Cabinet Committee on Foreign and Defence Policy, *Report on an Evaluation of Search and Rescue* (1982), pp. 359–412.
30. *Proceedings*, SCSNP, *supra* note 2, p. 30A:55.
31. Canada Employment and Immigration Commission, Fisheries and Oceans, and Transport Canada, *Marine Human Resources*: 1981 *Inventory and Training Requirements for* 1982–85 *and* 1986–90 (1982).
32. *Position Statement* TC/CMTA, *supra* note 3, pp. 39–40.
33. *Ibid.*, Appendix I, pp. 14–15.
34. *Ibid.*, p. 14.
35. *Supra* note 5.
36. Transport Canada, "Arctic Marine Transportation R & D Five-year Plan (1981/82 to 1985/86)", Discussion paper (1981), pp. 5–6.
37. TC/CMTA, *supra* note 3, p. 15; see also, *Report*, SCSNP, *supra* note 7, p. 55.
38. *Report*, SCSNP, *supra* note 7, p. 55.
39. *Ibid.*, p. 56.
40. See *Proceedings*, SCSNP, *supra* note 2, pp. 31A:15–59.
41. Transport Canada, *supra* note 36, pp. 21–28.
42. *Ibid.*, p. 19.
43. *Position Statement* TC/CMTA, *supra* note 3, p. 44.

6

Northern decision making: a drifting net in a restless sea

DAVID L. VANDERZWAAG
CYNTHIA LAMSON

I Introduction

Fish, nets, corporations and government. Although seemingly unrelated, the four entities have much in common. By instinct, corporations desire to roam the economic sea at will, free to pursue prey – minerals, oil and gas so as to maximize profits. Such "free-swimming" entails at least three dangers. Corporations may devour one another – the monopoly problem. Corporations may overexploit the flora and fauna–environmental and social values. Or corporations may themselves flounder because of outside predations such as inflation, high interest rates or trade restrictions. Government acts as a protective net, directing corporate movement into politically desired directions.

This chapter examines the Canadian decision-making net for controlling Northern development. Section II surveys the drifting floats, the convolutions in Northern policies. Section III discusses the administrative mesh, the threads of governmental bureaucracy operating in the North. Section IV reviews the legislative mesh, the threads of federal legislation and regulations controlling industrial development. Section V views the net in action by focusing on governmental reviews of ten major industrial proposals during the past decade – the Alaska Highway Gas Pipeline, Beaufort Sea hydrocarbon development, Norman Wells Oilfield expansion, the Arctic Pilot Project, Panarctic Arctic Islands exploration, Polaris Mine, Nanisivik Mine, North Davis Strait, South Davis Strait, and Lancaster Sound drilling. The role of Cabinet, while treated tangentially,[1] and the role of Parliament[2] are beyond the scope of the present paper.

No one study will ever capture all the intricacies of the decision-making net. Much will always remain hidden by the murky waters of informal consultations among government officials, industry leaders and public interest groups. Much will always remain shrouded by "back room" decision making of Cabinet and departmental officials. Nevertheless, the rough outline of the net may be viewed, however darkly.

II Government policies: the floats

Government policies – the goals upholding and guiding the decision-making net – show a drift between industrial promotion and environmental-social protection. In his comprehensive Northern policy statement in 1972 Jean Chretien, then Minister of Indian Affairs and Northern Development, attempted to set out a well-anchored string of policy floats. Improving Northern social conditions, enhancing the environment, and stimulating renewable resource development were clearly ranked above encouraging non-renewable resource projects.[3] Such an ordering seemed compatible with the societal current in the early 1970s for a clean environment and an improved quality of life.

However, the OPEC oil embargo in 1973 and the exciting discoveries of offshore oil in Prudhoe Bay and in the North Sea created a counter-current. Demands for secure energy supplies spurred other government officials to fish experimentally for new policies. On 12 July 1973 the Minister of State for Science and Technology announced a new Canadian oceans policy with a pro-development slant for the Arctic: "Canada must develop and control within its own borders the essential elements needed to exploit offshore resources by ... achieving within five years world recognized excellence in operating on and below ice-covered waters ...".[4] On 27 April 1976 the Minister of Energy, Mines and Resources, Alastair Gillespie, tabled "An Energy Strategy for Canada: Policies for Self-Reliance".[5] The document proposed to double expenditures (from $350 to $700 million per year) for exploration and development in Canadian frontier areas over the next three years.[6] On 28 October 1980 Allan MacEachen, Minister of Finance, announced a new National Energy Program, which further championed Canadian energy development.[7] The Program pledged to attain energy self-sufficiency by 1990,[8] and promised a great expansion of Northern oil and gas exploration by retaining a depletion allowance for frontier exploration and by granting operators up to 80 per cent of Northern exploration costs.[9]

Other policy statements have further complicated the drift in goals by re-emphasizing social or environmental concerns. In a National Energy Program *Update*, the Minister of Energy, Mines and Resources restated the need to develop resources at a rate "compatible with a delicate social and environmental balance".[10] In a still-being-formulated Northern environmental policy, statement, the Minister of the Environment seems ready to promote a comprehensive network of protected areas in the North, to promote environmentally sound technology, to encourage sustainable utilization of the North's renewable resources and to increase environmental information.[11] In a comprehensive land-use planning policy, the Minister of Indian Affairs and Northern Development has committed the Department to implementing a land-use planning programme.[12] In a land-claims policy, the Minister has promised cash payments, land consideration and the extension of local self-government to native groups.[13]

Perhaps the only firm verbal statement of the policy confusion is to describe government's overall policy as "the search for balanced development". Indeed, a statement in House debate by former Minister of Indian Affairs and Northern Development, John Munro, shows just such a position:

> I want to emphasize that we must strive to achieve balance in development; balance between renewable and non-renewable resource development; balance between conventional wage employment activities and ... the traditional native economy ... balance between using the land and resources and conserving them.[14]

The resultant uneasy policy tension between promotion and protection has spurred a mix of public responses. Some, like Mr Hetherington, President of Panarctic Oils Ltd, have called for a clear policy to develop and a concomitant streamlining in decision-making procedures:

> {W}e should have an understanding to supplement the national oil policy, to indicate that the government wants a program pursued to develop frontier oil and gas at the earliest date...
>
> The regulation process needs streamlining. "Single-window" ... procedures would help ... to the extent that we can be heard by a single agency and short cut these numerous other agencies, we would be better off.[15]

Others have despaired of gaining policy clarification or streamlined decision making. Senator Lucier questioned the possibility in these words:

> My question is: Regardless of who forms the government right now, the native people are saying: "We want our claims settled before there is any serious development in these areas". Industry is saying: "Let us go, we want to explore" ... Environmentalists are saying: "Go slow, let's take a look at it" ... You are saying government is unprepared to make many of the policy decisions. I wonder if it is unprepared or unable. I am being serious about this.[16]

Mr Chuck Cook, Member of Parliament on the Special House Committee on Regulatory Reform, questioned the chances for policy clarification this way:

> Realistically, can you see Cabinet being forced into the situation where they will give clear policy direction ... I have some doubts in a practical sense whether you can force government to operate in that kind of efficient manner ....[17]

Such seemingly pessimistic responses recognize the reality of the modern world. In a world of competing values, a world torn between worshipping the gods of man-made technology and the natural environment, political decision making tends to be dangerous to politicians' health, for most decisions create winners and losers. Almost any decision is precarious because of limited knowledge and information. The tendency is to seek a magical political formula – a policy for everyone – and then to leave it to non-elected bureaucrats to make the abstract concrete and to bear the brunt of informational uncertainty and social concern. Decision making tends to involve more negotiation than strict regulation, more fragmentation then centralization, and greater equity than efficiency. The "drift-netter" approach it might be called, for government tends to respond unsystematically, in small increments, rather than by tightly anchored, systematic, rational planning.[18]

Those urging clear policy and streamlined decision making, on the other hand, seem to envisage a different decision-making approach. The ideal is regulatory certainty, minimal discretion, centralization and efficiency.[19] The "set-netter" approach it might be labelled, for government is urged to anchor the decision-making process through systematic procedures and planning.[20]

As the following three sections will show, Canadian decision making has not uniformly followed one technique or the other, but has displayed a dynamic flux. Set-netting – the quest for centralization and rationalization – may be viewed in bureaucratic attempts to consolidate or coordinate the decision-making process and legislative attempts to regulate the Arctic environment. Drift-netting – the movement towards decentralized, *ad hoc* decision making and certainly the dominant approach in the North – may be seen in the constant struggling with fragmentation and jurisdictional overlaps, in legislation lacking policy direction and brimming with discretion, and in the hodge-podge of assessment reviews of major project proposals.

III Administrative mesh

The organization of government expresses the world view of decision-makers in several ways. The creation of a department, agency, or Crown corporation reflects the perceptions of government concerning what constitutes a priority, problem or special-needs area. Organizational structure also reveals how policy options will be evaluated and how decisions will be taken.[21] However, several factors inhibit synchronization between perceptions and realities of government operations. For example, persistent tension, created by competing ideas, ideologies and interest groups, places demands on government to achieve a judicious balance between the *status quo* and change.[22] In addition, bureaucratic self-interest also tends to obstruct synchrony: that is, it is difficult to dismantle an existing, although possibly outmoded or overburdened, structure because dismantlement may threaten established sets of power relations. However, administrative structures are vulnerable to dismantlement when their existence and activities are perceived as being obstacles to the realization of political goals, and politicians are adept at finding mechanisms to expedite the fulfilment of their agendas.

In this section, government reorganizations are viewed as indicators of consensus that new or different types of administrative structures are required to carry out the business of government effectively. The question of motivation for administrative change is also considered. Doern argues that the federal government experiments with a variety of policy instruments and structures to placate certain constituencies: "At times, the reorganization of agencies and departments or the creation of new ones constitutes a symbolic form of exhortation, demonstrating to a constituency that government is concerned".[23] Other observers imply that there may be Machiavellian motives for government organization restructuring.

To gain a better understanding of the complexity of Northern decision making, the evolution of several major federal departments is examined under the heading, "Knitting an Administrative Net". The departments involved include Indian Affairs and Northern Development (DIAND), Energy, Mines and Resources (EMR), Transport (DOT), and Environment (DOE). The section "Mending the Administrative Net" briefly surveys some of the alternative administrative and advisory mechanisms which government has created in response to criticisms concerning Northern policy coordination and consultation processes. In "Finding the Policy Floats", various categories of policy instruments are assessed in the context of Northern development. Finally, the relationship between administrative structure and the decision process is discussed under the heading "Northern Decision Making: Set- versus Drift-Netting".

A Knitting an administrative net

The evolutionary history of the Department of Indian and Northern Affairs is a case study of the shifts in government perception concerning Northern problems and opportunities. It also documents the conflict between interests promoting unregulated resource development and those favouring enhanced resource protection and management. At the heart of this conflict lies a fundamental, philosophical difference concerning how decisions should be made, and who should be given authority to make and carry out such decisions.

With respect to Northern decision making, a distinctive pattern has evolved. Periods of consolidation of responsibility and authority under "umbrella" or centralized agencies are interrupted when government philosophy shifts toward a preference for more widely distributed authority and specialization.

The origins of the Department of Indian Affairs and Northern Development, formally established by statute in 1966, extend back to Sir John A. Macdonald's first ministry in 1867. The process of department-building cannot be described as possessing any internal logic of its own. The route that eventually consolidated responsibility for the resources and administration of the Northern territories, as well as for Indian and Inuit affairs, has been circuitous.

Three examples demonstrate a reluctance by government to link specific legislative authority with a particular department or office. The post of Superintendent-General of Indian Affairs was established in 1867, but the office was occupied *ex officio* by ministers holding other portfolios. Although the Minister of the Department of the Interior was assigned the post most frequently, statutory provisions permitted the Governor in Council to designate any member of Cabinet to serve as Superintendent-General of Indian Affairs, until the post was abolished in 1936. Likewise, ministerial responsibility for overseeing matters relating to the Department of Mines (established in 1907) was not a full-time post until 1936, and the changes in portfolio linkages between Mines and other departments would make a fascinating study in its own right. Of equal significance is the fact that the post of Commissioner for the Northwest Territories was not a full-time job until 1964; previously, the office had been assigned to senior bureaucrats with additional departmental or agency responsibilities.

With the creation of the Department of the Interior in 1873, responsibility for overseeing matters relating to the Northern territories, inland resources, and Indian affairs was consolidated under a single ministry. By the turn of the century the Department was overextended and many responsibilities were being transferred to other ministries. Expansion and decentralization continued until 1936 when, as an economy measure, another umbrella department was established – the Department of Mines and Resources which consolidated the duties previously delegated to four ministries: Immigration and Colonization, Interior, Mines, and the Superintendent-General of Indian Affairs.

For many years, responsibility for Canada's Inuit (Eskimo) peoples shifted from department to department, and neither the Department of the Interior nor the Indian Affairs Branch was particularly enthusiastic about assuming responsibility for Inuit affairs, since Inuit status was uncertain. A memorandum from the Department of the Interior to the Department of External Affairs, dated 20 September 1932, illustrates this point:

> The doubt which exists as to the precise status of the Canadian Eskimo would appear to be due to the fact that parliament has not seen fit to enact any legislation indicating that the relationship between the Eskimo and the Government differs from that of the ordinary citizen. Generally speaking, the assumption would seem to be that because the Eskimo is an aboriginal inhabitant of the North American continent he must necessarily be a ward of the nation in the same way as the Indian.[24]

From 1936 until 1950 the Department of Mines and Resources discharged functions respecting Indian affairs, and the Department's Deputy Minister served as Commissioner of the Northwest Territories. In 1945 the Department of National Health and Welfare assumed responsibility for matters relating to Indian and Eskimo health, and with the extension of family allowance payments to Inuit families, the Canadian government recognized, for the first time, that the Inuit were Canadian citizens.[25]

Structural changes and administrative expansion occurred again in the post-World War II years. Two ministries (Mines and Resources and Reconstruction and Supply) were abolished, and three "new" departments emerged: the Department of Citizenship and Immigration (designated to oversee Indian Affairs), Mines and Technical Surveys, and Resources and Development.

With respect to Northern administration, perhaps the most significant structural change took place between 1950 and 1953 when the federal government relieved the Minister of Resources and Development of numerous duties to steer portfolio attention specifically towards Northern issues. When the Prime Minister, the Right Honourable L.S. St Laurent, moved for a second reading of Bill No. 6 respecting a departmental name change (from Resources and Development to the Department of Northern Affairs and National Resources (NANR)), he argued for a Northern focus:

> We think that the new name is rather important; it is indicative of the fact that the centre of gravity of the department is being moved north. That is the purpose of this measure. The Department of Resources and Development will become the Department of Northern Affairs and National Resources. It is quite proper that they should be together because the national resources subject to the jurisdiction of parliament and vested in Her Majesty in the right of Canada are the resources that are situated in the northern territories.[26]

The Opposition registered concern about what type of Northern development a new department might promote, and questioned who the benefactors of accelerated development activity might be:

> I suggest to you, Mr. Speaker, that if Canada is to play her part in that and the basis on which she will play her part is the development of our northern territories, then it is time that we and our government took account of these wide trends in world development today instead of relying, as they have done in the past, on that course, disastrous to the Canadian people, called free enterprise, which actually in many instances is merely a disguised form of contributions from the public treasury invested under private auspices.[27]

The Glassco Royal Commission on Government Organization reported in 1963. In addition to its recommendations concerning revised financial and personnel-management practices in government, the Commission also suggested ways to eliminate or reduce duplication, confusion, and inefficiency in departmental operations. For example, the inquiry revealed that branches in three separate departments, that is, the Indian Affairs Branch (Department of Citizenship and Immigration), the Northern Administration Branch (Northern Affairs and National Resources), and the Health Services Branch (National Health and Welfare), shared parallel responsibilities, and that it would make good sense to combine duties for Indian and Eskimo affairs under the aegis of the Department of Northern Affairs and National Resources.[28]

With the Government Organization Act, 1966, ministerial responsibility for Indian and Eskimo Affairs was assigned to the minister of a "new" department, the Department of Indian Affairs and Northern Development

(DIAND). The government also gave a new mandate to an expanded Mines and Technical Surveys Department, assigning responsibility for energy- and resource-development policy to the Energy, Mines and Resources (EMR) portfolio. From the start, Parliamentarians were at odds about the wisdom of dividing responsibility for resource policy and Northern resource development between two departments, that is, EMR and DIAND. The Honourable Edmund Davie Fulton commented in the House of Commons:

> Did you ever hear of such a half-baked idea of dividing administrative responsibility for areas having the same general problems between two different ministers?
>
> Perhaps even more important, Mr. Chairman, than the chaos created by this administrative division is the fact that the bill is completely silent as to the responsibility for planning and coordination of programs for development. It is important that we bear in mind this distinction between responsibility for administration and carrying out plans for physical development on the one hand and the responsibility for making those plans and coordinating national policy on the other.[29]

In 1968 the administrative boundary between EMR and DIAND was eventually clarified by an Order-in-Council as an amendment to the Canada Oil and Gas Lands Regulations.[30]

Although primary responsibility for Northern affairs is assigned to DIAND, numerous other departments have a legislative mandate to take part in Northern decision making. A recent government handbook listed 38 federal departments and agencies with responsibilities North of 60°. Energy decisions, although primarily the domain of the Department of Energy, Mines and Resources, likewise cannot be made unilaterally, and at least 43 other governmental departments, agencies and committees have energy interests.[31] EMR has been a rising star in the federal government since its creation in 1966, quickly assuming a dominant position at Cabinet level with authority rivalling that of Finance, Treasury Board, the Privy Council Office and the Prime Minister's Office. Because energy policy, economic policy and national unity have become virtually synonymous and now dominate the public agenda, EMR has been able to exert tremendous influence within the federal government through acceptance of its National Energy Program announced in 1980.[32]

In addition to DIAND and EMR, two other federal departments have a significant role in Northern decision making. The Department of the Environment, established in 1970, has not enjoyed the power or prestige of EMR, however, as environmental issues become more political, and as environmental assessment and monitoring procedures gain political legitimacy, DOE's role in the North is expected to expand. Transport, although a senior department in terms of longevity (it was established in 1936), has only recently begun to expand its Northern services because plans for year-round operations "require that the department increase its Arctic capabilities by a quantum step".[33] In January 1981, the Marine Administration of Transport Canada produced an Arctic Marine Service Policy to "foster development of an appropriate level of services in support of transport activities in the Arctic".[34]

In summary, administrative reorganizations and redistributions of responsibility represent attempts to rationalize government activity, but administrative revisions alone cannot remedy problems requiring policy direction. The *Glassco Report* (1963) addressed some of the problems associated with policy indecisiveness:

> Any programme dealing with underdeveloped areas and peoples must concern itself with three questions: what can be achieved? at what pace? and at what price? There must be, in short, a clear view of both the foreground to be traversed and the goals on the horizon, and a realistic schedule of progress. A casual disregard of the foreground can only cause the programme to founder before it is well begun; without goals there can be no sense of direction; impatience in the pace means that action is taken prematurely, and if the pace be laggard, the programme will be outstripped by events.[35]

In the context of Canada's North, fundamental questions such as Northern development for whom and at what price were not seriously considered for many years because official Ottawa agreed that the chief challenge facing the North was the attraction of business and opening up of the North.[36] But in the rush to create an environment favourable to Northern development, a number of important environmental and social issues were neglected.

B Mending the administrative net

After the Glassco Commission and public interest groups levied criticisms at government for *ad hoc* and backroom decision making during the late 1960s and early 1970s, the Canadian government initiated a number of administrative reforms. The introduction of a planning, programming, and budgeting system (PPB), the formation of the Cabinet committee system, the creation of major central advisory councils, and the emergence of a "management-by-objectives" philosophy (MBO) were among the most radical changes made to government operations.[37] In addition, to facilitate intergovernmental communications and consultation, committees and other advisory mechanisms proliferated. For example, more than 70 interdepartmental committees with Northern interests have been identified. A checklist of committees with a role in Beaufort Sea hydrocarbon exploration and production also illustrates how extensive decision and advisory networks have become.[38]

Intergovernmental committees serve a variety of purposes. Some committees are advisory while others are charged with the task of initiating or evaluating policies or programmes. Conflict resolution and consensus seeking are two important functions of intergovernmental committees.[39] Committees may be long-standing and comprehensive (the Advisory Committee on Northern Development (ACND)), or they may be project specific (the Senior Steering Committee on Lancaster Sound). Certain committees have great influence on policy (such as the Senior Policy Committee on Northern Resource Development Projects), while others are created to enhance communications (the Panel on Ocean Management, chaired by the Department of Fisheries and Oceans (DFO)). Some have very specific mandates (the Regional Ocean Dumping Advisory Committee), while others are more generalized, such as the External Affairs' Circumpolar Affairs Panel.

Since the Diefenbaker era, the federal government has turned frequently to non-departmental administrative structures to obtain advice, evaluate policy, or to expedite decision making on matters of special urgency. Non-departmental bodies, which have been labelled structural heretics, such as Crown corporations (for example, Petro-Canada), regulatory commissions and agencies (the National Energy Board and the Northern Pipeline Agency) and central advisory councils (Science Council of Canada), are particularly useful when autonomy from government is perceived as desirable.[40] Government also relies on mechanisms such as royal commissions and task forces to solicit public opinion on policy initiatives or to demonstrate government's responsiveness to politically charged issues (examples include the Berger Commission, the Lysyk Inquiry, and the Task Force on Northern Conservation). Coloured papers, such as the Lancaster Sound Green Paper, are anticipatory government-planning instruments which seek to ascertain political acceptability of policy in advance of proclamation or implementation.

C Finding the policy floats

Policy can be implemented through a variety of instruments available to government. G. Bruce Doern suggests that instrument choice is a reflection of dominant political thinking with respect to appropriate means (process), as well as to ends (policy).[41] Doern has constructed a typology of policy instruments which reflects degrees of coerciveness from one extreme, that is self-regulation (non-intervention), to the opposite extreme (complete public ownership). Exhortation (persuasion), expenditure, and regulation can be arranged along a continuum between self-regulation and public ownership (Figure 6.1). This section surveys the relationship between Northern policies and instrument choice in an effort to assess government commitment to expressed policies.

1 *Exhortation instruments*

Policy formulation and implementation is a delicate and politically risky business. Most politicians prefer to test the political waters before they introduce or implement new policies, and this testing is generally accomplished by using a variety of hortatory instruments.

Figure 6.1. Typology of policy instruments.

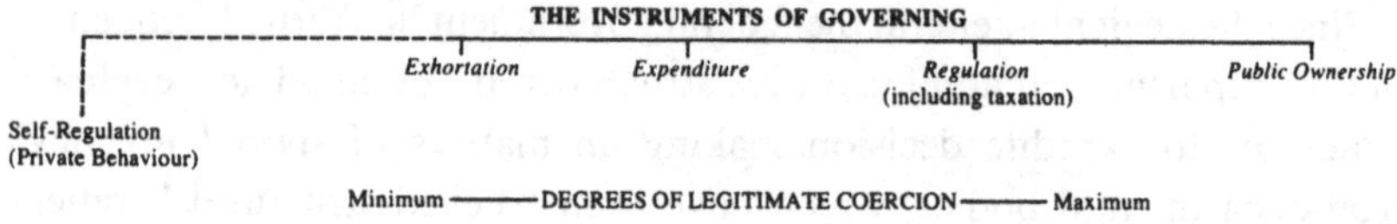

A SECONDARY CATEGORIZATION OF THE INSTRUMENTS OF GOVERNING: FINER GRADATIONS OF CHOICE

Exhortation	*Expenditure*	*Regulation*	*Public Ownership*
• Ministerial Speeches	• Grants	• Taxes	• Crown Corporations with Own Statute
• Conferences	• Subsidies	• Tariffs	• Crown Corporations under Companies Act
• Information	• Conditional Grants	• Guide-lines	• Purchase of Shares of Private Firm
• Advisory and Consultative Bodies	• Block Grants	• Rules	• Purchase of Assets
• Studies/Research	• Transfer Payments	• Fines	• Joint Ownership with a Private Firm
• Royal Commissions		• Penalties	• Purchase of Private Firm's Output by Long-Term Contract
• Reorganizing Agencies		• Imprisonment	

Source: Alan Tupper and G. Bruce Doern (eds), *Public Corporations and Public Policy in Canada* (1981), p. 17.

(*a*) *Speeches*:
Speeches are the most public and emphatic device to announce new policy directions. Because words are irretrievable, politicians are cautious when using speeches as a policy vehicle unless they are fully committed to a new programme regardless of cost. A good example was the February 12, 1958 "Vision" speech by Prime Minister John Diefenbaker:

> I think of a vast program on Frobisher Bay on Baffin Island in the Canadian Arctic, hiding resources that Canadians have little realization of ... We intend to start a vast roads program for the Yukon and the Northwest Territories which will open up for exploration vast new oil and mineral areas — thirty million acres. We will launch a seventy-five million dollar federal-provincial program to build access roads. THIS IS THE VISION[42]

Subsequently, the "Roads to Resources" programme was launched, and with it a Northern Pandora's box was opened.

(*b*) *Advisory and consultative bodies*:
Decision making in the pre-1968 era has been characterized as fragmented and highly decentralized. Visions of increased prosperity and wealth accruing from Northern resource development preoccupied the few who paid any attention to the North. The first interdepartmental committee, the Advisory Committee for Northern Development (ACND), was created by Cabinet in 1948 to provide "close and continuous interdepartmental coordination to ensure that all (civilian and military) responsibilities are discharged effectively and in acordance with overall government policy".[43] In January 1953 Cabinet further directed ACND to report immediately, and regularly thereafter, on all phases of Northern development and on "the means which might be employed to preserve or develop political, administrative, scientific and defence interests of Canada in that area".[44]

Political momentum for Northern development accelerated with the enactment of the Northern Affairs and National Resources Act in 1953. The new department (NANR) was charged with the responsibility to foster knowledge of the North, and to fulfil this mandate a Northern Coordination and Research Centre was established in 1954. The Centre was instructed to collect information and coordinate research on Northern subjects, and was empowered to issue licences and permits to scientists, explorers and archaeologists working in the Northwest Territories.[45]

ACND's visibility declined in the 1960s, and infrequent subcommittee meetings were poorly attended. The absence of crisis and the growth of multiple Northern administrative bureaucracies contributed to ACND's reduced role as a policy and programme coordinating instrument within the federal government.[46] Although ACND was partly revitalized after the discovery of oil in Prudhoe Bay in 1968, the Committee never regained its initial political influence. ACND's principal contributions have been the annual compilations of government activities in the North, and sponsorship of the "Science and the North" seminar on scientific activities in Northern Canada in 1972.[47]

Government's reaction to the discovery of Alaskan oil was to create an additional administrative advisory structure, a committee hybrid — the special Task Force on Northern Development. Jean-Luc Pepin, then Acting Minister of Energy, Mines and Resources, explained the three-fold role of the Task Force to members of the House of Commons in 1968:

> to bring together all information on the existing oil situation in the north, on transportation routes that might be used, and to coordinate all available information, from all federal agencies and departments, and then report and make proposals to government.[48]

The Task Force was chaired by the Deputy Minister of EMR; other members included senior officials representing DIAND, DOE, DOT, and the Chairman of the National Energy Board (NEB). Initially, four subcommittees — Pipelines, Economic Impact, Transport, and Marketing —were created (see Figure 6.2).

The Task Force was a powerful pro-development instrument due to its membership and accessibility to the Privy Council Office, Finance and Treasury Board. It could convene quickly if the need arose, and its deliberations were secret. As noted by Dosman, the Task Force served as a transmission belt for industry representatives requiring speedy approval by Cabinet:

> In theory at least, the Advisory Committee on Northern Development had a broad mandate and representation to address the broad multidimensional character of development in the North. In practice, the Advisory Committee was ineffective. But at the same time, it was too clumsy and exposed to serve as an adequate tool for pipeline decision making.
>
> *Something brand new, ad hoc, without commitments to existing jurisdictions* — a secret forum and transmission belt where senior officials could meet in confidence with business executives — this was required.
>
> *It could not be bureaucratized, or it would lose effectiveness.*[49]

Figure 6.2. Evolution of the Task Force on Northern Oil Development.

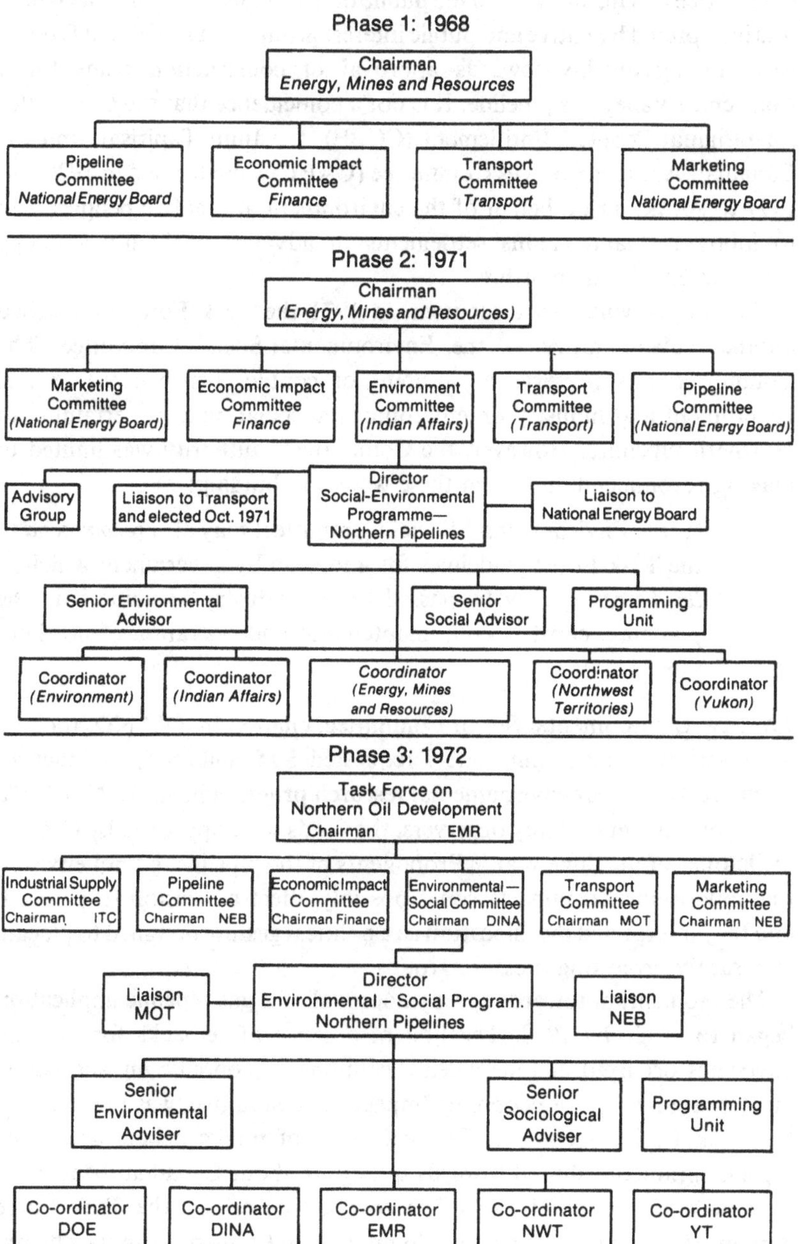

Sources: Edgar J. Dosman, *The National Interest*: *The Politics of Northern Development 1968–1975* (1975), pp. 22, 164; and Science Council of Canada, *Northern Development and Technology Assessment Systems* (1976), p. 194.

Northern Pipeline Guidelines were drafted during June and July 1970 by a Task Force working group in consultation with the Interdepartmental Committee on Oil, and were made public on 13 August. The announcement was interpreted by native and public interest groups as a signal that Ottawa was moving quickly towards approval of construction plans for a Mackenzie Valley gas pipeline. It is not a coincidence that the Committee for Original Peoples' Entitlement (COPE), the Inuit Tapirisat, and the Canadian Arctic Resources Committee (CARC) were formed in 1970 and 1971 to speak out on behalf of the environment and native peoples, and to lobby for land claims settlements in advance of high-technology developments in the North.[50]

To still the voices of discontent, in 1971 the Task Force established another subcommittee — the Environmental-Social Committee. The Committee was to serve as a point of contact between the federal government and industry on environmental-social matters pertaining to Northern pipelines. However, the Committee's authority was limited to making recommendations and these were non-binding:

> It {the Environmental-Social Committee} may also *recommend* to the Task Force guidelines for adoption by government which *at its discretion* may be passed to the industry to guide it in the planning, construction, maintenance and operation of northern pipelines.[51]

The new Environmental-Social Committee, chaired by DIAND, took its work seriously, and immediately requested $15 million to conduct an intensive, three-year environmental research programme in the North. To the astonishment of many observers, the funds were approved by Cabinet in October of the same year.[52] Proponents of the pipeline did not envision that environmental impacts could possibly affect a decision to proceed, and they interpreted the allocation as a political gesture designed to placate incessantly protesting pressure groups.

The momentum towards speedy approval of a gas pipeline application began to erode by 1972. The first of a series of setbacks for pipeline advocates occurred in June when a cautionary economic impact assessment, prepared by the Economic Impact Subcommittee, was presented to the Task Force. Secondly, the focal point of public debate about the pipeline gradually shifted towards questions about Canadian benefits of pipeline construction. In an effort to quell its critics, the Task Force appointed a sixth committee, the Committee on Industrial Supply, chaired by the federal Department of Industry, Trade and Commerce. Although the issue of *Revised Pipeline Guidelines*[53] suggested that progress towards pipeline approval was still being made, public debate began to focus on domestic economic and energy supply questions, in addition to environmental and social issues. The Quebec crisis, the formation of a minority government in the fall of 1972, the OPEC oil embargo, and EMR's new "Energy Policy for Canada" in 1973, further diverted Cabinet's enthusiasm for a transcontinental pipeline. After the appointment in 1974 of Justice Thomas R. Berger to chair a commission of inquiry to investigate the social, environmental and economic impacts of a Mackenzie Valley Pipeline, the Task Force lost its remaining control over project information and its exclusive access to Northern resource decision-makers.

Several other interdepartmental committees have been instrumental in Northern resource policy development and implementation since 1968. The Interdepartmental Committee on Oil enjoyed high-profile status for a short period, principally because its members were high-ranking Ottawa officials. For example, the Chairman of the National Energy Board was appointed the Committee's chairperson; other members were the Secretary of State for External Affairs, the Deputy Minister of EMR and representatives from the Privy Council Office. The Committee held chief responsibility for reviewing National Oil Policies in the 1970s, and participated in the drafting of pipeline guidelines.[54] In 1974 a Deputy-Minister level Interdepartmental Task Force on Energy Research and Development was created to design, implement, and coordinate a programme of federal energy research and development. Chaired by EMR, the Task Force released its report, *Science and Technology for Canada's Energy Needs*, in April 1975.[55] In 1980–1981, Transport Canada, the Department of the Environment and the Department of Fisheries and Oceans Interdepartmental Working Group reviewed the status of Canadian marine transportation research and development, and cited Arctic research and development (R and D) as a first priority.[56]

One final example of co-operative, interdepartmental policy and planning activity is provided by the Lancaster Sound regional study. The *Green Paper*, released in July 1982, was the product of two years' investigative work by five federal departments, the Government of the Northwest Territories, and Northern community representatives.[57] It is too early to know whether or not the exercise will set a precedent for future Northern decision situations, but clearly this strategy reduces vulnerability to criticism about central decision making and tends to diffuse confrontation and polarization.

(c) Royal commissions and task forces:
Royal commissions, commissions of inquiry, and task forces also fall under the rubric of hortatory policy instruments. Because Northern issues tend to be highly charged politically, this category of instruments has been relied upon frequently by the federal government to ascertain public opinion about development options. For example, the British Columbia-Yukon-Alaska Highway Commission came into being when national security concerns were aroused by highway construction to Alaska. The Commission, charged to investigate the "engineering, economic, financial and other aspects of the proposals to construct a highway through British Columbia and the Yukon Territories to Alaska", reported in 1941.[58]

The Berger, or Mackenzie Valley Pipeline, Inquiry is perhaps the best known and most significant Canadian inquiry. The Inquiry, established in 1974 and not reporting until 1977, generated considerable controversy because Justice Berger interpreted his mandate broadly and encouraged widespread public participation.[59] Subsequent Canadian inquiries are not likely to be given as much freedom or time as Justice Berger was granted, if the terms of reference of the Alaska Highway Pipeline Inquiry are an indication of future trends. Kenneth Lysyk, appointed chairman of the Inquiry in April 1977, was instructed by the federal government to present a preliminary report by August in order to meet a joint Canadian–US timetable for the project.[60]

The Task Force on Beaufort Sea Development, with representatives from DIAND, EMR, DOE, DFO, Transport, and the territorial governments, provides an example of an internal, intergovernmental advisory and planning mechanism. The Task Force, which reported to the Senior Policy Committee, Northern Development Projects, in April 1981, was requested to prepare a "situation report" concerning government and industry activities in relation to current policy and policy initiatives, overall planning and coordination, review and assessment, and research. The Committee was also asked to assess government's preparedness and to formulate a planning framework so government would be able to meet the demands of Beaufort Sea development.[61]

Commissions and inquiries have been used to test the waters of political change and have often become embroiled in controversy. A Northern example of this phenomenon is the *Report of the Advisory Commission on the Development of Government in the Northwest Territories* (Carrothers Commission), which was severely criticized for its paternalistic attitude towards native peoples and bias towards Southern development models.[62]

Reaction to the *Report of the Special Representative on Constitutional Development in the Northwest Territories* (Drury Report), released in 1980, was generally more favourable.[63] C.M. Drury, appointed by the Prime Minister, recommended changes which would bring the territorial government more in line with provincial governments. This would be accomplished by eliminating DIAND's "middleman" role as financial negotiator with Treasury Board and the Federal-Provincial Relations Office. Drury also proposed the transfer of ownership of onshore non-renewable resources and responsibility for land-use planning from Ottawa to the Government of the Northwest Territories, with the condition that overriding authority would remain with the federal government. A paramount concern, according to the Report, was the development of experience and expertise in government practice, and to this end, Drury advocated a strengthened civil service: "only by being given meaningful power will the members of government – particularly the elected Assembly and Executive Committee – come to develop a sense of responsibility".[64]

(*d*) *Conferences, seminars and workshops*:
Conferences, seminars, and workshops may be useful instruments of exhortation. The Canadian government has frequently used such forums as an opportunity to demonstrate its commitment to participatory planning for Northern development. One such event, the 1961 federal-provincial "Resources for Tomorrow" Conference, marked a significant departure in government planning philosophy. The need for a comprehensive, national resources policy to replace sector-based resource planning, and the importance of consultation and participation were recognized by Conference sponsors and participants as important components of the policy process: "resource development goals, no less than other objectives of national life, must find expression through national participation and discussion".[65] Unfortunately, such ideas were not generally incorporated into resource policy making in the 1970s, and resource decisions remained the domain of the highest echelons of the federal government.

Conferences have been useful forums for the public and interested groups to express their concerns, exchange information and engage in co-operative planning. For example, in 1972, the Advisory Committee on Northern Development convened a seminar for the purpose of developing guidelines and setting priorities for Northern scientific activities.[66] A.E. Pallister, Vice-Chairman of the Science Council of Canada, expressed his concerns about the timing and pace of large-scale resource developments:

"science does not exist in a political vacuum; political and economic decisions are necessary to provide a framework for scientific and technological activities".[67] As a follow-up to ACND's "Science and the North" seminar, the Science Council supported a systematic analysis of Northern development issues, and commissioned a number of studies to achieve this objective.[68]

A series of "National Northern Development Conferences" have convened triennially since 1958 to stimulate further the orderly development of Canada's Northland. Each conference has a central theme, and these themes reflect how perceptions of Northern opportunities and problems have changed in the last 25 years. Specifically, the list of conference themes illustrates a moving away from the perception that the North is a treasure trove of wealth, towards a recognition that there are social, economic and environmental costs associated with development: 1958 – "The Last Frontier in North America"; 1961 – "Canada's New Role in Resource Development"; 1964 – "Canada's Northland and World Markets"; 1967 – "Man and the North"; 1970 – "Oil and Northern Development"; 1973 – "Mining and Canada's North"; 1976 – "Energy and Northern Development"; 1979 – "At the Turning Point"; and 1982 – "Partners in Progress".

The Canadian Arctic Resources Committee (CARC) has been a catalyst for public and academic scrutiny of Northern decision making. CARC has sponsored three *National Workshops on People, Resources, and the Environment North of* 60° (1972, 1978 and 1983), and several symposia on topics of special interest, such as the symposium *Marine Transportation and High Arctic Development* in 1979. Such forums bring Canadian experts together to evaluate existing Northern policy frameworks, and to recommend priorities for improving Northern resource management and regulatory regimes.[69]

Ministers' conferences also function as important consultative and policy-initiating mechanisms. A recent example was the Consultative Task Force on Industrial and Regional Benefits from Major Canadian Projects (Major Projects Task Force) following a federal-provincial industry ministers' conference in 1978.[70] The Task Force reported in 1981, and Cabinet responded by announcing new federal industrial-benefit objectives and by creating the Office of Industrial and Regional Benefits. At the same time, a new interdepartmental committee, the Committee on Megaproject Industrial and Regional Benefits (C-MIRB) was established. In 1982 C-MIRB's mandate was expanded to review plans submitted to the Canada Oil and Gas Lands Administration as a prerequisite to project approval. In 1983 a First Ministers' Conference discussed native rights and constitutional issues, and agreed to continue deliberations in subsequent years.

(*e*) *Studies and research*:

As national interest and attention slowly shifted Northwards, Canadians gradually discovered their almost total ignorance of the North and its peoples. The Northern Coordination and Research Centre was established in 1954 to promote and oversee Northern research, and special emphasis was given to social science research. From 1978 to 1981, the Northern Research Information and Documentation Service (Northern Affairs Program, DIAND) issued annual compilations of current and recent Northern social research projects, providing a valuable service as a clearinghouse and network for academics, government, and the public.[71]

Concurrent with the International Geophysical Year (1957–1958), and in response to concerns about national security, the federal government inaugurated the Polar Continental Shelf Project (PCSP). The Project's mission was to conduct geological, geophysical and oceanographic surveys of the polar continental shelf. Under the auspices of the Department of Energy, Mines and Resources, the project continues to promote co-operative Northern research by government and academic interests.

In the early 1970s, when proposed pipeline routes were being assessed, industry, government and the public recognized the lack of reliable environmental data which could be used to make sound decisions. As a result, the federal government became an active environmental research partner with industry and academic institutions. Federal involvement was justified on the grounds that the task was urgent, and that Canada lagged in the acquisition of Arctic expertise and experience. The Mackenzie River Basin study, one of the earliest government-industry co-operative research projects, sought to obtain comprehensive data on a regional basis. Other examples of joint government-industry research programmes include the Baffin Island Oil Spill Project (BIOS) and the Eastern Arctic Marine Environmental Studies (EAMES) programme. Both projects collected data for environmental-impact statements, and acquired information for oilspill contingency planning. The Arctic Petroleum Operators' Association (APOA) was formed in 1970, and has sponsored, or been associated with, approximately 170 co-operative research projects with total costs in excess of $ 34 million.[72]

(*f*) *Reorganizing ministries and agencies*:

The most powerful exhortation instrument available to government is the authority to reorganize itself and to create new administrative agencies, structures, and regulatory boards. For example, in 1959 the National Energy Board (NEB) was created after national energy issues and policy processes had been scrutinized by two royal commissions, the Gordon Royal Commission on Canada's Economic Prospects (1955), and the Borden Royal Commission on Energy (1957).[73] The NEB, a quasi-judicial body, was established, following the political storm created by government's financial aid to TransCanada Pipeline Ltd in 1956, in an effort to remove decisions on pipeline construction and natural gas exports from the political arena.[74]

Another example of government's ability to create new structures, when gaps are perceived, is the passage of legislation in 1971 enabling the appointment of Ministers and Ministries of State. The legislation grants the Prime Minister and Cabinet flexibility to tackle problems in new or emerging areas where appropriate policy apparatus is lacking. For example, a Minister of State for Science and Technology (MOSST) was appointed in 1971, but the Ministry's effectiveness has been undermined by frequent reorganizations and intergovernmental rivalry.[75] Two additional Ministries have been created in response to provincial gaps in existing government agencies and departments — the Minister of State for Social Development (MSSD) and the Minister of State for Economic and Regional Development (MSERD) — and both may play significant roles in Northern development planning.

Three further examples of administrative innovations are relevant to Northern development. In 1973 a Cabinet directive created the Federal Environmental Assessment Review Office (FEARO), an independent arm of the Department of the Environment, to facilitate environmental assessment of major federal projects.[76] To date, environmental assessment panels have reported on seven Northern projects — the Alaska Highway Gas Pipeline, the Shakwak Highway Project, South Davis Strait offshore drilling, Lancaster Sound offshore drilling, the Arctic Pilot Project, Norman Wells Oilfield development and Beaufort Sea hydrocarbon development.

The Northern Pipeline Agency (NPA) was created by an Act of Parliament in April 1978 "to provide a single regulatory body to undertake federal responsibilities for planning and monitoring construction of the 2028-mile, main-line portion in Canada of the joint Canadian–US system".[77] NPA's mandate requires the project be "carried forward in a way that will yield the maximum economic, energy and industrial benefit for Canadians with the least possible social and environmental disruption".[78] Although the Agency represents a "single-window" approach to project planning, Parliament authorized two committees to maintain surveillance of the Act and its implementation, that is the House Standing Committee on Northern Pipelines and the Senate Special Committee on the Northern Pipeline.

The Canada Oil and Gas Lands Administration (COGLA) was created in 1981 by a Memorandum of Understanding between the Ministers of DIAND and EMR to manage oil and gas exploration, production and development activity in the Canada Lands:

> {T}he principal purpose for the creation of COGLA was to concentrate, within a single body, the oil and gas management functions exercised by DIAND, with respect to Canada lands situated north of the line of administrative convenience defined in Schedule IV of the Canada Oil and Gas Lands Regulations and by EMR with respect to Canada lands located south of that line.[79]

COGLA is an unique administrative structure within the framework of the Canadian government: "it is an administrative body with dual functional responsibility to Northern Policy (DIAND) and Energy Policy (EMR) and whose authority, derived from the Ministers of both parent departments, is exercised to the extent that ministerial delegation is made".[80] COGLA consists of eight branches — Land Management, Engineering, Resource Evaluation, Environmental Protection, Canada Benefits, Policy Analysis and Coordination, Nova Scotia, and Special Investigations (see Figure 6.3).

2 *Expenditure instruments*

Expenditures provide clues about government's commitment to policy. Unfortunately, statistics and figures are easily manipulated, and so it is difficult to obtain precise figures which can be used for comparative purposes. Since 1973, ACND's annual report of *Government Activities in the North* has tabled figures to indicate how government funds have been distributed to meet the seven Northern objectives outlined by DIAND in 1972 (see Figure 6.4 and Table 6.1).

Although the Northern expenditure record looks impressive, the expenditures of a single department such as EMR (Table 6.2) suggest that government is relying more heavily on expenditure instruments to realize energy policy objectives than Northern policy objectives. EMR expenditures for 1980–1981 were $3 billion, while total Northern expenditures were registered at $885 million. From 1980–1983, forecast expenditures for energy-industry incentives under the National Energy Program totalled $2.5 billion. However, it could be argued that spinoff effects from

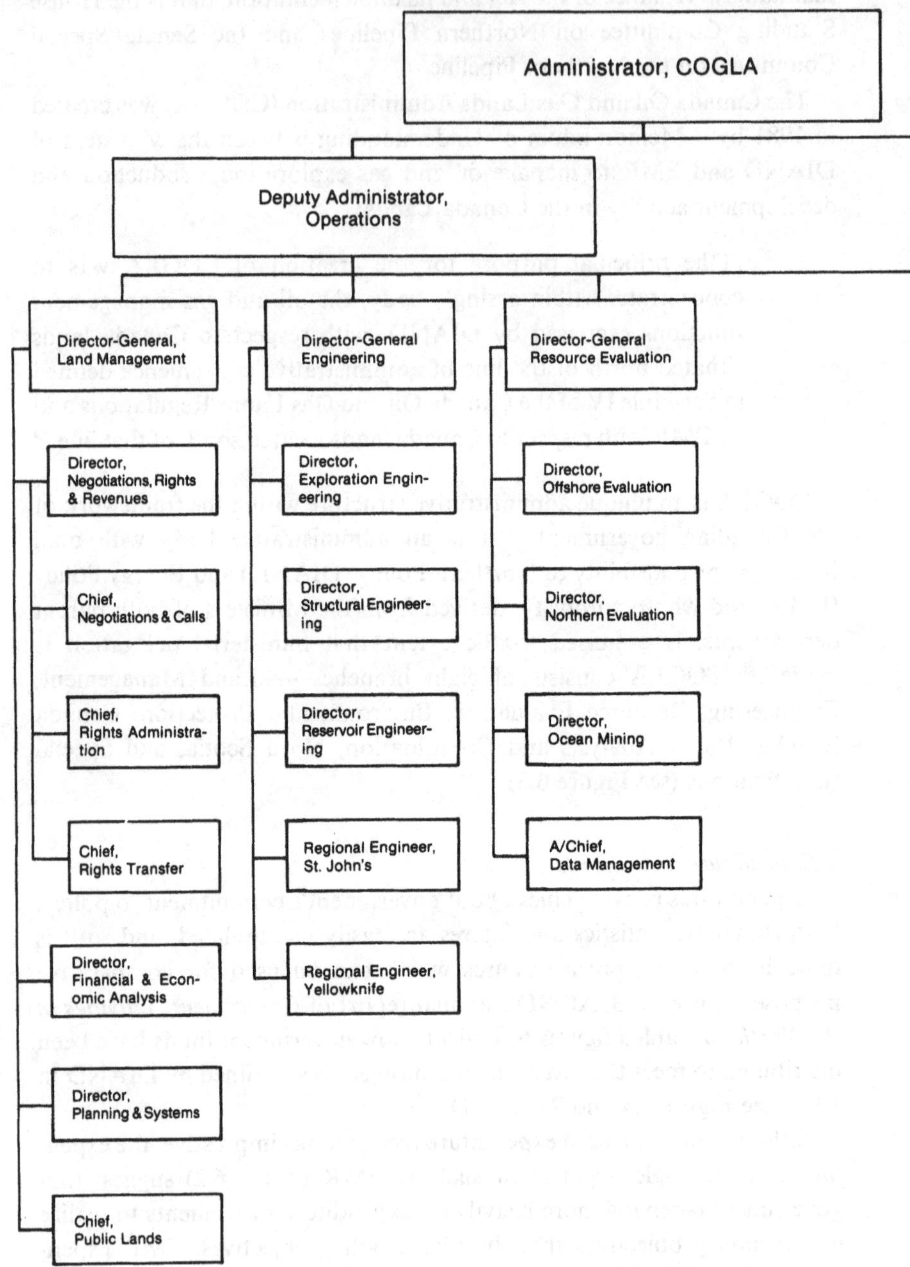

Figure 6.3. Canada Oil and Gas Lands Administration (COGLA).

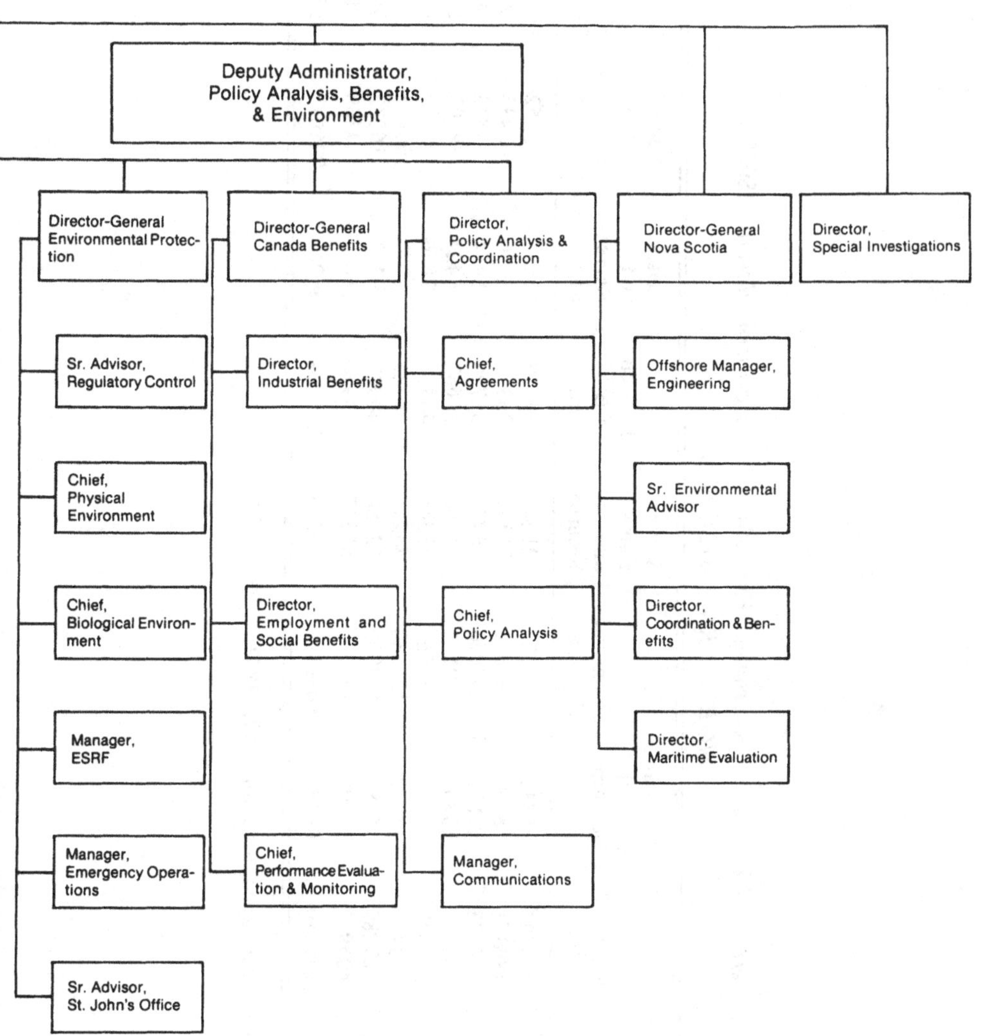
Deputy Administrator, Policy Analysis, Benefits, & Environment
Director-General Environmental Protection
Director-General Canada Benefits
Director, Policy Analysis & Coordination
Director-General Nova Scotia
Director, Special Investigations
Sr. Advisor, Regulatory Control
Chief, Physical Environment
Chief, Biological Environment
Manager, ESRF
Manager, Emergency Operations
Sr. Advisor, St. John's Office
Director, Industrial Benefits
Director, Employment and Social Benefits
Chief, Performance Evaluation & Monitoring
Chief, Agreements
Chief, Policy Analysis
Manager, Communications
Offshore Manager, Engineering
Sr. Environmental Advisor
Director, Coordination & Benefits
Director, Maritime Evaluation

Table 6.1. *Distribution of Federal and Territorial Government Expenditures by Northern Objective* ($000'*s*)

	NORTHERN OBJECTIVE								
	Evolution of Government	Quality of Life	Social and cultural Development	Economic Growth	Sovereignty and Security	Protection of Environment	Leisure and Recreation	Administration and Support	Total
1973/74	9227	150032	5831	90522	21244	19459	3254	45252	344811
1974/75	14383	180171	8059	90018	21137	26351	4056	63435	407610
1975/76	19567	225928	11831	109076	26731	31706	5454	75833	506126
1976/77	22757	268731	13851	129886	29661	31662	6424	88503	591485
1977/78	25715	306547	16142	137568	33346	40488	7846	96235	663887
1978/79	30945	333497	17772	138530	44244	38226	8970	106425	718609
1979/80	38513	369289	16289	150959	52880	40922	9192	116232	794276
1980/81	30003	406915	19408	173257	56277	53910	11059	134168	884997
Totals	382220	2241110	109183	1019816	285520	282724	56265	726073	4911801

(Source: ACND. Annual Northern Expediture Plans, 1975–1982, Table V, VII).

Table 6.2. *Department of Energy, Mines and Resources Program Budgetary Expenditures as a Percentage of Total Departmental Budgetary Expenditures (millions of dollars)*

Year	Total Department	Administration	%	Minerals, Energy Resources	%	Earth Sciences	%	Energy[1]	%
1970–71	70.4	7.6	12	41.8	59	20.9	30	—	—
1971–72	83.3	7.8	9	45.2	54	30.2	36	—	—
1972–73	80.5	6.7	8	41.9	52	31.0	39	—	—
1973–74	240.5	7.2	3	197.8	82	35.5	15	—	—
1974–75	565.0	10.2	2	514.6	91	40.2	7	—	—
1975–76	127.0	11.1	9	72.0	57	46.1	36	—	—
1976–77	188.5	14.1	7	124.4	66	50.0	27	—	—
1977–78	1319.4	14.6	1	19.8	2	86.0	7	1198.9	91
1978–79	836.9	14.9	2	22.1	3	89.2	11	710.7	85
1979–80*	1816.1	15.9	0.1	22.6	1	94.0	5	1683.7	93
1980–81*	3047.1	17.0	0.03	23.8	0.7	101.2	3	2905.1	95

Source: *Estimates* 1970–71 to 1980–81 (from G. Bruce Doern: *How Ottawa spends your tax dollars*, 1981, p. 70).
[1] Minerals and Energy became separate programs in 1977.
* Estimates rather than actual expenditures.

government-supported exploration activity will also accelerate realization of the economic-growth objective of Canada's Northern policy.

3 *Regulation*:
The regulatory framework of Northern decision making is examined in detail in Section IV, "Legislative Mesh".

4 *Public ownership*:
Crown corporations and equity participation are at the extreme end of the policy instrument continuum in terms of degree of government involvement or control. The Canadian government has utilized this type of instrument to achieve diverse political and economic objectives, but because of the varied histories of Crown corporations, it is difficult to generalize about them: "the goals of Crown corporations, far from being carved in stone, are dynamic and often change under the influence of technological imperatives, market circumstances, and political philosophies".[81]

A number of Crown corporations have been created to facilitate Northern development because private industry has not been able to raise sufficient capital, or because the federal government is obligated to provide services to remote regions. The Northern Transportation Agency (NTA) has operated in the Mackenzie River Basin since 1934, and along the Western Arctic coast since 1957. The company's primary purpose is to provide economic, reliable, and comprehensive transportation and related services to the North. In 1980, the company transported 305,952 tonnes of deck and bulk cargo.[82]

Figure 6.4. Total direct Northern expenditures (in millions of dollars).

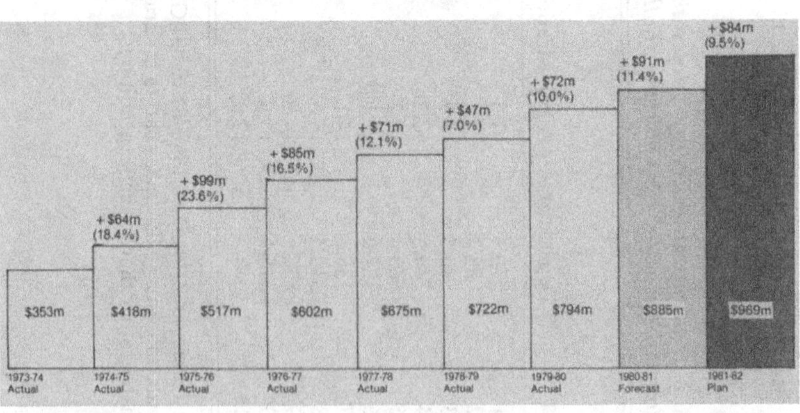

Source: Department of Indian Affairs and Northern Development, *Annual Northern Expenditure Plan* (1981–1982), p. 12.

The Northern Canada Power Commission (NCPC) is a federal Crown corporation which operates under the authority of the Northern Canada Power Corporation Act. The Commission is the principal producer of electrical energy in the North, and operates main transmission networks in the Northwest Territories and Yukon. The Act requires that operations be self-financing within each zone, so rates can cover expenses and operating allowances. In 1980–1981, the NCPC provided retail electric utility service to 53 Northern communities, as well as heating, water and sewage systems in several locations.[83]

The decision-in-principle to create a national oil company was proposed by the Minister of Energy, Mines and Resources in October 1973. The idea was adopted by Cabinet in November, and Prime Minister Trudeau announced, in December, the decision to proceed to Parliament. The Petro-Canada Act was proclaimed on 30 July 1975, and, by promoting accelerated domestic hydrocarbon exploration and production, was heralded as a bold response to the OPEC oil embargo.[84] A recent analysis of Petro-Canada's performance since 1976 suggests that its role has changed since the Act was first drafted:

> Petro-Canada was originally established ... with a developmental orientation. Its principal function was to increase Canadian energy supply, especially from frontier sources, not to collect rents or Canadianize the industry. The 1980 National Energy Program and the new Canada Oil and Gas Act imply, however, that the state corporation is increasingly to have a redistributive purpose — shifting industry activity (and future economic rent) from the provinces to federal lands, and redirecting the benefits from new energy development from foreign investors to Canadians.[85]

Petro-Canada, as a member of the Arctic Islands Exploration Group (AIEG) and as the principal shareholder of Panarctic Oils Ltd, is a participant in Arctic Islands and Beaufort Sea drilling. Petro-Canada is also involved with East Coast offshore operations and has several international interests.

A final example of government exerting policy through public financing occurred on 30 September 1982 when the federal government announced a $ 1 billion refinancing package to avert the collapse of Dome Petroleum Ltd. The federal government denied the move was a direct bail-out, and insisted that it would gain effective control of the company through agreement respecting rights to nominate a new board of directors.[86]

D *Northern decision making: set- versus drift-netting*

Activities of the Canadian government demonstrate a persistent uncertainty about its appropriate role: that is whether government should act as an initiator of policy and programmes, or whether its principal function is to respond and assist initiatives of the private sector. Government is well aware of its own shortcomings, as the appointment of the Glassco Royal Commission and subsequent investigations indicates. That policy is most appropriately formulated through widespread discussion and consultation was not a common notion in the 1970s, although government appears to be taking such recommendations more seriously in recent years. The appearance of policy branches in most federal departments in the 1970s indicates that policy initiation is no longer viewed as an exclusive function of Cabinet or the Prime Minister's Office, but has become the domain of the federal civil service as well. Finally, the release of public consultation policies by several departments suggests a willingness to extend policy discussions beyond the arena of the federal bureaucracy, although such documents may also be viewed as mere exhortations camouflaging a reluctance to extend substantive policy-making powers to the public.

The rapid proliferation of intergovernmental, governmental-industry, and citizen advisory committees concerned with Northern, environmental, and energy issues is a reflection both of government's recognition of its own complexity (and the need for improved coordination mechanisms), and of the need to broaden public participation. Environmental Assessment and Review Panels, the Berger and Lysyk inquiries, as well as the appointment of the Carrothers and Drury Commissions, are examples of government seeking to be (or appearing to be) more open and willing to solicit public input prior to reaching decisions with far-reaching consequences. However, initial experimentation with public participation proved to be frustrating for government, industry, and public interest groups alike and, as a consequence, confrontation became a characteristic feature of such forums. Citizens' expectations of having a greater role in the policy process failed to materialize, and development proponents were stymied by delays and what they perceived as government indecisiveness and-or bureaucratic inefficiency. By 1980, attitudes towards public participation were changing:

> {R}eformers of all kinds, including those who want better science and technology assessment processes, are often properly viewed as part missionaries and part used-car salesmen. Governments increasingly feel overloaded with, and sceptical about, the latest reforms and the latest reformers.[87]

From a structural perspective, the Canadian government might be viewed as an innovator, a type of administrative entrepreneur willing to experiment with new structures and a variety of instruments to improve both policy-making and information-dissemination procedures. Another interpretation of why such a variety of structures and instruments have been used to make and implement Northern policies rests on the premise that politicians have short time horizons. What counts is performance, not procedure. When structure, process, or instrument becomes a liability to the realization of political goals, change is embraced as a necessary remedy. A good example of this process is provided by the Task Force on Northern Oil Development, whose demise was a consequence of its failure to obtain unequivocal approval for pipeline construction. In a process not dissimilar to ecological succession, the vacant niche of the Task Force in the decision "system" was filled by numerous competing bureaucratic communities. From 1973–1979 intergovernmental committees proliferated and Northern decision making was characterized by an incremental, *ad hoc* regulatory approach.[88]

The succession and proliferation of Northern administrative structures since the 1950s must also be viewed within the context of state behaviour, as argued by Pratt in his Petro-Canada case study:

> The origins of Petro-Canada must therefore be traced to the behaviour of the state itself, and especially the tendency of the executive branch, to react to external threats and instability by expanding its own domain of bureaucratic control. The concept of a national oil company found its strongest advocates within the government, notably in an Energy Department attempting to extend its control over the petroleum industry. Large permanent bureaucracies crave predictability and stability ...[89]

Although the Department of Energy, Mines and Resources has rapidly ascended to major-actor status in Northern decision making, other federal departments and agencies have managed to get a foothold, albeit less firm, on central policy and priority-setting processes. The foothold is sustained, in part, by governmental emphasis on consensus decision making and the Cabinet committee system. The proliferation of environmental, social and economic regulations has been another factor contributing to the distribution of policy responsibility through numerous federal bureaucracies.

Growing frustration with the regulatory maze gave rise to reassessments of both policy-formulation and project-approval procedures. In terms of Northern policy formulation, the appointment of commissions of inquiry, circulation of coloured papers, and the creation of multiple advisory committees, suggest that the federal government is currently displaying a preference for anticipatory planning.[90] Nevertheless, however laudable planning exercises appear to be, the fact remains that consultation further delays major policy decisions:

> If a major concern of contemporary governments is to develop a capacity to anticipate problems and gain lead-time in their policy-making activities, these techniques will serve to create delay or lag rather than anticipation.[91]

The exemption of energy policies and megaproject planning from serious public debate may be explained in terms of political power relations in Ottawa. As noted by Gurston Dacks, government may view energy and Northern megaproject development as too important to be entrusted to DIAND:

> Having been bitten once by the Mackenzie Valley Pipeline Inquiry, Ottawa is very shy about creating any institution around which opposition to megaprojects can organize. Megaprojects are seriously scrutinized but in relation to an artificially narrow definition of the national interest applied in a setting which limits effective participation to those who share a pro-development bias.[92]

In the absence of a systematic Northern policy planning framework, decisions are made by high-level intergovernmental or project-specific committees. Abele and Dosman have recommended a number of possible structural innovations to rectify the current *ad hoc* approach to Northern policy planning, including: creation of a Northern Research Secretariat (similar to the Science Council or the National Research Council); the appointment of a Minister of State for Northern Affairs; or the creation of a central Northern Policy Office (similar to the Federal-Provincial Relations Office).[93] Adoption of any one of these options would serve to focus national attention on Northern policy issues in a way that DIAND, because of its history and structure, is incapable of doing.

Without systematic and formal procedures for Northern policy formulation, industry has gained a slight advantage over other Northern interest groups, whose attention is now divided between concern for improved policy mechanisms and concern about the impacts of specific projects. Industry, being less concerned about policy formation than project approval mechanisms, tends to favour what has popularly come to be known as the "single-window" approach to project approval. The Northern Pipeline Agency (NPA), created in 1978 to facilitate and oversee post-certificate construction of the Canadian portion of the Alaska Highway Gas Pipeline, was the first such agency in government. Although the NPA is generally cited by industry as a useful mechanism to coordinate government-industry activities, there is some reluctance to adopt the NPA as a model for other projects:

> Although there was considerable justification for adopting this approach in the case of a project of such magnitude and involving such important public interest considerations, it is questionable whether it would be either desirable or realistic to contemplate the adoption of the same approach in the case of other prospective frontier energy developments. Quite apart from the practical difficulty of obtaining the necessary legislative authority required, it is open to question whether the establishment of such an extensive new regulatory mechanism to deal with an individual project is necessary or desirable.[94]

Esso Resources Canada, Ltd, in a presentation to the Special Senate Committee on the Northern Pipeline, advocated streamlining the project approvals process and argued that delays caused by the existing system reduced the reliability of benefit-cost estimates.[95] Senator H.A. Olson, Minister of State for Economic and Regional Development (MSERD) and the Minister responsible for the Northern Pipeline Agency, suggested to the same committee that the federal government was moving, admittedly in an *ad hoc* fashion, towards acceptance of the single-window concept. The decision to appoint project directors — appointed by Cabinet and attached to MSERD — was taken to facilitate coordination of future large-scale projects through a single point of contact, parallel to the Northern Pipeline Agency. When queried whether or not this step was indicative of government philosophy to retain a highly centralized administration, the Minister replied: "We hope that it will not be centralized. We hope it will be decentralized, to a large extent. There are two concepts of 'centralized' and 'decentralized'. We call it 'regionalized'."[96]

Government's desire to rationalize the project approvals process was made clear with the creation of the Canada Oil and Gas Lands Administration (COGLA). Although COGLA is not a policy-making forum, some fear it will expedite Northern decision making to the extent that industrial development projects will be given a green light before native claims are resolved, or prior to approval of DIAND's land-use management plans. Fears that alternative use options will not be evaluated, or that certain interest groups will be excluded from project decision making, were expressed to the Senate Committee on the Northern Pipeline by representatives of the territorial governments (Yukon and the Northwest Territories have no direct input into COGLA).[97]

The future status of COGLA within the Northern decision-making framework remains uncertain. COGLA may represent the creation of a powerful new wheel of decision making, or it may, in fact, represent a bureaucratic reinvention of the wheel. The organization of COGLA has a striking similarity to the 1973 structure of the Task Force on Northern Oil Development, an earlier attempt to centralize decision making for Northern resource development. Both organizations display similar committee structures, for example committees concerned with Canada benefits, environmental-social impacts and engineering technology (see Figures 6.2 and 6.3). Whether COGLA's evolution matches the history of the Task Force is likely to depend on numerous factors such as the ability of the other departments to protect their realms of authority, the ability of Northerners to gain an autonomous hand in decision making, and public acceptability of COGLA's largely non-participatory approvals process.

The debate between advocates of systematic, rational (anticipatory) planning procedures and those who advocate flexible, project-specific (*ad hoc*) decision making is unlikely to subside as long as "frontier" exploration remains a cornerstone of the National Energy Program. Before a balanced approach can be achieved, fundamental alterations — at the perceptual level — will have to occur within central agencies at the highest levels of government. Until perceptions of "national interest" are expanded to include Northern needs, as well as energy needs, the prognosis for achieving balance or equity is not very positive.

IV Legislative mesh

Four types of legislative mesh may snag or restrict the movement of industrial development: Land-Use Controls; Water-Use Controls; Special-Use Provisions governing specific resource uses such as oil and gas or mineral exploration and development; and Public Review Mechanisms such as the National Energy Board and the Environmental Assessment Review Process, which are potentially applicable to both land and water uses.[98] This section, while examining all four types of mesh, will not lay bare every minute stitch, for such an unravelling would fill many volumes. Rather, this section emphasizes key federal legislative-regulatory provisions[99] and highlights, where applicable, the potentiality for drift-netting, that is, loosely knit, open-ended language that allows great official discretion and legal gaps, permitting the avoidance of decision making.

A Land-use controls

Land use control in the North is essentially of two types: land-use permitting under the Territorial Lands Act;[100] and land-use withdrawal through the Territorial Lands Act, the Canada Wildlife Act,[101] the National Parks Act,[102] and the Migratory Birds Convention Act.[103]

1 Land-use permitting

The Territorial Lands Act is the fountainhead of power in the control of land uses in the Northwest Territories and Yukon.[104] The Act allows Cabinet to set apart any territorial lands as land management zones[105] and allows Cabinet to make regulations for protecting surface lands in land management zones and for the issue of land-use permits.[106]

Through the Territorial Land Use Regulations,[107] Cabinet has exercised such powers by declaring all of the Northwest Territories and Yukon as land management zones[108] and by implementing a two-class permit system. Class A permits, which could be subject to up to 12 months of preliminary studies by the Land Use Engineer, Department of Indian Affairs and Northern Development,[109] are required for major undertakings such as using a vehicle over 22,046 pounds, using drilling equipment over 5511.5 pounds, and using earth-moving machinery.[110] Class B permits, subject to a maximum 10 days of government consideration,[111] are required for smaller operations such as using any vehicle between 11,023 and 22,046 pounds.[112]

Such a land-use permitting system brims with discretion in three ways. First, the Land Use Engineer is not bound by any criteria for granting or denying a permit. For example, there is no guiding, legislated policy such as: "If the Engineer finds a land use entails a reasonable opportunity to cause significant environmental impairment, the permit shall be denied." Second, the Engineer is free to impose 13 classes of conditions into land-use permits, including the timing of work, erosion prevention, disposal of chemical or toxic materials, wildlife and fisheries protection, ecological protection, security deposit, debris and brush disposal, and whatever else is necessary for protecting the physical or biological character of the land.[113] Third, Cabinet retains the latent power to order special inquiries, unbounded by pre-established procedures or time-frames, to investigate land use requests.[114]

The land-use permitting system also contains major gaps. It does not clearly apply to mining operations in Yukon,[115] which are covered by the less stringent Yukon Quartz Mining Act[116] and Yukon Placer Mining Act.[117] It does not apply to lands whose rights have been disposed of by the Minister,[118] namely lands leased or sold to private interests or lands transferred to the Territories.[119] It does not provide for dealing with social problems.[120] The permit powers appear to be too narrow to impose any social conditions such as: "Work crews shall refrain from entering X native community"; or "Z Company should pay Y community $ 200 000 to cover increased service costs as a result of construction". Nor does it stipulate a comprehensive administrative review process. An applicant is not guaranteed the right to confront evidence adduced by public interest groups or government officials against the application.[121] Public interest groups or local residents are not guaranteed input or a public hearing. The composition of the Land Use Advisory Committee, charged with making recommendations, is left to official discretion.

Special protection may be given to archaeological, ethnological or historical sites through permits issued by the Department of Indian Affairs and Northern Development pursuant to the Northwest Territories Archaeological Sites Regulations[122] and the Yukon Archaeological Sites Regulations.[123]

2 *Land-use withdrawal*

Three statutory nets may be cast over Northern areas to grant special land-use protection.[124] The National Parks Act[125] allows Cabinet, after consultation with the Territories, to proclaim National Park Reserves,[126] which could invoke strict environmental protection, including a zoning system. A Special Preservation Zone would exclude all vehicles, buildings, and possibly human visitation. A Wilderness Zone would limit the number of users and restrict development to "primitive visitor facilities".[127] The Canada Wildlife Act[128] allows Cabinet to set aside areas for wildlife conservation[129] which could prohibit any commercial or industrial activities.[130] The Migratory Birds Convention Act[131] empowers Cabinet to set aside migratory bird sanctuaries, wherein disturbing migratory birds without a permit would be unlawful.[132]

The mesh of all three, however, is subject to discretionary drift. The National Parks Act allows Cabinet to authorize within parks right-of-ways for oil or gas pipelines and facilities, right-of-ways for telephone or electrical transmission lines,[133] and various developments such as buildings[134] and businesses.[135] Canada Wildlife Act regulations empower the Minister of the Environment to allow prohibited activities such as industrial activity or removal of sand and gravel through a special permit, provided the "activity will not interfere with the conservation of wildlife".[136] The Migratory Birds Convention Act's regulations allow the Minister of the Environment to permit development activities, if *in his opinion* the activities are not likely to be harmful to migratory birds or their habitat.[137] Given such loosely knit provisions, many critics have urged new park legislation, perhaps a Northern Wilderness Act with bolstered wilderness protection or perhaps a new Canada Wilderness Act with nation-wide expansion of land protection.[138]

The Northern areas actually set aside pursuant to the Acts are not extensive. Five national park areas, Wood Buffalo National Park,[139] Kluane,[140] Nahanni[141] Auyuittuq,[142] and Northern Yukon[143] have been established. Former Environment Minister, John Roberts, announced a commitment to establish Polar Bear Pass, a rich wildlife oasis on Bathurst Island, as the first National Wildlife Area in the Arctic.[144] The number of migratory bird sanctuaries stands at 16.[145]

Parks Canada, however, is considering additions. Northern Ellesmere Island, East Arm of Great Slave Lake and Northern Baffin are currently rated highest as potential national parks. The ideal has been set at 10 new park reserves, several with marine components, which would cover about five per cent of the territorial land area.[146]

B Water-use controls

Four federal departments may play a part in deciding Northern water uses: Indian Affairs and Northern Development; Environment; Fisheries and Oceans; and Transport.

1 *Department of Indian Affairs and Northern Development*

The Department of Indian Affairs and Northern Development may control water use in the North through two major vehicles:[147] The Northern Inland Waters Act[148] and The Arctic Waters Pollution Prevention Act.[149]

(*a*) *The Northern Inland Waters Act*:

The Northern Inland Waters Act stands as a potential bulwark for regulating inland waters of Yukon and the Northwest Territories.[150]

Broad powers are delegated to Cabinet, including: the power to enter into co-operative water management agreements with the provinces and Territories;[151] the power to withdraw lands to protect water resources;[152] the power to set priorities for water uses;[153] and the power to prescribe water-quality standards.[154] Broad authority is also placed in the hands of Territorial Water Boards, the Yukon Water Board and the Northwest Territories Water Board appointed by the Minister of Indian Affairs and Northern Development.[155] The Boards are empowered to impose strict conditions on water use through the issue of licences.[156]

The bulwark appearance, however, displays at least six cracks. First, Cabinet has not exercised the broad powers mentioned above. It has not withdrawn areas to protect water quality; it has not established water-use priorities; nor has it established water-quality standards. Second, Cabinet has exempted three substantial water uses from the licensing requirement: use for municipal purposes by an unincorporated settlement or construction camp;[157] use for water engineering purposes;[158] and use at a rate less than 50,000 gallons per day.[159] Third, the Water Boards enjoy wide discretion in issuing licences. No objective criteria limit the Boards, just the subjective open door of: "A board may attach to any licence ... any conditions that it considers appropriate".[160] Fourth, Board procedures tend to be uncertain. The Northwest Territories Water Board has issued no rules of procedure as regulations. The Yukon Water Board, while having issued rules with regulatory status,[161] still operates with discretionary gaps. No time limit is placed on the Water Board's decision making. No public hearing or public input is required for water authorizations granted without licences (for example, where the water use is less than 50,000 gallons per day). Fifth, DIAND officials, at least on occasion, have not strictly enforced licence conditions, but have preferred a negotiation process. For example, when United Keno Hill Mines discharged toxic chemicals from 1.5 to 17 times the licence limits into Yukon waters, DIAND inspectors carried on negotiations for nearly three years before finally pushing for prosecution.[162] Sixth, courts have tended to levy small fines on violators.[163]

(b) The Arctic Waters Pollution Prevention Act:

The Arctic Waters Pollution Prevention Act[164] grants the Department of Indian Affairs and Northern Development a potential mega-net for controlling offshore and coastal activities. The Act, while prohibiting all non-domestic water disposals[165] in all Arctic waters[166] and coastal lands,[167] allows DIAND to continue regulating industrial waste-disposals through land-use permits, land leases and oil and gas drilling permits.[168] The Department may also require any person about to undertake a work[169] likely to result in waste deposit into Arctic waters to submit work plans for review.[170] Such plans would then be subject to modification or rejection.[171]

As in the case of inland water management, the Department has been granted broad discretionary powers. No pre-established procedures, such as a limited time-frame, bridle departmental review of work plans:

> If after reviewing any plans ... and affording ... the person who provided those plans ... a reasonable opportunity to be heard, the

> Governor in Council is of the opinion that the deposit of waste ... will ... or is likely to occur ... he may, by order, either a) require ... modification ... or b) prohibit the carrying out of the construction, alteration or extension.[172]

Another uncertainty lingers in the legislation as well. The Act expressly exempts inland waters from coverage[173] but fails to define inland waters. At least two definitions are possible. Inland waters might be defined as those water areas landward of the baselines from which the territorial sea is measured, or inland waters might be defined more narrowly as only those water areas considered inland at common law, namely waters *inter fauces terrae* (between the jaws of land).[174] Occasions thus might occur, where the regulatory requirements would be unclear. Would the Northern Inland Waters Act, with its water licence requirement, apply? Or would the Arctic Waters Pollution Prevention Act, with its strict waste prohibitions and work-plan review, apply?

2 *Department of the Environment*

The Department of the Environment may seek to control industrial water uses through three legislative webbings: the Ocean Dumping Control Act,[175] the Fisheries Act,[176] and the Canada Water Act.[177]

(*a*) *The Ocean Dumping Control Act*:

Parliament giveth and Parliament taketh away. That theme partly describes the Ocean Dumping Control Act. The legislation gives the Minister of the Environment power to control, through permits, any deliberate disposal of substances from ships, aircraft or other man-made structures at sea,[178] any loading of disposal substances,[179] any placing of disposal substances on the ice,[180] and any disposal of ships, aircraft or other man-made structures at sea.[181] Much power is taken away, however, through three exceptions. Permit powers do not extend to disposals from the normal operation of ships and aircraft,[182] to emergency disposals[183] or to discharges associated with the exploration/exploitation of seabed mineral resources.[184] The latter exception arguably clips Environment's control strings over oil and gas operations, including disposal of drilling effluents, dumping of debris such as barrels or pipes, and construction of artificial islands, in so far as those activities are "incidental to or derived from" mineral exploration.[185] Artificial island construction might also be excepted on the ground that such islands are not deliberate disposals but constructive placements. The Department has, however, regulated some oil and gas activities, including harbour dredgings,[186] placement of coal dust on the ice to hasten the spring thaw[187] and experimental oil deposits.[188]

The statute also leaves many open-ended threads of ministerial discretion. Although Parliament included a list of prohibited pollutants, such as mercury, high-level radioactive waste and crude oil,[189] the Minister of the Environment is still left with freedom to permit dumping if he believes one of four conditions exist:

(a) the substance is rapidly rendered harmless by physical, chemical or biological process of the sea and does not render normally edible marine organisms inedible or endanger human or animal health;

(b) the substance is contained in another substance and not above a prescribed maximum;[190]

(c) the substance must be dumped to avert an emergency posing an unacceptable risk to human health and there are no other feasible solutions; or

(d) the substance may be transformed by burning into a non-prohibited substance.[191]

In deciding whether to issue a permit, the Minister is not bound by any set procedures, such as a time-frame, and is allowed to consider all factors "he deems necessary".[192] In issuing the permit, the Minister is free to impose conditions "as the Minister deems necessary in the interests of human life, marine life or any legitimate uses of the sea ..."[193] Although a permit applicant, faced with a permit refusal, amendment, or disagreeable provision, has the right to a Board of Review,[194] the Minister retains discretion to order a review in case of public complaints.[195]

Uncertainty over the Act's applicability to estuarine waters looms in the legislation as well. The Act specifically states that it governs internal waters but not inland waters.[196] Inland waters are defined as "all the rivers, lakes and other fresh waters in Canada".[197] Fresh waters are not defined, therefore a jurisdictional grey area is latent for those waters containing a mix of fresh and salt water. Does the Act apply all the way up river estuaries to the last iota of salt water intrusion? Does the Act just apply to the point where salinity is over half of fresh water salinity? Or does the Act only apply where water salinity matches the open sea salinity? The question remains debatable.

(b) The Fisheries Act:

Through a memorandum of understanding,[198] the Minister of Fisheries has delegated the enforcement powers of the Fisheries Act[199] Section 33 to the Minister of the Environment. At first glance, the delegation appears to grant the Minister of the Environment enough legal clout to forbid almost any waste disposal in Arctic waters. Section 33 forbids all deposits of deleterious substances from ships,[200] and from land into waters frequented by fish.[201] Deleterious substance is defined so broadly as to include almost any disposal: "any substance, that if added to water would ... alter ... the quality of that water so that it is rendered or is likely to be rendered deleterious to fish or fish habitat or to the use by man of fish ...".[202]

This power is limited, however, by two exceptions. Persons may deposit pollutants if authorized by regulations made by Cabinet under other Acts[203] or if authorized by Cabinet under Fisheries Act regulations.[204]

The first exception's language is clouded with some uncertainty. The exact language is:

> No person contravenes ... by depositing ... of waste or pollutant of a type, in a quantity and under conditions authorized by regulations applicable to that water or place made by the Governor in Council under any Act other than this Act ....[205]

At least two interpretations seem possible. One could read the language to mean that only Cabinet pollution-related regulations passed under other Acts override the provisions of the Fisheries Act. Cabinet has authorized few waste disposals through regulation under other Acts. For example, pursuant to the Ocean Dumping Control Act, Cabinet has approved deposits of prohibited agents if contained within another substance at a low level;[206] pursuant to the Arctic Waters Pollution Prevention Act, Cabinet has approved experimental deposits of oil.[207] One could also interpret the language to mean that departmental permits such as land-use permits, water licences and oil and gas drilling permits issued pursuant to regulations under other Acts are also paramount. However, the impact of such an interpretation may be lessened by provisions in or under other Acts. Section 3(3) of the Northern Inland Waters Act provides that water licences must not contravene any provision of any other Act while Section 10(3)(b) states that a Water Board may not include conditions contrary to pollution regulation under the Fisheries Act. Section 13 of the Territorial Land Use Regulations indicates that land use permits are not to allow deposits of any material or debris contrary to the Fisheries Act.

The second exception — pollution regulations issued by Cabinet under the Fisheries Act — is straightforward. To date, Cabinet has authorized waste disposals, subject to maximum effluent levels, for six industrial uses: Pulp and Paper;[208] Petroleum Refineries;[209] Potato Processing;[210] Metal Mining;[211] Chlor-Alkali Plants;[212] and Meat Processing.[213]

(c) The Canada Water Act:

The Canada Water Act[214] might be labelled more a net in the hold rather than a net on the deck. The Act's accomplishments have been limited to forcing provinces into enacting their own water management legislation, to fostering basin management studies of the Yukon and Mackenzie Rivers,[215] and to imposing a maximum limit on phosphorus in laundry detergents.[216]

3 *Department of Fisheries and Oceans*

The Department of Fisheries and Oceans may play two roles in the North related to water management. The Department of Fisheries and Oceans Act[217] endows the Department with the administrative role of coordinating ocean policies and programmes,[218] of carrying on hydrographic and marine science research[219] and of managing fisheries.[220] The Fisheries Act,[221] meanwhile, gives the Department a regulatory role through three key provisions. Section 20 allows the Minister to require fish passages for any obstructions of streams. Section 30 allows the Minister to prosecute anyone destroying fish unless authorized to do so. Section 33.1 grants by far the greatest power: the Minister may require a full review of any work or undertaking likely to deposit deleterious substances into water or likely to alter or destroy fish habitat. Subject to Cabinet approval, the Minister could thereafter modify, restrict or prevent the work or undertaking.[222]

Such a review process is open to official discretion. No procedures are set forth, thus the Minister would be free to set his own time-frame and his own informational requirements. The applicant would have no pre-established right to confront (for example, through cross-examination) those opposed to the application but would have to rely on the right of "a reasonable opportunity to make representation".[223]

4 *Department of Transport*

The Department of Transport plays a double-barrelled function in controlling Northern waters. The Department controls obstruction to navigation through the Navigable Waters Protection Act,[224] and controls shipping and ship-generated pollution through the Arctic Waters Pollution Prevention Act and the Canada Shipping Act.[225]

(a) The Navigable Waters Protection Act:

The Navigable Waters Protection Act grants the Minister of Transport three powers over obstructions to navigation. He must approve and may condition the construction of structures, such as wharves or pipelines, in the water.[226] He must approve the dumping of fill and excavation of materials from the waterbed,[227] which would encompass such activities as artificial island construction and harbour dredging, and he may require removal of debris, such as tools and vehicles, left in the water by work-crews.[228]

Ministerial approval could be bypassed, however, in two situations. No approval would be needed for a work constructed under the authority of an Act of Parliament[229] and no approval would be needed for a work, not a bridge, boom, dam or causeway, which "in the opinion of the Minister, does not interfere substantially with navigation".[230]

Discretion is once again a prominent theme. No set procedures or fixed time-frame limit ministerial review of construction proposals. The Minister is free to impose conditions in permits "upon such terms ... *as he deems fit*."[231]

For marine terminal proposals, a special review lurks beneath the legislative surface, namely the TERMPOL review process. If an inter-departmental TERMPOL Coordinating Committee, chaired by the Ship Safety Branch of the Canadian Coast Guard, believes a proposed terminal could have an adverse environmental impact, the Committee could ask the proponent to undergo an extensive TERMPOL review, including preparation of a detailed environmental impact statement and various studies such as vessel traffic patterns, location of fishing grounds, navigation conditions and underkeel clearances in the site area.[232] Recommendations of the Committee could then be incorporated into permits under the Navigable Waters Protection Act or other Acts such as the Fisheries Act.

(*b*) *The Arctic Waters Pollution Prevention Act*:

The Arctic Waters Pollution Prevention Act, where enforcement powers over shipping have been delegated to the Department of Transport,[233] seeks to curtail shipping pollution, in a 100-nautical-mile zone adjacent to the Arctic mainland and islands, in three ways. First, the Act prohibits all waste disposals from ships into Arctic waters[234] with two minor exceptions. Ships may dump human and animal wastes generated aboard,[235] and ships may deposit oil if necessary to save life or the ship itself.[236] Second, the Act establishes 16 shipping safety zones[237] (see Figure 6.5), wherein 14 classes of vessel are restricted in movement according to ice thickness and construction design.[238] For example, the strongest and most powerful vessel, an Arctic Class 10 icebreaker, could navigate any Arctic zone at any time of the year, whereas the least powerful vessel, a Type E ship, could never enter six of the zones and only enter the other 10 zones during summer or fall months. Inside the zones, pollution prevention officers have wide powers. They may board ships and undertake inspection for compliance with Canadian standards.[239] They may also order ships out of a zone if there are reasonable grounds to suspect a contravention of Canadian regulations[240] or if weather, sea conditions, ship condition or cargo condition justify an order "in the interests of safety".[241]

A third way the Act seeks to curtail shipping pollution is through special equipping and manning requirements. Some of the special equipments include gyro-compasses,[242] marine radars,[243] echo-sounding devices[244] and radio-telephones.[245] Some of the special manning requirements include two deck watches[246] and an ice navigator.[247]

Since the Department of Transport has been delegated enforcement powers and not decision-making powers under the Act, discretionary open-endedness is not a problem. Instead, the major problem has been the Department's capability to keep up with regulating new ship-designs. Innovations have come so fast and furiously that Transport officials have publicly acknowledged the difficulty, as exemplified in the Departmental brief to the Special Committee of the Senate on the Northern Pipeline:

Figure 6.5. Shipping safety control zones.

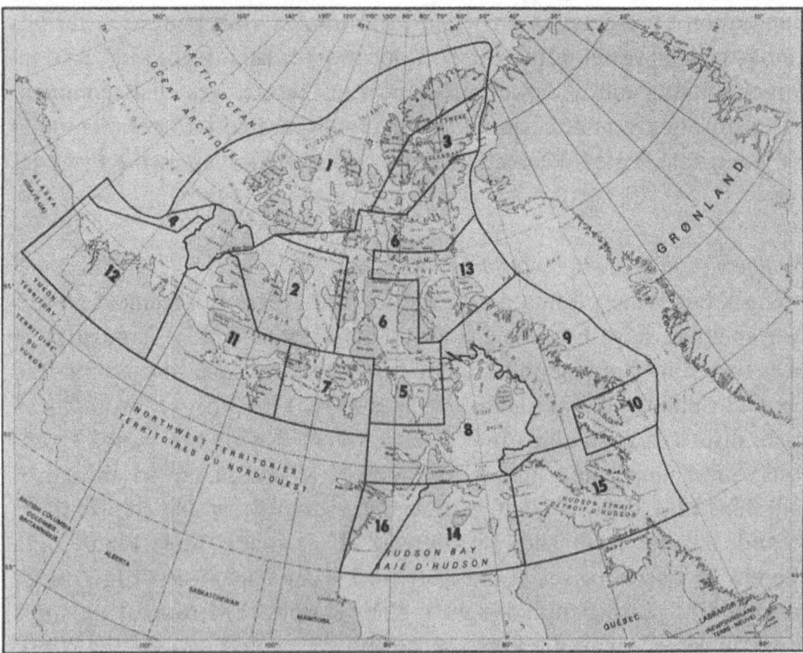

Source: Department of Fisheries and Oceans, *Sailing Directions-Arctic Canada*, V. 1 (1982), p. 100.

> New technology in the field of Arctic marine transportation and resource extraction is now developing at such a rate that there is not sufficient time to assimilate and translate data into objective regulatory provisions. As a result, decisions are now being made with less than a satisfactory level of review and assessment ... Coast Guard leadership capability in designing ships for operations in the Arctic is falling behind the private sector. As a consequence, the private sector ... can be expected to construct cargo carriers that ... challenge the design and operating aspects of our regulatory regime.[248]

(*c*) *The Canada Shipping Act*:

The Department of Transport also administers the Canada Shipping Act, which regulates Arctic shipping pollution between 100 and 200 nautical miles offshore, as shown by statutory logic. The Canada Shipping Act expressly revokes coverage of the 100-nautical-mile safety control zones under the Arctic Waters Pollution Prevention Act[249] but extends coverage to any fishing zones of Canada.[250] In 1977, Cabinet declared a 200-nautical-mile fishing zone in the Arctic.[251] Therefore, the Shipping Act applies between 100 and 200 nautical miles offshore.

The Shipping Act imposes less stringent pollution requirements than the Arctic Waters Pollution Prevention Act (AWPPA) because of a different regulatory philosophy. While the AWPPA adopts the philosophy, "No pollution until excepted by regulation", the Shipping Act embraces a contrary approach of "Pollute until excepted".[252] To date, the Shipping Act has excepted, that is prohibited, the disposal of oil and oily mixtures[253] and approximately 400 other pollutants such as mercury, arsenic, phosphorus and cyanide compounds.[254]

C Special-use provisions

Besides the special permits required for timbering,[255] quarrying[256] and coal mining,[257] special regulatory requirements also apply to the two major resource uses in the North, mineral mining and oil and gas exploration and exploitation.

1 *Mineral mining*

Harnessing water uses by the mining industry, for example tailings disposal or soil washings, would almost totally depend on three of the water mesh statutes discussed above. The Northern Inland Waters Act would require a water licence or water authorization for mining activities impinging on inland waters.[258] The Arctic Waters Pollution Prevention

Act, covering offshore waters, could impose an environmental review or prosecution by the Department of Indian Affairs and Northern Development. Or the Fisheries Act could empower the initiation of an environmental assessment or prosecution.

Specialized regulations do apply to river mineral dredgings,[259] but the regulations deal primarily with administrative matters, such as staking and royalties, and only tangentially with water protection. The regulations merely prohibit dredging lessees from interfering with the general right of the public "to use the river for navigation or other purposes"[260] and, more specifically, from obstructing streams with tailings accumulations.[261]

In Yukon, two statutes brimming with administrative provisions, but limited in environmental controls, govern mining. The Yukon Placer Mining Act[262] sets out the basic staking, surveying, recording and royalty requirements for mining with water, usually for gold. The Yukon Quartz Mining Act[263] establishes similar requirements for the mining of mineral veins such as zinc, lead or tin. Both statutes limit environmental consideration to two narrow areas. The statutes allow mining inspectors to order changes in mining practices to protect the safety of the public or employees (for example, by ordering an abandoned pit to be filled).[264] They also prohibit the dumping of earth or stones onto neighbouring claims.[265]

Land-use impacts of mining in the Northwest Territories, meanwhile, are governed by two regulatory enactments flowing from the Territorial Lands Act. The Canada Mining Regulations,[266] while mainly administrative, covering such details as recording and surveying, grants the Minister of Indian Affairs and Northern Development discretionary power to order waste treatment and even mine closures where discharges are "likely to be harmful to humans, animals or vegetation".[267] The Territorial Land Use Regulations[268] would likely require a Class A land-use permit, which could include such environmental conditions as debris disposal, wildlife protection and equipment limitations.[269]

2 *Oil and gas*

Special controls governing oil and gas exploration and exploitation may be grouped into six categories: legislated approvals; administrative approvals; ministerial controls; Cabinet powers; general regulations; and pipeline provisions.[270]

(*a*) *Legislated approvals*:

Legislation and regulations, relating to oil and gas development, set forth six special hoops through which corporations may have to manoeuvre. Each hoop — an operating licence, an exploration agreement, a drilling programme approval, an authority to drill, a production licence and a subsurface storage licence — remains uncertain due to discretionary open-endedness.

(i) An *operating licence*, required by the Oil and Gas Production and Conservation Act,[271] merely allows the holder to enter and use the surface of Canada Lands. No right to drill or right to oil and gas is thereby conveyed. The licence, renewable annually, is open to official discretion through the statutory words: "{T}he Minister ... may issue an operating licence ... subject to such requirements as he determines ...".[272]

(ii) An *exploration agreement*, required by the Canada Oil and Gas Act[273] and negotiated by the Canada Oil and Gas Lands Administration (COGLA) on behalf of the Minister of Indian Affairs and Northern Development,[274] confers four rights. The holder receives the right to explore, the exclusive right to drill for oil or gas, the exclusive right to develop land for production, and the exclusive right to obtain a production licence.

The Minister is allowed to insert numerous conditions into the exploration agreement. Besides requiring a strict work schedule (for example, the Minister might require two wells to be drilled within one year), the Minister may provide for "any other relevant matter", including rental payments, informational requirements and equity participation by Canadians.[275] He must require a Canadianization plan from the proponent, guaranteeing the opportunity for Canadian employment and Canadian business involvement. Such a plan need only be "satisfactory to the Minister".[276] The Act also allows the Minister to include provisions ensuring that disadvantaged persons have access to training and employment.[277]

Procedural discretion is also granted to the Minister in considering exploration proposals. The Minister need give no notice calling for proposals, if he believes foregoing notice would be in the public interest, or if he needs to act expeditiously.[278] In selecting among proposals, he need only consider "the factors he considers appropriate in the public interest".[279] While the Minister must publish a notice 30 days before entering an exploration agreement, he is not required to consider any public inputs.[280]

In case an interest holder under previous regulations or a permittee eligible for renewal does not wish to be burdened with work requirements for the first year, they may secure a provisional lease instead of an exploration agreement. Unlike the exploration agreement, the lease would entail an annual rental fee of $ 30 per hectare to compensate for loss of drilling productivity, and would not be renewable after a five-year term.[281]

(iii) A *drilling programme approval*, required by the Canada Oil and Gas Drilling Regulations,[282] may be described as COGLA's checklist for assuring an applicant's overall drilling plans meet technical requirements for safety. The pre-conditions for approval include adequate life-saving equipment, adequate rig design, sufficient medical and rescue facilities, blowout prevention and a contingency plan for foreseeable emergencies such as an oilspill.

(iv) An *authority to drill*, required by the Oil and Gas Production and Conservation Act[283] and the Canada Oil and Gas Drilling Regulations,[284] is a licence to drill a particular well. The licence requires much site-specific information such as the name of the well, the geographical coordinates, the proposed depth, the water depth (if offshore) and the procedures to protect the environment in the well vicinity.

Like an exploration agreement, the authority to drill contains some uncertainty. The Minister, not limited by any time-frame or hearing procedure, may impose "requirements as he determines", including liability requirements, environmental studies requirements and reimbursement requirements for governmental costs in approving production facilities.[285] A mandated Canadianization plan is left open-ended with the subjective criterion of "satisfactory to the Minister".[286] The required financial responsibility, such as an indemnity bond or a letter of credit, is left open-ended with the words "in any form satisfactory to the Minister, in an amount satisfactory to the Minister".[287]

(v) A *production licence* confers the exclusive right to produce oil or gas and, upon payment of royalties, title to the oil or gas.[288] The only special preconditions for corporations seeking a licence are two in number. A corporation must be incorporated in Canada and must have a Canadian

Figure 6.6. Typical COGLA approvals process for a fixed hydrocarbon system.

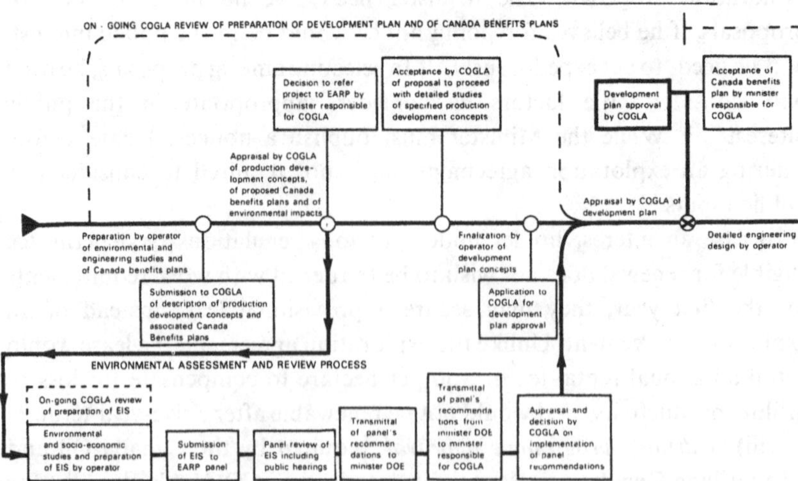

ownership rate of 50 per cent or more.[289] If the ownership falls below 50 per cent, either before or after the licence grant, the Minister may reserve a share in production equal to the difference between the actual Canadian ownership rate and 50 per cent.[290] For example, if the Canadian ownership rate were only 40 per cent, the Minister could reserve a 10 per cent share in production.

(vi) A *subsurface storage licence* subject to any terms and conditions considered appropriate by COGLA would be required by anyone wishing to store oil or gas, at depths greater than 20 metres, in federal lands.[291]

(*b*) *Administrative approvals*:

Hidden beneath the legislated surface are four administrative hoops constructed by COGLA to control oil or gas production. A Development Plan Approval may require a detailed production field appraisal, including environmental, geological and economic studies, and would require proofs of financial responsibility "satisfactory to the Minister" (for example, a performance bond and an indemnity bond).[292] A Production Structures Approval and a Production Facilities Approval would harness actual constructions such as production platforms and offshore terminals, respectively, by requiring detailed information and possibly imposing special conditions of design and operation.[293] A Transportation Approval would cover the design and operation of a pipeline production system, including intraterritorial and offshore pipelines[294] (Figure 6.6).

Since COGLA has only implemented the approval hoops once — for the Norman Wells Expansion Project — the hoops are admittedly in a

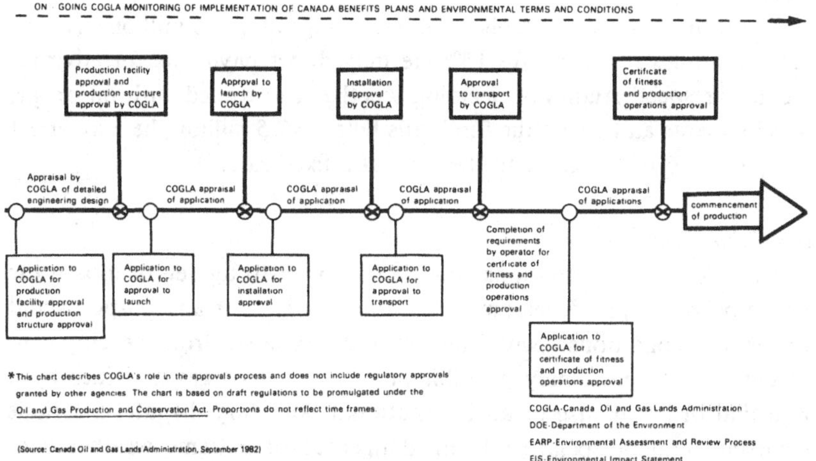

formative stage. COGLA, in a presentation to the Special Senate Committee on the Northern Pipeline, described the rather *ad hoc* process in these words:

> It must be emphasized that COGLA views the approvals process as an iterative one, in which COGLA and the concerned applicants will co-operate to the greatest extent possible ... to obtain timely and satisfactory results ... many of the details of an approvals process, as well as its related policy issue, not only will be further refined as the need arises, but also vary from one project to the next.[295]

(c) Ministerial controls:

Besides the powers exercised by the Minister of Indian Affairs and Northern Development through the legislative and administrative hoops, the Minister retains three other legislative draw-strings for influencing oil or gas development. First, the Minister has broad power to hasten along drilling programmes. By declaring an area a "significant discovery", he may order each interest owner to drill up to three wells at a time.[296] By declaring an area "a commercial discovery", he may order each interest owner to drill even more than three wells, for the Canada Oil and Gas Act places no upper limit on his powers.[297]

Second, the Minister may act as the "Czar of oil and gas production". He may order a non-producer, having the capability of producing, to begin production and to deliver the oil or gas as specified. The order may name the time, place and quantity of delivery and the purchaser.[298] He may also order producers to increase, decrease or cease production and associated works.[299]

Third, the Minister controls the purse-strings of a $ 15 million environmental studies revolving fund.[300] He may direct payments for environmental or social studies in deciding whether to proceed with oil or gas developments and where the fund falls below $ 7.5 million, he may direct companies to recontribute to the fund at a fixed rate.[301]

(d) Cabinet powers:

Three legislative provisions empower Cabinet to tighten or loosen the noose on oil or gas developments. Cabinet may "for any purposes and under any conditions" withdraw oil and gas lands from development activities.[302] Cabinet may prohibit or restrict oil or gas activities by a remedial order in case of an international boundary dispute, a serious environmental or social problem, dangerous or extreme weather conditions, or "other special circumstances".[303] Cabinet also enjoys regulatory power over almost every facet of oil and gas development through section 12 of the Oil and Gas Production and Conservation Act. That section bestows 23 classes of powers, including the right to enact measures necessary to prevent air, land or water pollution as a result of any phase of oil or gas development.[304]

(*e*) *General regulations*:

Prior to 1983, three sets of regulations applied to oil and gas development — the Canada Oil and Gas Drilling Regulations,[305] the Canada Oil and Gas Drilling and Production Regulations,[306] and the Canada Oil and Gas Land Regulations.[307]

The latter two sets of regulations are relatively minor and have been largely superseded. The Canada Oil and Gas Drilling and Production Regulations, in so far as its drilling regulations are concerned, have likely been totally superseded by the newer Canada Oil and Gas Drilling Regulations.[308] Production regulations merely cover a few areas such as testing and metering gas wells.[309] The Canada Oil and Gas Land Regulations, meanwhile, which used to control the issuing of oil and gas rights, have been almost totally superseded by the Canada Oil and Gas Act which provides for a renegotiation of oil leases or permits. Only a few regulatory remnants remain, such as the grid system covering Canada Lands and surveying requirements for well locations.

The major set, the Canada Oil and Gas Drilling Regulations, besides setting out multitudinous requirements for attaining a drilling programme approval and an authority to drill, is also a storehouse of regulations covering drilling from start to finish. For example, at the start-up stage, the regulations admonish drillers to minimize ground and vegetation disturbance "to the extent practicable".[310] At the operational stage, the regulations cover such matters as: suspending offshore drilling in case of iceberg threat or severe weather;[311] disposing of wastes such as drilling fluids and non-combustibles;[312] maintaining personnel safety through mandatory rest periods[313] and protective clothing;[314] and reporting information to government.[315] At the termination stage, the regulations cover such matters as plugging and marking abandoned wells[316] and final well reporting.[317]

Although the Drilling Regulations do show some tightly-anchored provisions, such as "every drilling unit shall ... carry at least ten approved life-buoys",[318] many provisions harbour room for discretionary drift as shown by two examples. While operators are told to suspend offshore drilling upon certain conditions, including where there is an excessive motion of the drilling unit because of sea or weather conditions,[319] or where there is a serious and imminent threat of ice or icebergs,[320] the undefined language "excessive motion" and "serious and imminent threat" leaves great leeway for individual judgment. Disposals of drilling fluids, sewage and acids, meanwhile, depend on the open-ended words, "in a manner approved by the Chief".[321]

Three additional sets of regulations were issued in 1983. Canada Oil and Gas Operations Regulations establish conditions for obtaining an operating licence, a work authorization and for reporting an oilspill.[322] Canada Oil and Gas Interests Regulations set out procedures for surrendering and transferring an interest in Canada Lands and a fee schedule for various transactions (for example $10 to grant a production licence).[323] Environmental Studies Revolving Fund Regions Regulations prescribe 14 regions for sub-funding purposes from a revolving fund.[324]

Additional regulations wait in the wings. COGLA has proposed at least seven new sets of regulations: Canada Oil and Gas Royalties Regulations, Canada Oil and Gas Production Regulations, Canada Oil and Gas Geophysical Regulations, Canada Oil and Gas Structures Regulations, Canada Oil and Gas Regulations — Diving, Canada Oil and Gas Pipelines Regulations and Offshore Mining Regulations.[325]

(f) Pipeline provisions:

Special controls over pipeline development may be described as a few threads of set regulations but a potential spool of drift. Only two major sets of regulations presently govern pipeline construction, Gas Pipeline Regulations[326] and Oil Pipeline Regulations[327] passed pursuant to the National Energy Board Act.[328] Both sets of regulations cover such details as pipeline specifications[329] (for example, notch strengths), welding requirements,[330] field pressure testing[331] and environmental constraints. For example, companies are urged to work on frozen ground, to use vehicles with low ground bearing pressure and to stabilize disturbed areas by seeding or planting.[332] Waste burial sites must not contaminate ground water entering streams, lakes or other surface waters.[333] Fish-spawning beds must be protected from construction sediments.[334] Animal crossings must be maintained in passable condition.[335]

Potential for drift looms in two sources. As discussed in the following section on public review mechanisms, the National Energy Board, through a certificate of public convenience and necessity, may impose numerous technical, social, or environmental conditions on any interprovincial or international pipeline. Cabinet may manipulate construction requirements of offshore and intraterritorial pipelines through two broad regulatory powers granted by the Oil and Gas Production and Conservation Act. Section 12(k) of the Act allows Cabinet to authorize the Minister of Indian Affairs and Northern Development or another suitable person to exercise powers necessary for pipeline construction. Section 12(j) expands the power even further to allow Cabinet to authorize the Minister of Indian Affairs and Northern Development or another suitable person to exercise powers necessary for the removal of oil or gas from the offshore or Territories.

A preview of what such powers could mean in practice may be glimpsed in the Northern Pipeline Act,[336] passed in 1978. The Act attempted to hasten construction of the Alaska Highway Natural Gas Pipeline, not only by declaring a certificate of public convenience and necessity in force[337] and by trimming hearings before the Yukon Territory Water Board,[338] but by authorizing Cabinet to transfer powers of other departments to the Minister of the Northern Pipeline Agency.[339] Cabinet subsequently crowned that Minister as "King of Decision-Makers" by passing to him the power sceptres of the Departments of Fisheries and Oceans, Environment, and Indian Affairs and Northern Development.[340] As a result, the Minister could unilaterally make almost all environmental decisions as shown by a few excerpts from the Socio-Economic and Environmental Terms and Conditions for operations by Foothills Pipe Lines Ltd in Southern British Columbia:

> Foothills shall, in the planning and construction of the pipeline, minimize *to the satisfaction of the designated officer*, any adverse environmental impact on land ....[341]
>
> Foothills shall take such measures as are *satisfactory to the designated officer* to ensure that significant quantities of deleterious substances, as defined in the Fisheries Act, ... are not left in an area where they are likely to enter into any waterbody.[342]
>
> Where the construction or operation of the pipeline takes place within a waterbody, Foothills shall take such measures as are *satisfactory to the designated officer* to protect the quality of the water in that waterbody.[343]

D *Public review mechanisms*

Industrial proposals for the North, besides having to run the regulatory gauntlet discussed above, may also have to manoeuvre through two mega-review mechanisms — the National Energy Board and the Environmental Assessment Review Process.

1 *National Energy Board*

Four types of industrial activities need the National Energy Board's regulatory blessing before proceeding. Oil or gas pipelines, if just interprovincial, require a certificate of public convenience and necessity before construction commencement.[344] Oil or gas pipelines, if international, would require not only a certificate but also a licence to export or import.[345] Oil or gas exports or imports, if transported by means other than pipeline (for example, by tanker), would just require an export or import licence.[346] Construction of an international power line and the export of power would require a certificate[347] and an export licence[348] respectively.

Such certifying and licensing functions contain potential for official discretion in at least five ways. First, the National Energy Board, unanchored by any taut legislative criteria, may roam widely in the factors considered for issuing certificates of public convenience and necessity. The Board is merely mandated to "take into account all such matters as to it appear to be relevant", including:

(i) availability of oil, gas or power;
(ii) the existence of markets, actual or potential;
(iii) economic feasibility;
(iv) financial responsibility of the applicant and the extent of Canadian participation; and
(v) any public interest.[349]

The Board's Rules of Practice and Procedure[350] have expanded "public interest" to include environmental impacts. The Rules establish some seven pages of informational requirements including resource inventories, an environmental impact statement (EIS) and methods for mitigating environmental effects.[351]

Second, the Board is given almost unbridled leeway to place conditions into certificates and licences. Language governing certificate issue states:

> The Board may issue a certificate subject to such terms and conditions as the Board considers necessary or desirable in the public interest.[352]

Language governing licence issue states:

> Subject to the regulations, the Board may, on such terms and conditions as it may impose, issue licences ....[353]

Third, although the Board in licensing must satisfy itself that "the quantity of oil, gas or power to be exported does not exceed" reasonably foreseeable

Canadian requirements,[354] the Board is free to consider all factors it considers relevant.[355]

A fourth factor of drift is the power retained by Cabinet. Even if the National Energy Board approves a certificate or licence, Cabinet may still impose additional conditions or reject the application.[356] Cabinet may also direct the Board to supervise and control oil or gas movements, not otherwise subject to National Energy Board Review, such as interprovincial oil and gas transportation via tanker or aircraft.[357]

A fifth factor is meagreness of Board procedures. No set time-frame binds the Board in holding hearings or in reaching decisions. Rendering reasons for decision is not generally mandated.[358] Rules of evidence remain flexible.[359] Whatever procedures are mandated may summarily be taken away by the regulatory scalpel:

> The Board may, in any proceeding before the Board upon an application, direct either orally or in writing that ... these Rules ... shall not apply ... for the purpose of ensuring the expeditious conduct of the business of the Board ....[360]

The tendency of much legislation to drift has resulted in inconsistent administrative practice. In the Northern Pipelines decision, where the Board recommended a $200 million socio-economic fund and a coordinating agency to facilitate the Alaska Highway Pipeline, the Board excused environmental deficiencies by granting a conditional certificate pending further environmental studies.[361] In the TransCanada Pipelines Ltd: Q & M Pipelines Ltd decision, the Board refused the Q & M application due to insufficiency of environmental information.[362] In the Trans Mountain Pipe Line Company Ltd decision the Board, in deciding whether to certify a pipeline carrying oil from a proposed terminal in Washington through Canada to landlocked northern states, considered the environmental implications of tankering and refining, even though the terminal and tanker movements were beyond its jurisdiction.[363] In an earlier decision concerning an LNG import-export licence application, the Board refused to consider the safety aspects in the Bay of Fundy, as such a matter was outside the Board's jurisdiction.[364]

Courts have shied away from anchoring the National Energy Board's wide discretion as shown by the case, *Committee for Justice and Liberty Foundation* v. *Inter-provincial Pipeline (NW) Ltd.* There various environmental groups challenged the National Energy Board's grant of a certificate to construct a pipeline from Norman Wells, Northwest Territories, to Zama, Alberta. The Court responded to claims of "socio-environmental deficiency" with a hands-off attitude:

> The Court is not entitled to substitute its opinion as to whether or not these facts justified a finding of public convenience and necessity for the opinion of the Board. In a situation of this kind, the determination of public convenience and necessity is not a question of fact, but is, rather, a formulation of an opinion by the Board, and by the Board only.[365]

2 *Federal Environmental Assessment Review Process*

In 1979, when the Department of Fisheries and the Environment was split into two separate departments — Fisheries and Oceans and Environment — the new Department of the Environment gained a legislative base for launching environmental assessments. The Government Organization Act, 1979[366] provided:

> The Minister of the Environment ... shall ... ensure that new federal projects, programs and activities are assessed early in the planning process for potential adverse effects on the quality of the natural environment and that a further review is carried out of those projects ... found to have probable significant adverse effects, and the results thereof taken into account ....[367]

Such a mandate legislatively solidified an environmental review process already initiated by the Canadian Cabinet through previous policy memoranda. On 18 December 1973 the Cabinet Committee on Science, Culture and Information recommended an environmental review process for all projects initiated by a federal department or agency or "for which federal funds were solicited or for which federal property was required". Proprietary Crown corporations and regulatory agencies were merely invited to participate and environmental study panels were limited to Department of Environment personnel.[368] On 8 February 1977 the Cabinet Committee on Government Operations recommended opening panel membership to those outside the Public Service and adopting a financial policy of sharing environmental assessment costs between the federal government and non-federal government proponent(s).[369]

The Environmental Assessment Review Process (EARP) still follows the Cabinet directives and has evolved into essentially a three-stage process:

(*a*) *Environmental screening*:

The initiating federal department (that is the department with the lead involvement in a development project) undertakes an initial environmental screening to determine if a project carries significant environmental effects.[370] If the department concludes there are no adverse effects or no significant adverse effects, the project would proceed subject to mitigative measures.

(*b*) *Initial environmental evaluation*:
If the department decides the potential for significant environmental effects is not certain and requires further study, a more detailed study takes place called an initial environmental evaluation.[371]

(*c*) *Environmental Assessment Review Panel*:
If the department, based on preliminary screening or a subsequent initial environmental evaluation, decides the environmental effects are potentially significant, the department would refer the matter to the Executive Chairman of the Federal Environmental Assessment Review Office (FEARO), an independent office of the Department of the Environment. The Chairman would establish an environmental assessment panel, usually consisting of public servants and independent experts, which would draw up guidelines for an environmental impact statement (EIS) to be prepared by the proponent. The Panel would review the environmental impact statement, hold public hearings,[372] and make recommendations to the Minister of the Environment and the Minister of the initiating department.

Many have criticized the process[373] on at least four fronts. First, since the initial environmental screening and the ultimate decision whether to refer a project to FEARO lies largely within the discretion of the sponsoring or initiating department and since "significant adverse effects" are not defined in legislation, the opportunity exists for a department, with an urge to develop without delay, to avoid the process. As an example, the Department of Energy, Mines and Resources has never submitted the extensive East Coast exploration drilling programmes off Nova Scotia and Newfoundland to the EARP process.[374] Second, since the slim legislative base does not impose a duty on the departments to submit proposals to EARP, the public has had almost no recourse to judicial review as shown by the case of *Esso* v. *Atomic Energy Control Board*.[375] There the Township of Tosorontio had requested a court to require an EARP hearing for a proposed radioactive storage site. The court bailed out with the words:

> While the apparent failure to follow the procedures so glowingly advertised by the Minister of the Environment in his May, 1979, publication, *Revised Guide to the Federal Environmental Assessment and Review Process*, may be regarded as peculiar in such a conspicuous situation, the Guide created no legal obligations ....[376]

Third, the EARP process is advisory. No assurance exists that the independent assessment will be implemented in practice.

Fourth, an Environmental Assessment Review Panel may be prevented from undertaking a full, systematic review by its narrow terms of reference. For example, the Beaufort Sea Environmental Assessment Panel has been limited to considering environmental effects North of 60°.[377] Such a reference could preclude consideration of major issues like the effects of tanker traffic off the Labrador coast.

New Cabinet guidelines issued in June 1984 could substantially strengthen the process, however.[378] Initiating departments are required to develop, in co-operation with FEARO, a list of the types of proposals to be automatically referred to the Minister of Environment for public review. Initiating departments are also required to establish written procedures for assessing project proposals. Since the guidelines were issued as an Order-in-Council, they could provide a legal foundation for expanded judicial review of the environmental assessment process.

V The decision-making net in action

The potential for drift in Canadian law and administration has been borne out in practice. A brief comparison of ten Arctic industrial proposals, which have attempted to navigate through the tricky mesh of government decision making in the past decade, reveals sharp contrasts. Three proposals — the Nanisivik Mine, the Polaris Mine and Panarctic Arctic Island exploration — avoided the formal environmental review process altogether. One proposal — South Davis Strait drilling — underwent a formal environmental assessment but managed to manoeuvre through the mesh quickly. Two projects — the Alaska Highway Gas Pipeline and Norman Wells Oilfield expansion — ran into extensive and delaying reviews. The Arctic Pilot Project, while slipping through the environmental assessment process, became entrapped in National Energy Board hearings, and proponents finally shelved the project. Three proposals — Lancaster Sound drilling, Beaufort Sea hydrocarbon development, and North Davis Strait drilling — are still enmeshed in the decision-making net. This section provides an overview of the decision-making process for the ten projects and concludes by analysing the reasons for varied reviews. (For the location of the various projects, see Figure 6.7).

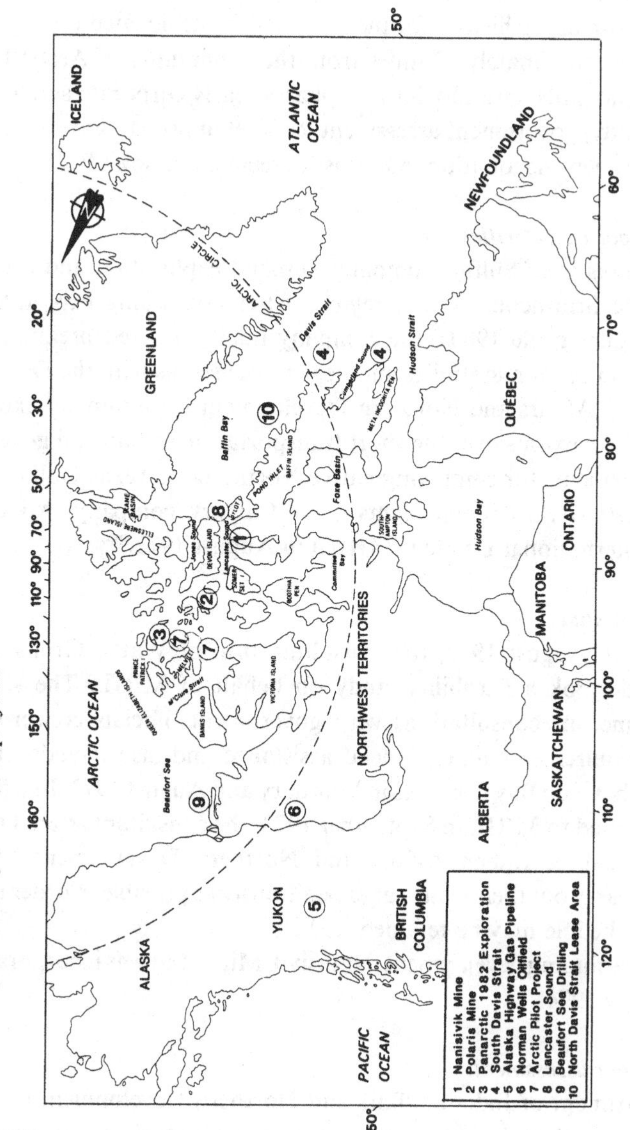

Figure 6.7. Major Northern projects.

A *Project reviews*

1 *Nanisivik mine*

The decision-making history for the Nanisivik mine, located on Strathcona Sound, approximately 17 miles from the community of Arctic Bay on Baffin Island, falls naturally into six phases: early corporate activities, consultant study, government assessment, cabinet approval, government-industry agreement negotiation, and post-agreement reviews.[379]

(*a*) *Early corporate activities*:

In 1957 Texas Gulf Sulfur Company initiated exploration and claim-staking in the Strathcona Sound region. After continuing exploration sporadically during the 1960s, the company finally ordered preliminary engineering and economic studies and began discussions with the Department of Indian Affairs and Northern Development for a mineral export permit in 1970. Anxious over the short transportation season and governmental requirements for employing native Northerners, Texas Gulf eventually transferred Strathcona rights to a Calgary company, Mineral Resources International Limited (MRIL) in August 1972.[380]

(*b*) *Consultant study*:

Beginning in August 1972, the consulting firm of Watts, Griffis and McQuat undertook a feasibility study on behalf of MRIL. The study included numerous consultations with government officials concerning regulatory requirements and financial assistance and also included two meetings with Arctic Bay residents in February and August 1973. In a final report, submitted to MRIL in September 1973, the consultants stated that the Department of Indian Affairs and Northern Development "had promised to carry out their review as expeditiously as possible in order that the project schedule may be retained".[381]

To facilitate mining development, Nanisivik Mines Ltd was incorporated on 17 January 1974.

(*c*) *Government assessment*:

The Department of Indian Affairs and Northern Development coordinated a five-month review of the mining proposal by various government departments-agencies including Energy, Mines and Resources, Transport, Treasury Branch Secretariat, Finance, Manpower and Immigration, Environment and Regional Economic Expansion, and the government of the Northwest Territories. Despite recommendation by several DIAND officials for additional social studies and recommendation by DOE experts for up to a year of environmental studies, the Department of Indian Affairs and Northern Development recommended project approval to Cabinet on 8 March 1974.

(*d*) *Cabinet approval*:

DIAND's memorandum emphasized the potential for native employment benefits and government financial gain, and emphasized the project's urgency due to a private financing deadline and the risk of losing Inuit support. Cabinet quickly granted approval-in-principle on 28 March 1974 and charged DIAND and the Department of Industry, Trade and Commerce to negotiate a development agreement with MRIL.

(*e*) *Government-Industry agreement negotiation*:

Negotiation of a development agreement was handled by a Special Interdepartmental Working Group on the Strathcona Sound Project, composed of representatives from the Department of Indian Affairs and Northern Development, the government of the Northwest Territories and 14 other federal departments or agencies.[382]

For environmental terms and conditions, DIAND and DOE requested the Northwest Territories Water Board to coordinate departmental views and to make recommendations to the Working Group.

The Government-Industry agreement,[383] reached on 18 June 1974, pledged the federal government: to supply $ 16.7 million for infrastructure such as roads, an airport, a wharf and town construction; to grant a surface land lease in return for corporate agreement to give the federal government an 18 per cent interest in Nanisivik;[384] to extend production from eight to twelve years; to pursue the goal of a 60 per cent native work force; to favour Canadian goods and services; and to undertake environmental studies "as required by appropriate government agencies". In addition, the agreement named the Northwest Territories Water Board as the initial point of contact on environmental matters and approved land disposal of tailings subject to further studies. Marine disposal, meanwhile, would have required detailed environmental studies and DIAND-DOE approval.

(*f*) *Post-agreement reviews*:

Although MRIL undertook environmental studies to justify the less-costly marine disposal of tailings, government scientists and an independent consultant remained concerned over the potential for heavy metal contamination of Strathcona Sound. As a result, Judd Buchanan, Minister of Indian Affairs and Northern Development, denied the company's request for marine disposal on 13 November 1975.

In the spring of 1976 the Northwest Territories Water Board held a public hearing on the company application to dispose of tailings into Twin Lakes and agreed to issue a one-year interim licence subject to terms and conditions. The Water Board retained the right to re-evaluate tailings disposal within one year and to impose other conditions at that time.[385]

Two government committees have subsequently monitored company compliance with the 1974 agreement. The Nanisivik Monitoring Committee, the successor to the Strathcona Sound Working Group, has overseen several matters such as use of Canadian shipping, rate of exploitation and continuance of exploration. The Training and Employment Advisory Committee has guarded the standards for training and employing Northerners.[386]

Nanisivik production commenced in 1976.

2 *Polaris mine*

Approval of the Polaris lead-zinc mine, located on Little Cornwallis Island approximately 90 miles from the magnetic North Pole, involved essentially four major movements: getting off the mark, skipping the EARP, wading the Water Board, and dealing with DIAND.[387]

(*a*) *Getting off the mark*:

In 1964 Cominco Ltd obtained a 75 per cent interest in the lead-zinc property of Bakeno Mines Ltd and in 1971 formed a joint venture company, Arvik Mines Ltd, to facilitate mining development (Arvik would later become Polaris when Cominco purchased the remaining 25 per cent in January 1979). In 1973–1974 various consultations occurred among Cominco, the Department of Indian Affairs and Northern Development, the Department of the Environment, and the Northwest Territories Water Board, concerning the need for, and the adequacy of preliminary environmental studies.

(*b*) *Skipping the EARP*:

In February 1975 DOE concluded Cominco should undergo a formal environmental assessment before issue of a water licence and directed the chairman of the Federal Environmental Assessment Review Office (FEARO) to draw up guidelines. In January 1976 the Department of Indian Affairs and Northern Development officially requested the formation of an EARP panel to review the Arvik proposal. On 18 March 1976 the chairman of FEARO wrote to the Assistant Deputy Minister of

DIAND and asked for strong reasons why Arvik should undergo the EARP process. DOE's position on the need for a formal environmental assessment had changed for at least three reasons. DOE officials felt Cominco's environmental studies, submitted since February 1975, answered many original questions. DOE officials were calmed by Cominco's decision to forgo marine disposal of tailings in favour of disposal into Garrow Lake, a highly saline and unproductive body of water. Officials also believed the Water Board was sufficient for overseeing environmental requirements. On 5 April 1976 DIAND agreed a Panel review was not required.

After Cominco announced, on 5 November 1979, a definite decision to proceed with Polaris, interest groups, including the Canadian Arctic Resources Committee and the Inuit Tapirisat of Canada, petitioned the Ministers of DOE and DIAND to refer the proposal to EARP review. No EARP referral occurred, however, due to official concerns that EARP could lose credibility, since construction had already begun and the "no-go" option had been eliminated.

(*c*) *Wading the Water Board*:

Cominco first submitted a licence application to the Northwest Territories Water Board on 9 October 1974, and, by May 1980, had submitted sixteen separate reports to the Board. On 22 May 1980 the Water Board held a public hearing on the application in the community of Resolute and on 9 October 1981 the Chairman of the Water Board forwarded a signed licence to the Minister of DIAND for final approval.

(*d*) *Dealing with DIAND*:

After years of dickering over conditions for mine development, Cominco and DIAND reached a letter of understanding in February 1980. The letter pledged Cominco to comply with existing regulations, to undertake a smelter feasibility study within five years of production, to ship one-half of each year's ore via the Canadian vessel, the *MV Arctic*, to give Canadian vessels preference in transporting remaining ores, and to enter a socio-economic action plan with the government of the Northwest Territories.

After construction of a barge-mounted processing plant in Quebec and a 4800 kilometre towage of the plant to Little Cornwallis Island, sustained production began in February 1982.[388]

3 *Panarctic Oils Ltd/Arctic Islands exploration*

The first exploratory well was drilled on Melville Island in the Sverdrup Basin in 1961, and the first hydrocarbons were discovered by Panarctic Oils in 1969 at Drake Point.[389] By 1981, 158 test wells had been drilled, 19 of which had been drilled through artificially thickened ice platforms.[390] Current estimates of gas reserves in the Arctic Islands are quoted to be somewhere in the 10–16 tcf range.[391] Panarctic is the operator of the Arctic Islands Exploration Group (AIEG), which is composed of Esso Resources Canada Ltd, Gulf Canada Resources Ltd, Panarctic Oils Ltd, and Petro-Canada.[392]

A major constraint on drilling in the Arctic Islands is the presence of sea ice, often for 11 to 12 months of the year. Operators have developed a technique to reinforce natural ice by spraying sea water on the surface, allowing it to freeze up to seven metres in thickness. A demonstration project at Drake F-76 in 45 metres of water, 1000 metres offshore, proved the feasibility of hydrocarbon production and transportation under sea ice.[393] Despite continuous exploration from the 1960s, Panarctic activities received scant public notice until the consortium made its own announcement of a gas discovery in 1974. Although approval-in-principle was required to pursue annual drilling, neither DIAND nor Cabinet has raised strong objections to proposed drilling plans — despite the pioneering aspects of ice island technology — and, in general, both have accepted the information supplied by Panarctic as being adequate for approval. CARC has charged the government with turning a blind eye to the potential environmental risks associated with Panarctic's offshore drilling for reasons of self-interest, namely a 45 per cent ownership interest in Panarctic.[394]

In the 1982–1983 winter drilling season Panarctic had approval to drill four wells.[395] On 15 November 1982 John Munro (Minister of DIAND) and Jean Chretien (EMR Minister) announced that 20 exploration agreements spanning a five-year period had been concluded with Panarctic on behalf of 67 companies. The package was the largest ever negotiated by COGLA to date and, according to Chretien, the agreements represented significant progress in establishing Canada's new resource management regime in Canada Lands. The COGLA approval process required submission of a Canada Benefits Plan to the Departments of Industry Trade and Commerce, Regional Economic Expansion, Indian Affairs and Northern Development and the Canada Employment and Immigration Commission, in consultation with the government of the Northwest Territories. Panarctic proposed an all-Canadian workforce of 500 and promised to undertake a $ 1.6-million training programme to increase native employment opportunities. Under the terms of the agreement, specific approval by COGLA would be required for each well to be drilled.[396] Two transportation modes, pipelines and LNG tankers, have been proposed for transporting gas from the Arctic Islands, but as yet final decisions have not been reached.

In early 1984 Panarctic unveiled plans for a new pilot project to transport oil from the Bent Horn field, discovered in 1974 on Cameron Island. The Bent Horn Project would produce and store oil in tanks over the winter and ship moderate amounts of oil (approximately 100000 barrels) during the summer to eastern markets. In order to demonstrate the feasibility of the plan, Panarctic's ship *MV Lady Franklin* carried one barrel of crude from the Arctic Islands to the South during the summer of 1984.

Concerning governmental and public review, DIAND, together with Environment Canada, COGLA, EMR, Fisheries and Oceans and Transport Canada, drew up guidelines for Panarctic's preparation of an Initial Environmental Evaluation.[397] Canadian native groups formed the High Arctic Development Review Committee, consisting of two persons each from Grise Fiord, Resolute Bay, Arctic Bay and Pond Inlet and a staff coordinator, to provide community input into the decision-making process.[398] Panarctic received conditional approval from the Minister of DIAND on 5 February 1985 and no formal EARP review was required.[399]

The *MV Arctic*, assisted by *MV Imperial Bedford*, transported the first large shipment of Northern crude oil (100000 barrels) to market in 1985 and a second voyage was completed in 1986.

4 *South Davis Strait*

In the summer of 1976 a consortium of oil companies (Imperial Oil Ltd, Aquitaine Company of Canada Ltd, and Canada Cities Service Ltd) submitted a proposal to undertake a two- to three-year exploratory drilling programme in South Davis Strait to evaluate hydrocarbon potential using drillships or semi-submersible platforms in water depths ranging to 6000 ft.[400] DIAND presented initial guidelines for preparing an EIS to companies in July, and the proponents initiated a series of environmental studies in the same year. When the Minister of DIAND formally submitted the proposal to EARP in 1977, revised EIS Guidelines were issued to reflect Panel requirements.[401]

In early 1978, after reviewing the proponents' EIS, the EARP Panel identified a number of information deficiencies and, shortly thereafter, the Eastern Arctic Marine Environmental Studies (EAMES) programme was integrated into the EIS data-collection process.[402] The Panel reported to the Minister in November 1978, and concluded that the environmental risks associated with the proposal were acceptable, provided that certain conditions were met: submission and approval of a detailed industry contingency plan six months prior to drilling; preparation of a government contingency plan to delineate the response of government agencies to the occurrence of an oilspill; continuation of environmental studies and monitoring programmes; and development of a mechanism to ensure that compensation would be paid out for damages incurred and clean-up costs.[403]

Drilling in South Davis Strait in 1979 (Esso), 1980 (Aquitaine), and 1982 (Canterra Energy Ltd) produced disappointing results. However, Canterra Energy, acting as operator for the Baffin Labrador Group, negotiated a five-year exploration agreement with COGLA in July 1982, with a commitment to drill two wells off Southeastern Baffin Island.[404]

5 *The Alaska Highway Natural Gas Pipeline*

The Alaska Highway Natural Gas Pipeline Project, often referred to as the Alcan Project, was approved by the US and Canadian governments in September 1977. The Project proposed to pipe natural gas from Prudhoe Bay, Alaska, through the Southern Yukon along the Alaska Highway corridor into Alberta. Near Edmonton a western leg, continuing down through Alberta and British Columbia, would carry the gas into the Western United States. An eastern leg, continuing down through Alberta and Saskatchewan, would carry gas to the Eastern United States.

Four tangled twists of the governmental decision-making net led to project approval. First, the Mackenzie Valley Pipeline Inquiry, appointed by Cabinet pursuant to the Territorial Lands Act to suggest environmental, social and economic conditions for granting a pipeline right-of-way, defused two initial pipeline proposals. The Arctic Gas Project had proposed to build a gas pipeline from Prudhoe Bay, Alaska, across the Northern Yukon to the Mackenzie Delta, then south through the Mackenzie Valley into Alberta and the United States. The Maple Leaf Project, a proposal led by Foothills Pipe Lines Ltd, offered an all-Canadian pipeline from the Mackenzie Delta along the Mackenzie Valley and into Alberta. In a report tabled in the House of Commons on 9 May 1977, Justice Thomas Berger, head of the Inquiry, quashed both proposals by recommending a permanent ban on pipelines across the Northern Yukon to protect wildlife and a 10-year moratorium on pipeline construction through the Mackenzie Valley to allow sufficient time for native land claims settlement and social adaptations.[405]

Second, after holding extensive hearings, the National Energy Board, on 4 July 1977, also rejected the two initial proposals. As an alternative, the Board recommended approval of the "last-minute" Alcan proposal[406] subject to three major conditions. Foothills Pipe Lines (Yukon) Ltd, parent company of the Canadian portion of the pipeline,[407] would have to undertake a feasibility study for a Dempster Lateral line from the Mackenzie Delta, would have to seek a certificate from the NEB for the Dempster line before 1 July 1979 (or later if ordered by Cabinet), and would have to provide Government with $200 million to mitigate social and economic costs.[408]

Third, an Environmental Assessment Panel, formed soon after a request by the Minister of Indian Affairs and Northern Development on 21 March 1977, granted a provisional OK to the project. Although the Panel was unable to undertake a full environmental review due to a government imposed deadline of August 1977, the Panel's Interim Report of 27 July 1977 concluded the pipeline could be built in an environmentally acceptable manner if subjected to further environmental and technical studies.

Fourth, the Alaska Highway Pipeline Inquiry,[409] established by the Minister of Indian Affairs and Northern Development on 19 April 1977 to investigate the socio-economic implications of the Alcan Project, was not totally opposed to the Southern Yukon route. In an 1 August 1977 report the Inquiry recommended approval, subject to a governmental advancement of $50 million towards settlement of Yukon Indian land claims, a company compensation payment of $200 million, establishment of a single planning and control agency, and deferral of actual construction for four years.[410]

Following US and Canadian government approvals of the Project in September 1977, three administrative reviews provided additional approval conditions. In December 1977 the Alaska Highway Gas Pipeline Environmental Assessment Panel issued guidelines for the preparation of an EIS by Foothills. After the company submitted its EIS to the Panel in January 1979, the Panel held public meetings in seven Yukon communities and issued a report in August 1979.[411] The report requested further environmental-technical studies, including studies as to why the proponent should be allowed to detour from the Alaska Highway in four areas — Kluane Lake, Mt Michie-Squanga Lake, the Rancheria River Valley and Ibex Pass. On 1 March 1981 Foothills submitted a study supporting the Ibex Pass Route.

In July 1981 the Panel recommended an alternative Whitehorse route for two reasons.[412] The Whitehorse route would be closer to a future Dempster Lateral line and the Ibex Pass was a sensitive wildlife area needing protection against increased highway access. In September 1982 the Panel issued a final report approving the other three detour proposals.[413] In March 1983, the Northern Pipeline Agency approved the

Ibex Pass route, after the Yukon government had undertaken studies concerning ways to lessen environmental impacts.[414]

The Northern Pipeline Agency, created in 1978,[415] has played the central role in regulating pipeline conditions. For example, by March 1982 it had reviewed four out of 20 environmental protection plans and ten out of 12 socio-economic plans required for the Yukon segment.[416]

The National Energy Board, meanwhile, was active in opening the door for Foothills to build the southern pipeline portions first and to export Alberta gas, a boost which could allow Foothills to weather difficulties in financing the northern portion of the pipeline.[417] The Board approved both gas exports by Pan-Alberta (a sister company of Foothills) and a financing plan for the southern portions.[418]

Cabinet approved the southern construction on 17 July 1980. Construction began on the western leg in December, 1980, and on the eastern leg on 4 May 1981. Although the eastern and western legs have been constructed and are transporting surplus Canadian natural gas to markets in the United States, construction of the northern section through Alaska and Yukon is on indefinite hold due to depressed markets and prices for gas.[419]

6 *Norman Wells Oilfield expansion*

Norman Wells oil production and transport is not a wholly new project. In 1919 Imperial Oil Ltd crews first drilled exploratory wells along the Mackenzie River. In 1939 the company built a small refinery to supply local markets in the Western Northwest Territories. During World War II the oilfield was expanded and the Canol pipeline was built to Whitehorse, Yukon, to provide fuel to American forces building the Alaska Highway.[420]

In the 1970s Esso proposed a major expansion. Through 200 new wells, six artificial islands for drilling in the Mackenzie River and a central processing plant Esso proposed to increase crude production from 3000 barrels per day to over 25000 barrels per day. The added volume would be transported through a new 12.75 inch diameter buried pipeline, operated by Interprovincial Pipe Line (NW) Ltd, to Zama, Alberta, where the pipeline would connect with an existing pipeline system.

To gain approval for the expansion, Esso had to manoeuvre through three major reviews. The Canada Oil and Gas Lands Administration had to grant a drilling-programme approval, development-plan approval, production-structure approval and production facilities approval.[421] The National Energy Board, after receiving an application for a certificate of public convenience and necessity from Interprovincial (NW) in March

1980, recommended approval to Cabinet in April 1981 and issued a certificate in November 1981. The certificate required further environmental assessments, mitigation measures for wildlife and fish, and monitoring of pipeline construction and operation.[422]

An Environmental Assessment Panel, formed by request of the Minister of Indian Affairs and Northern Development in February 1980, held public hearings in 12 communities during August 1980. In January 1981 the Panel recommended project approval, subject to a two-year construction delay, so as to allow government time to expand social programmes and industry time to further refine know-how on artificial island construction, oilspill contingency planning, and environmental mitigations.[423]

Cabinet approved the pipeline proposal in July 1981, and, following the advice of the Assessment Panel, pushed back the start-up date from 1981 to 1983. Thus, actual construction could not begin until January 1983.

To coordinate communications among the federal government, the Territorial government, industry and residents, and to oversee funding of various social programmes, the Department of Indian Affairs and Northern Development has established a Project Coordination Office in Yellowknife. Mr Scallion, the Coordinator, summarized the innovation in these words:

> I sometimes think it is more of an ombudsman than anything else ... One of the nice things about the job is that I do not assume any regulatory authority ... I am acting as a one-window agency where I will appear as a delivery boy, if you want, from the general public to the companies or to the governments, and the other way around ... in order to stop confusion and to make sure that the program in support of the projects occurs on time and with a minimum of fuss.[424]

The Norman Wells pipeline began transporting oil to Zama, Alberta in March 1985.

7 *Arctic Pilot Project*

The Arctic Pilot (APP), conceived by Petro-Canada in 1976, was designed to test the feasibility of producing natural gas from wells in the Arctic Islands. The project envisioned production of 9.0 million m^3 of natural gas per day from the Drake Point field on the Northern Sabine Peninsula of Melville Island, and transportation of gas by a 160 km buried pipeline across the island to a liquefaction plant at Bridport Inlet. Two Arctic Class 7 icebreakers would deliver liquefied natural gas to a terminal site in Eastern Canada, with distribution to Southern markets by conventional pipelines. Principal partners of the APP included: Petro-Canada Exploration Inc. (37.5 per cent); NOVA, an Alberta Corporation (25.0 per cent); Dome Petroleum Ltd (20.0 per cent); and Melville Shipping Ltd (17.5 per cent). Partners in Melville Shipping are Federal Commerce and Navigation Ltd, Upper Lakes Shipping Ltd, and the CSL Group, Inc.[425] An application, filed with the National Energy Board, requested a licence to export natural gas. The Department of Transport was requested to approve construction of a work in Bridport Inlet and operation of LNG carriers in Arctic zones 6 and 13.[426]

In 1977 APP proponents initiated discussions and information sessions in Northern communities, and Petro-Canada, on behalf of the APP, met with Danish officials in Greenland.[427] In November 1977 the Minister of DIAND and Petro-Canada made a joint referral of the APP to the Federal Environmental Assessment and Review Office (FEARO), which subsequently formed an Environmental Assessment Review Panel. In January 1979 the proponents issued an environmental impact statement. The Panel responded by issuing "Draft Guidelines for the Completion of the Assessment for the APP" in June and "Final Guidelines" in November.

In April 1980 the Panel convened public hearings in four Northern communities: Resolute, Arctic Bay, Grise Fiord, and Pond Inlet. Presentations were made by EMR, DFO, DOE, DIAND, DOT, GNWT, as well as by community representatives, the Baffin Regional Council (BRC), the Baffin Region Inuit Association (BRIA), the Inuit Tapirisat of Canada (ITC), the Canadian Nature Federation, CARC, and a number of non-affiliated individuals.[428]

In October 1980 the Environmental Assessment Review Panel reported that the Northern component of the APP was environmentally acceptable, subject to certain conditions:

> The Panel believes that it is essential that ships be routed to avoid environmentally sensitive areas in Parry Channel and that advantage be taken of the "pilot" nature of this project to monitor and research the effects of year-round shipping in the Arctic. The panel concludes that this can only be achieved through the formation of a control authority to monitor ship movements, and enforce good seamanship and appropriate environmental regulations. Without further research on marine mammals, guided by the advice of Inuit, and of government scientists and without a monitoring and control mechanism for the selection of the shipping routes, the Panel is unable to recommend that the project is environmentally acceptable.[429]

The government responded to the Panel's recommendations by delegating responsibility for implementation of an Arctic Shipping Control Authority to the Coast Guard. An Environmental Advisory Committee, co-chaired by DFO and DOE, with representatives from a number of other government departments and the territorial governments, was also created: to assess environmental information; to identify data gaps; and to make recommendations to the Control Authority concerning environmental protection measures, R and D needs, and monitoring activities.[430] Inuit groups declined an invitation to sit on the Committee until land claims were settled.

At the provincial level, public hearings were conducted in Quebec in January and February 1981, and acceptance of the Gros Cacouna terminal site was given in June. In Nova Scotia, public hearings were conducted in April and June 1981, with approval granted for the Melford Point terminal in September.[431]

The Arctic Pilot Project was referred to the Advisory Committee on Industrial Benefits to advise and evaluate the project in terms of benefits for Canadians, such as increased employment and use of domestic goods and services. Project proponents also submitted the APP to the voluntary TERMPOL Code Assessment conducted by DOT. Three TERMPOL Committees reviewed proposed terminal sites at Bridport Inlet, Gros Cacouna, and Melford Point. Preliminary assessments, based on available information, indicated that each site was acceptable, but a final assessment and report would not be filed until more specific information was available.[432]

On 2 February 1982 the National Energy Board began hearings on the APP application for the export of natural gas transported from the Arctic in a liquefied state, and on an application by TransCanada Pipelines for a Southern terminal and re-gasification facilities. In July TransCanada requested an adjournment of the hearings since sales to Europe might eliminate an LNG terminal site in Southern Canada. On 31 August 1982 the NEB adjourned the hearings until the sponsors could provide more detailed market information. The APP appealed the suspension but the application was denied on 4 November 1982.[433] On 3 August 1984 the NEB finally dismissed the APP application after the project sponsors decided not to extend agreements with American buyers.[434]

8 *Lancaster Sound*

DIAND approved initial oil exploration permits to Magnorth Petroleum Ltd in 1968, and in 1971 Magnorth began conducting seismic studies in Lancaster Sound.[435] In 1974 DIAND gave approval-in-principle to

Norlands Petroleum (a subsidiary of Magnorth) to drill a single exploratory well, conditional on Norlands completing a series of environmental studies. The study requirements were presented to Norlands in March 1975, but when the approval-in-principle expired in August 1977 Norlands had not fulfilled its obligations in providing sufficient data on ocean currents or an oilspill contingency plan.[436]

In 1976 two scientists from the former Department of Fisheries and Environment, Allan Milne and Brian Smiley, offered to review the Norlands' proposal, which had not received much public attention, and the Arctic Waters Oil and Gas Advisory Committee (AWAC) accepted the offer. The Milne and Smiley report, *Offshore Drilling in Lancaster Sound*, released in February 1978, emphasized the fragility of the Lancaster Sound ecosystem and its vulnerability to an oilspill hazard.[437] In the same year, 1976, the National Energy Program was unveiled, and oil companies, anxious to take advantage of the lucrative incentives offered through the NEP, were quick to propose exploration programmes. In addition to the Norlands' proposal, Imperial Oil, Aquitaine, and Canada Cities Service submitted a proposal for Davis Strait, and Petro-Canada began developing plans for drilling in Baffin Bay.

In 1977 the Minister of DIAND announced that comprehensive studies of the Baffin Region would be conducted prior to future drilling approvals, and that all Eastern Arctic offshore development proposals would be referred to EARP. The Norlands' proposal was referred to EARP in the autumn, and in March 1978 EIS guidelines were released. In November 1977 DIAND Minister Hugh Faulkner announced a four-year $13 million Eastern Arctic Marine Environmental Studies (EAMES) programme for Lancaster Sound, Davis Strait and Baffin Bay: "EAMES will enable us to determine the environmental restraints on a systemic basis before any offshore drilling for petroleum might be allowed in the eastern Arctic."[438]

The Norlands' proposal involved the use of the *Pelerin*, a dynamically-positioned drillship, to drill at Dundas K-56 in 770 metres (the maximum depth previously drilled in the Arctic was 600 metres). For reasons which are not clear, Norlands apparently thought itself to be exempt from the EAMES study, and went forward with preparing and submitting an EIS to the EARP panel in early summer 1978.[439] Public hearings on the EIS were held in Arctic Bay, Resolute Bay, Grise Fiord and Pond Inlet in October and November. In 1979 the Panel reported that a meaningful asssessment of exploratory drilling could not be made in isolation from broader issues concerning the region:

> This fundamental question of whether there should be hydrocarbon development at all in Lancaster Sound is clearly a matter

> for government to decide. For these reasons the panel concludes that for it to make a recommendation in favour (or not in favour) of exploratory drilling at this time would be *arbitrary*.
>
> The panel recommends that the responsible federal coordinating and planning body (DINA) use the time available from a deferment of drilling to address, on an urgent basis, with adequate national and regional public input and taking into account the various forces at work, the best use(s) of the Lancaster Sound Region. The Panel stresses that socio-economic considerations must be included as a major factor in this determination.[440]

In the fall of 1979 the Lancaster Regional Study (established by DIAND in conjunction with the GNWT and the federal departments of EMR, DOE, DFO, and Transport) was announced to the Baffin Regional Council. From 1979–1980 a thorough compilation of the existing data base was assembled. Using a regional planning approach to identify and analyse significant issues bearing on future use, an interdepartmental working group produced a draft green paper which was released to the public in February 1981 in Frobisher Bay. Public review of the draft green paper was initiated in April 1981, with community meetings in Pond Inlet, Arctic Bay, Resolute and Grise Fiord. Native and public groups held workshops on the draft paper in the spring and in July 1981 the Minister of DIAND announced Cabinet approval of the Northern Land Use Planning Policy.[441]

In January 1982 the green paper entitled, *The Lancaster Sound Region: 1980–2000: Issues and Options on the Use and Management of the Region* was released. It identified six possible options: (i) no new development; (ii) environmental protection; (iii) renewable resource economy; (iv) Northwest Passage shipping; (v) balanced development; and (vi) non-renewable resource economy. Following the release of the document, the Lancaster Sound Regional Study Committee solicited comments and conducted workshops to receive public input on the six options. A final report issued in May 1984 recommended that an integrated land-use strategy be formulated for the Lancaster Sound region before any final development decisions are made.[442]

9 *Beaufort Sea-Mackenzie Delta hydrocarbon development*

Imperial Oil drilled the first exploratory offshore well from an artificial island in the Beaufort Sea in 1973 and by 1981 eighteen such islands had been constructed.[443] In 1973 Dome Petroleum obtained Cabinet approval-in-principle for drilling offshore from floating platforms (drillships), but Cabinet stipulated that a regional assessment must be conducted prior to the initiation of a drilling programme. Subsequently, a jointly funded (government-industry) impact studies programme known as the Beaufort Sea Project was inaugurated. By completion of the programme in 1976, 46 technical reports had been produced on various aspects of marine biology, oceanography, meteorology, sea ice, marine geology, oilspill technology, and wildlife resource use.[444] Despite the continuing assessment project, the Minister of DIAND approved Dome's initial drilling proposal in May 1974, and in April 1976 Cabinet approved permits for Dome-Canmar drillships to commence Beaufort operations in the summer.[445] In May 1977 Cabinet approved an exploration programme for the years 1977–1979, and stipulated that operators were to submit annual post-season drilling reports. Subsequent exploratory drilling was contingent on Cabinet acceptance of these reports, and conditions were often attached to drilling approvals.[446]

Beginning in 1978, a Beaufort Sea Steering Committee, chaired by DIAND with representation from EMR, DOE, DFO, COGLA, Industry, Trade and Commerce, Transport Canada, External Affairs and the government of the Northwest Territories, reviewed hydrocarbon activities with respect to technical operations, marine management, environmental impacts and socio-economic-cultural matters.[447] In February 1981 the Senior Policy Committee, Northern Development Projects, created an intergovernmental Task Force on Beaufort Sea Development to prepare a situation report concerning government's preparedness for anticipated hydrocarbon development in the Beaufort.[448] In June 1981 Esso Resources Canada Ltd, Dome Petroleum Ltd, and Gulf Canada Resources Inc released their proposal for *Hydrocarbon Development in the Beaufort Sea–Mackenzie Delta Region*.[449] The proposed programme included: the drilling of 150 exploratory and delineatory wells, and 575 production wells; construction of a variety of production systems; producing petroleum from these systems; construction of pipelines or oil tankers and liquid natural gas (LNG) tankers; and building harbours to support operations. Proponents envisaged a three-phased development strategy to the year 2000: pre-production, 1982–1985; early production, 1986–1990; and final development, 1991–2000.[450]

The Minister of DIAND referred preliminary proposals for the production and transportation of oil and gas from the Beaufort Sea to FEARO in July 1980, and an EARP panel was formed to draft EIS guidelines. Public hearings on the draft guidelines were held in November and December 1981, and final guidelines were released in March 1982.[451]

Dome, Esso, and Gulf filed their EIS with FEARO in November 1982. Following a ninety-day review period, the Panel issued a Deficiency

Statement in March 1983 which cited information deficiencies with the EIS in four categories: (i) assessment of socio-economic effects; (ii) assessment of environment effects; (iii) oilspills; and (iv) zone summaries.[452] In August 1983, the Panel decided EIS Supplementary Information was sufficient and proceeded with public sessions.

In a final report issued in July 1984 the Panel recommended a cautious, phased approach to developing offshore hydrocarbon reserves in the Beaufort Sea-Mackenzie Delta region. The Panel advised that the transport of oil should only be authorized to begin through a single, small-diameter (for example, 400 mm), buried pipeline. If proponents should desire a future large-diameter oil pipeline (for example 1000 mm), the Panel recommended a comprehensive public review on socio-economic grounds as a precondition. The Panel recommended the Government of Canada withhold approval of the tanker option until after two evaluation stages. A Research and Preparation Stage would focus on such questions as the effect of tanker traffic on marine mammals and the need for increased government support systems such as ice detection and hydrographic surveys. A Two Tanker Stage would involve actual field trials and performance studies of two Arctic Class 10 oil-carrying tankers.[453]

10 *North Davis Strait*

In July 1977 a Petro-Canada proposal to drill in North Davis Strait was referred by DIAND to FEARO, and subsequently an EARP Panel was struck and initial EIS Guidelines were formulated.[454] Petro-Canada has worked steadily since 1978 on the preparation of an EIS for Baffin Bay, and has integrated data from the EAMES North Program to produce numerous field-data reports and to develop computer simulation models to assess the risks associated with operating offshore in ice-infested waters.[455] However, the EIS review process has been on hold pending a completed review of the Lancaster Sound Regional Study Committee report. Given the long delay between referral of the proposal to EARP and the initiation of a formal review, the chairman of the North Davis Strait EARP Panel, R.G. Connelly, indicated that revisions in Panel membership, as well as in EIS Guidelines, would be made when or if the proposal is reactivated through the Environmental Assessment Review Process.[456]

B Reasons for variation

The experiences of each of the ten projects with the governmental decision-making net has been unique. It is simply not possible to identify any one administrative or legal mechanism which can stymie or expedite project approvals, nor is it possible to trace a predictable decision channel that proponents must successfully navigate in order to obtain approval for their plans. Decision-makers appear to pull the draw-strings of the approval net in a somewhat random or arbitrary manner. How a proposal is perceived and subsequently reviewed depends in large measure on the timing of submission, the degree of technological innovation required, and the location of the project.

Timing is clearly one of the most critical factors influencing government response to Northern industrial projects. The mid-1970s were characterized by heightened public awareness, activism, and demand for public participation in decision-making processes. EARP panels, commissions of inquiry, and, more recently, regional and land-use planning exercises have given interested parties the opportunity to listen and give testimony before government representatives.

However, as yet, no provisions make public consent a prerequisite to project approval. The Alaska Highway Pipeline decision was hastened by the course of international diplomacy between the US and Canadian interests. The National Energy Program, introduced in 1976, gave a green light to companies interested in promoting Northern energy projects. By giving energy self-sufficiency a high priority, government began to evaluate hydrocarbon exploration and production proposals by "how much and for whose benefit" criteria, while potentially negative environmental consequences were perceived as being a matter of designing appropriate monitoring systems and mitigation procedures. With the creation of COGLA, a formal, albeit non-legislated, approval process has been articulated. Although it may be premature to speculate on the function of COGLA, the number of exploration permits issued since May 1982 suggests that it will serve as a mechanism to expedite hydrocarbon exploration.[457]

Distinctions may be drawn between the approval process for hydrocarbon proposals and those for production and transportation of Northern hard minerals. Northern mining proposals have not been submitted to EARP inquiries, and one may speculate that this is a consequence of several factors including: accessibility (land as opposed to water); the boundedness of the projects; and technology (given the greater experience with mining, there is greater confidence in mitigation measures designed to respond to potential hazards). Whereas mining proponents were able to provide a fairly accurate assessment of projected new employment opportunities, Canadian experience with offshore hydrocarbon exploration and production is limited.

Proposals for marine projects are particularly vulnerable because of the complexity of ecosystem interactions, making it difficult to assess with accuracy the consequences of an oilspill or blow out, the impact of artificial island construction, or the potential hazards of new transportation technologies such as icebreaking LNG tankers, submarines and airships. Although the scientific database is rapidly expanding, several proposals have already been delayed on the grounds of insufficient data and unreliable modelling.

Finally, domestic and international economic circumstances tend to influence how government manipulates the decision-strings. Fluctuating and uncertain market demand has been an issue in pipeline and APP debates, while the extraordinary costs associated with the design and implementation of pioneer technologies in remote and hostile environments is a topic of current debate. The extent of government investment in Northern energy projects, as well as the apparent economic vulnerability of several key corporate actors, has also become a critical factor in the decision process.

VI Conclusions

Everyone is likely to agree: Canadian decision making for the North has shown great drift. Northern policies have floated a disordered course between the poles of industrial development and environmental protection. Administrative responsibilities continue to be fragmented and administrative committees continue to proliferate. Legislation and regulations brim with discretionary open-endedness. Project reviews have varied greatly, depending on such factors as the strength of political protest and economic urgency.

Many persons may disagree, however, over the "goodness" or "badness" of such *ad hoc*, fragmented decision making. Some may argue Canada has fared pretty well. Major projects, such as Polaris Mine, Nanisivik Mine, the Alaska Highway Gas Pipeline and Norman Wells Oilfield expansion, have survived the regulatory swim. When lengthy delays have occurred, the extra time was essential in order to develop infrastructure, institutions, technology and environmental mitigations. *Ad hoc* and fragmented decision making is perhaps necessary in order to deal adequately with modern-day complexities. A balance among competing interests — fisheries, the environment, industrial development — is perhaps best achieved by informal negotiations among many departments with special expertise, rather than by the fiats of a central agency. Others may argue that the decision-making process has fared badly.

Companies have faced great financial drains in trying to meet many repetitive review processes and waiting years for a final decision. Public interest groups and Northern residents have often been unable to have full input into decision making due to *ad hoc* meetings and *ad hoc* procedures.

One thing is certain: Canadian decision making for the North is in the process of evolution. Acknowledging the need for "a more considered and integrated approach" to reaching decisions, the Department of Indian Affairs and Northern Development is in the process of developing a land-use planning system.[458] Settlement of native land claims could result in more local control over resource developments. Both Yukon and the Northwest Territories continue to push for more autonomy on the road to full provincehood.

How far the federal bureaucracy will be willing to divest powers to the territorial, regional and local levels remains uncertain. Perhaps the new Canadian Charter of Rights and Freedoms offers a guiding light. While government agencies, worshipping the hallowed principle of ministerial responsibility, may be unwilling to sacrifice final decision-making powers, the Charter beckons officials to assure all concerned individuals receive a fair and just hearing. Major legislative and regulatory renovations will have to be carried out — setting maximum and minimum time-frames for reviews, establishing the sequence and interrelationship of reviews, and guaranteeing the right to confront opposing views and opposing data. A true concern for Northern individual rights also beckons government to reconsider the adequacy of Southern decision-making processes such as EARP and NEB for the North. New administrative mechanisms, spawned by resource sharing and management agreements, land-use planning, or coastal zone co-operation agreements, could provide Northerners a stronger legal hand in controlling or at least directing their personal destinies. The challenge for Canada in the 1980s is to establish clear, reasonable and workable procedures and administrative structures, so that all interested parties — companies, native groups, individuals, and public-interest groups — will know how to proceed without the quirks of unseen currents.[459]

Notes

1. For discussions of Cabinet's role in decision making, see G. Bruce Doern and Peter Aucoin (eds), *Structures of Policy-Making in Canada* (1971); Richard W. Phidd and G. Bruce Doern, *The Politics and Management of Canadian Economic Policy* (1978); Audrey D. Doerr, *The Machinery of Government in Canada* (1981); and Government of Canada, Treasury Board, *A Guide to the Policy and Expenditure Management System* (1980).

2. Suffice it to say, the role of Parliament is primarily three-fold, amending legislation, passing legislation and providing oversight of governmental programmes through committees. Of some 21 standing committees of the House of Commons, three in particular, the Committee on Indian Affairs and Northern Development, the Committee on Northern Pipelines and the Committee on National Resources and Public Works, have concentrated on Northern issues. For example, the Committee on National Resources and Public Works held over six months of hearings on the Canada Oil and Gas Act, a major stimulant for Northern development, and a series of hearings on LNG transport from the Arctic. See Minutes of the Proceedings and Evidence of the Standing Committee on National Resources and Public Works, 1st Sess., 32nd Parliament, Issues 16–69 (6 November 1980–25 June 1981) (Canada Oil and Gas Act hearings) and Issues No. 13–14 (27 October 1980–29 October 1980) (LNG hearings). Of some 14 Senate committees, one in particular has devoted special concern to Arctic development. The Special Committee of the Senate on the Northern Pipeline held over seven months of hearings on Arctic offshore transportation. See Proceedings of the Special Committee of the Senate on the Northern Pipeline, 1st Sess., 32nd Parliament, Issues 16–37 (9 February 1982–16 September 1982).
3. Minutes of Proceedings and Evidence of the Standing Committee on Indian Affairs and Northern Development, 4th Sess., 28th Parliament, Issue 3 (28 March 1972), pp. 7–8. Other goals, such as furthering governmental evolution in the Territories and maintaining Canadian sovereignty were also mentioned but not included in the listing of priorities. The same ranking of priorities – social improvement, environmental enhancement, renewable resource development and non-renewable resource development – was reiterated by the Science Council of Canada in its 1977 Science Policy for Northern Development. See Science Council of Canada, *Northward Looking: A Strategy and a Science Policy for Northern Development* (August 1977), pp. 45–6.
4. Ministry of State for Science and Technology, *New Oceans Policy* (1973), quoted in E. Dosman, "Arctic Seas: Environmental Policy and Natural Resource Development" in O.P. Dwivedi (ed.), *Resources and the Environment: Policy Perspectives for Canada* (1980), p. 205.
5. Debates of the House of Commons, 1st Sess., 30th Parliament (27 April 1976), p. 12910.
6. Such a proposal caused an Opposition member to state: "{T}he government has all but made the final decision to go ahead with Arctic oil and gas development". *Ibid.*, p. 12912 (statement of Mr Gillies).
7. The announcement was made on behalf of the Minister of Energy, Mines and Resources. Debates of the House of Commons, 1st Sess., 32nd Parliament (28 October 1980), p. 4186.
8. *Ibid.*, p. 4190.
9. *Ibid.*, p. 4188. For the full text of the Program, *see* Department of Energy, Mines and Resources, *The National Energy Program* 1980 (1980).
10. Department of Energy, Mines and Resources, *The National Energy Program Update* 1982 (1982).
11. Environment Canada, *Environment Canada and the North* Discussion Paper (July 1983).
12. "Northern Land-Use Planning Policy Announced", Department of Indian Affairs and Northern Development Communique No. 1–8124 (July 1981).
13. Minutes of Proceedings and Evidence of the Standing Committee on Indian Affairs and Northern Development, 1st Sess., 32nd Parliament, Issue 22 (24 March 1981), p. 7 (statement of Honourable John C. Munro).

14. "Notes for Remarks by the Honourable John C. Munro, Minister of Indian Affairs and Northern Development, in Debate of Bill C-48 in the House of Commons", Department of Indian Affairs and Northern Development Communique No. 101–1–3 (19 October 1981), p. 4. For a policy statement with a similar slant, see "Notes for Remarks by the Honourable John C. Munro, Minister of Indian Affairs and Northern Development", Speech delivered to the Canadian Club, Edmonton, Alberta on behalf of the Minister by Senator Jack Austin, Department of Indian Affairs and Northern Development Communique No. 3–8205 (17 June 1982).
15. Proceedings of the Special Committee of the Senate on the Northern Pipeline, 1st Sess., 32nd Parliament, Issue 28 (9 June 1982), pp.46–47. Also see Statement of Ian R. Smyth, Executive Director, Canadian Petroleum Association in *Frontier Oil and Gas Development: The Decade Ahead* (Proceedings of the Tenth Environmental Workshop, Montebello, Quebec, 28 April–1 May 1981), p. 130.
16. Proceedings of the Special Committee of the Senate on the Northern Pipeline, 1st Sess., 32nd Parliament, Issue 19 (16 March 1982), p. 27.
17. Minutes of Proceedings and Evidence of the Special Committee on Regulatory Reform, 1st Sess., 32nd Parliament, Issue 21 (27 October 1980), p. 7.
18. In public administration parlance, this approach, first formulated by Charles Lindblom, might be labelled disjointed incrementalism or strategic decision making. For a general discussion of the approach, see Charles E. Lindblom, "The Science of Muddling Through", in Edward V. Schneier (ed.), *Policy-Making in American Government* (1969), pp. 24–37; Charles E. Lindblom, *The Policy-Making Process* (1968); and Timothy M. Hennessey, "Evaluating Marine Policy: Criteria from Two Models and a Comparative Study" in Center for Ocean Management Studies, University of Rhode Island, *Comparative Marine Policy* (1981), pp. 227–28.
19. The underlying ideal, of course, might also be a lot less romantic and noble, namely, "Stop trying to appease other interest groups and let's get on with development."
20. In public administration terminology this approach might be called comprehensive rationality or synoptic decision making. The approach advocates systematic decision making through a five-step process: clarifying and ranking goals; listing means for achieving goals; investigating all important consequences of the various means; comparing consequences of each means with goals; and choosing the means most closely matching the goals. Charles E. Lindblom, *The Policy-Making Process* (1968), p. 13. Such an approach might be facilitated by a centralized institutional structure. For example, a centralized oceans agency would likely seek to coordinate all ocean uses in a highly rationalized, comprehensive manner. Timothy M. Hennessey, *supra* note 18, p. 224.
21. For a more detailed discussion of these points, see Annmarie H. Walsh, in Alan Tupper and G. Bruce Doern (eds) *Public Corporations and Public Policy in Canada* (1981), p. 13.
22. The tension within government to balance conflicting ideas and interests is described by G. Bruce Doern, see *The Peripheral Nature of Scientific and Technological Controversy in Federal Policy Formation* (1981), p. 14.
23. *Ibid.*, p. 24.
24. Diamond Jenness, *Eskimo Administration: II* (1964), p. 32.
25. *Ibid.*, p. 77.
26. Debates of the House of Commons, 1st Sess., 22nd Parliament (8 December 1953), pp. 696–697.

27. Statement of Mr Colin Cameron (Nanaimo); *ibid.*, p. 705.
28. See *Royal Commission on Government Organization*, J. Grant Glassco, Chairman (1963), V. 4.
29. Debates of the House of Commons, 1st Sess., 27th Parliament (31 May 1966), p. 5751.
30. Order in Council, P.C. 1968—1574 (14 August 1968).
31. Proceedings of the Special Committee of the Senate on the Northern Pipeline, 1st Sess., 32nd Parliament, Issue 32 (29 June 1982), pp. 42–43.
32. See G. Bruce Doern, *How Ottawa Spends Your Tax Dollars*: *Federal Priorities* (1981).
33. Transport Canada, Position Statement submitted to Beaufort Sea Environmental Assessment Review Panel (16 August 1982). See also John W. Langford, *Transport in Transition*: *The Reorganization of the Federal Transport Portfolio* (1976).
34. Transport Canada, Position Statement to the Beaufort Sea Environmental Assessment Review Panel (16 August 1982), pp. 11–14, and Appendix 1, pp. 1–55.
35. *Royal Commission on Government Organization*, *supra* note 28, p. 157.
36. Edgar J. Dosman, *The National Interest*: *The Politics of Northern Development*, 1968–1975 (1975), p. 11.
37. Richard W. Phidd and G. Bruce Doern, *The Politics and Management of Canadian Economic Policy* (1978), p. 86.
38. Frances Abele and E.J. Dosman, "Interdepartmental Coordination and Northern Development", *Canadian Public Administration* V.24 No. 3 (1981), p. 438.
39. Audrey D. Doerr, *The Machinery of Government in Canada* (1981), pp. 136–164.
40. J.E. Hodgetts, quoted in *ibid.*, p. 106.
41. See Tupper and Doern, *supra* note 21, p. 17.
42. Peter C. Newman, *Renegade in Power*: *The Diefenbaker Years* (1963), p. 218.
43. Proceedings of the Special Committee of the Senate on Science Policy, 2nd Sess., 30th Parliament, Issue 14 (June 1977), pp. 8–11.
44. *Ibid.*, p. 8.
45. Department of Northern Affairs and National Resources, *Annual Report* 1957–1958, p. 77.
46. Dosman, *supra* note 36, p. 8.
47. Special Committee of the Senate on Science Policy, *supra* note 43, pp. 8–11.
48. Debates of the House of Commons, 1st Sess., 28th Parliament (20 December 1968), p. 4221. The Task Force on Northern Oil Development was unusual in the sense that it lacked a specific mandate, did not have instructions to report within a given time-frame, and representation was not a consideration. See V.S. Wilson, "The Role of Royal Commissions and Task Forces," in G. Bruce Doern and Peter Aucoin (eds), *The Structures of Policy-Making in Canada* (1971), pp. 113–129.
49. Dosman, *supra* note 36, pp. 24–25 (emphasis added).
50. The Committee for Original Peoples' Entitlement (COPE) was formed in 1970 to provide a united voice for all native Northerners. In 1973 COPE became a regional organization representing the Inuit of the Western Arctic (the Inuvialuit). COPE headquarters are located in Inuvik, Northwest Territories (NWT). The Inuit Tapirisat of Canada (ITC) represents approximately 25000 Inuit of the NWT, Northern Quebec and Labrador. The organization has six regional affiliates: the Baffin Region Inuit Association, the Keewatin Inuit Association, the Kitikmeot (Central Arctic) Inuit Association, COPE, the

Makivik Corporation and the Labrador Inuit Association. ITC offices are located in Ottawa and Frobisher Bay. The Ottawa-based Canadian Arctic Resources Committee (CARC), was founded in 1971 as an *ad hoc* citizens' information committee. For a detailed account of CARC's origins, see Douglas M. Pimlott, "People and the North: Motivation, Objectives and Approach of the Canadian Arctic Resources Committee", in *Arctic Alternatives*, Douglas H. Pimlott, Kitson M. Vincent, and Christine E. McKnight (eds), (Proceedings of a National Workshop on People, Resources and the Environment North of 60°, Ottawa, 24–26 May 1972), pp. 3–24.

51. Task Force on Northern Pipeline Development, *Pipeline North: The Challenge of Arctic Oil and Gas* (1972) (emphasis added).
52. Dosman, *supra* note 36, pp. 163–66.
53. Department of Indian Affairs and Northern Development, *Expanded Guidelines for Northern Pipelines* (28 June 1972).
54. Dosman, *supra* note 36, pp. 18, 181–82.
55. Department of Energy, Mines, and Resources, *Annual Report* 1975–1976, p. 23.
56. Transport Canada, "Arctic Marine Transportation R & D Five Year Plan (1981/82 to 1985/86)", Discussion paper (1981).
57. Department of Indian Affairs and Northern Development, *The Lancaster Sound Region*: 1980–2000 (January 1982).
58. Erik Madsen, "Federal Commissions and Northern Development", in Nils Orvik and Kirk R. Patterson (eds), *The North in Transition* (1976), pp. 93–111.
59. *Northern Frontier, Northern Homeland: The Report of the Mackenzie Valley Pipeline Inquiry*, Mr Justice Thomas R. Berger, Chairman (1977), Vol. 2.
60. *Alaska Highway Pipeline Inquiry*, Kenneth M. Lysyk, Chairman (1977).
61. See *Report of the Task Force on Beaufort Sea Developments*, submitted to Senior Policy Committee, Northern Development Projects (1981).
62. *Advisory Commission on the Development of Government in the Northwest Territories*, A.W.R. Carrothers, Chairman (1966).
63. *Constitutional Development in the Northwest Territories: Report of the Special Representative*, C.M. Drury (1980).
64. See also Gurston Dacks, *A Choice of Futures: Politics in the Canadian North* (1981).
65. Department of Northern Affairs and National Resources, *Annual Report* 1961–1962, p. 15.
66. *Science and the North* (A Seminar on Guidelines for Scientific Activities in Northern Canada, Mont Gabriel, Quebec, 15–18 October 1972), sponsored by the Subcommittee on Science and Technology, Advisory Committee on Northern Development.
67. See Science Council of Canada, *Northward Looking: A Strategy and a Science Policy for Northern Development* (1977).
68. For a list of commissioned studies, see *Northward Looking*, *ibid.*, p. 6.
69. See *Arctic Alternatives*, *supra* note 50; R.F. Keith and J.B. Wright (eds), *Northern Transitions* (Proceedings of The Second National Workshop on People, Resources and The Environment North of 60°, Edmonton, 20–22 February 1978), Canadian Arctic Resources Committee (1978); and *Marine Transportation and High Arctic Development: Policy Framework and Priorities* (Symposium Proceedings, Montebello, Quebec, 21–23 March 1979) (hereinafter referred to as Symposium Proceedings).
70. *Major Canadian Projects, Major Canadian Opportunities*, Report of the Consultative Task Force on Industrial and Regional Benefits from Major Canadian Projects (June 1981).

71. Department of Indian Affairs and Northern Development, *Northern Affairs Program, Current and Recent Research and Studies Relating to Northern Social Concerns* (1978–1981). See also Department of Northern Affairs and National Resources, Northern Coordination and Research Centre, *Government Research and Surveys in the Canadian North*, 1956–1961, and *Social Science Abstracts* (1959–1965). For a list of the Arctic Land Use Research (ALUR) programme projects, see Department of Indian Affairs and Northern Development, *Toward a Northern Balance* (1973).
72. See Arctic Petroleum Operators' Association, *Catalogue of Research Project Reports*, (May 1982); and Peter N. Duinker and Gordon E. Beanlands, *Environmental Impact Assessment Procedures in Canada: Improvements from a Scientific Perspective*, (paper delivered at the International Conference on Oil and the Environment, Halifax, Nova Scotia, 16–19 August 1982).
73. See *Royal Commission on Canada's Economic Prospects*, W.L. Gordon, Chairman (1955), and the *Royal Commission on Energy*, Henry Borden, Chairman (1957).
74. Proceedings of the Special Committee of the Senate on the Northern Pipeline, *supra* note 31, p. 115.
75. Peter Aucoin and Richard French, "Ministry of State for Science and Technology," *Canadian Public Administration* V.17 (1974), pp. 461–81.
76. Federal Environmental Assessment and Review Office, "Detailed Outline of Contents of the Cabinet Memoranda Establishing the Federal Environmental Assessment and Review Process" (1 April 1978). For a detailed discussion of the environmental process, see notes 366–372, and accompanying text, *infra*.
77. *Organization of the Government of Canada* 1978–1979, p. 392.
78. Northern Pipeline Agency, *Annual Report* 1980–1981, p. 25.
79. Proceedings of the Special Committee of the Senate on the Northern Pipeline, 1st Sess., 32nd Parliament, Issue 35 (14 September 1982), p. 35A:6.
80. *Ibid.*, p. 7.
81. Tupper and Doern, *supra* note 21, p. 21.
82. Advisory Committee on Northern Development, *Government Activities in the North* 1980–1981, pp. 140–141.
83. *Ibid.*, pp. 136–137; see also Northern Canada Power Commission, *Annual Report* (1982).
84. S.C. 1974–75–76, c. 61.
85. Larry Pratt, "Petro-Canada", in Tupper and Doern, *supra* note 21, pp. 143–44.
86. "$ 1.5b bailout agreed to for Dome", (Toronto) *Globe and Mail* (30 September 1982), pp. 1–2; plans for refinancing were vehemently opposed by a number of parties including a group of Dome shareholders; after a year of negotiations, final arrangements were still pending, see "Dome Pete to unveil refinancing plan", *Globe and Mail* (2 November 1983), p. 31. See also, Dome Petroleum Limited, *News Release* (1 December 1983; 16 April 1984; 30 April 1984).
87. Doern, *supra* note 22, p. 31.
88. Abele and Dosman, *supra* note 38, p. 439.
89. Pratt, *supra* note 85, p. 109.
90. Department of Indian Affairs and Northern Development, Position statement to Beaufort Sea Assessment Panel (October 29, 1982).
91. Doerr, *supra* note 39, p. 153.
92. Gurston Dacks, "Serving Notice on the North", *Canadian Forum* V. LXI, No. 713 (1981), p. 8.
93. Abele and Dosman, *supra* note 38, pp. 443–47.

94. Proceedings of the Special Committee of the Senate on the Northern Pipeline, 1st Sess., 32nd Parliament, Issue 36 (15 September 1982), p. 36A:123.
95. Proceedings of the Special Committee of the Senate on the Northern Pipeline, 1st Sess., 32nd Parliament, Issue 17 (16 February 1982), p. 17A:94.
96. Proceedings of the Special Committee of the Senate on the Northern Pipeline, 1st Sess., 32nd Parliament, Issue 16 (9 February 1982), pp. 18–19.
97. Proceedings of the Special Committee of the Senate on the Northern Pipeline, 1st Sess., 32nd Parliament, Issue 22 (27 April 1982), p. 22.
98. Regulation of air quality in the North is beyond the scope of the present paper. Suffice it to say, protection would depend on the general provisions of the Clean Air Act, S.C. 1970–71–72, c. 47 and on the special air-pollution regulations for ships under the Canada Shipping Act, Air Pollution Regulations, C.R.C. 1978, c. 1404.
99. Territorial regulatory regimes are beyond the scope of this chapter. Suffice it to say, territorial control over industrial development is relatively minor, primarily limited to socio-economic conditions and wildlife management. Territorial legislative powers are set forth in the Territorial Acts, the Northwest Territories Act, R.S.C. 1970, C. N-22, s. 13 and the Yukon Act, R.S.C. 1970, c. Y-2, s. 16.
100. R.S.C. 1970, c. T-6.
101. S.C. 1973–74, c. 21.
102. R.S.C. 1970, c. N-13.
103. R.S.C. 1970, c. M-12.
104. For a detailed discussion of land use legislation, see K. Beauchamp, *Land Management in the Canadian North* (1976).
105. R.S.C. 1970, c. T-6, s. 3.1 as amended by R.S.C. 1970, c. 48 (1st Supp.), s. 24.
106. R.S.C. 1970, c. T-6, s. 3.2 as amended by R.S.C. 1970, c. 48 (1st Supp.), s. 24.
107. C.R.C. 1978, c. 1524.
108. *Ibid.*, at s. 3.
109. Class A permits could be issued within 10 days of application receipt but may undergo an extensive study and consultation process of up to 12 months. *Ibid.*, at s. 25. Applications would be forwarded to the Land Use Advisory Committee, the Ottawa Land Use Office, the Department of Indian Affairs and Northern Development District Office, and possibly community and special-interest groups for recommendations about environmental concerns. Department of Indian Affairs and Northern Development, *Northern Natural Resource Development: Requirements, Procedures, and Legislation* (1981), pp. 12–13. Also see Department of Indian Affairs and Northern Development, *A Guide to Territorial Land Use Regulations* (1981), pp. 4–5.
110. C.R.C. 1978, c. 1524, s. 8. Other uses requiring a permit are the use of over 330 pounds of explosives within any 30-day period, the use of any campsite for over 400 person-days, the establishment of petroleum storage over 17,597 gallons or the use of a single petroleum container over 800 gallons, and the preparation of trails or rights-of-way over 4.92 feet wide and over 9.88 acres in area. *Ibid.*
111. *Ibid.*, at s. 27.
112. *Ibid.*, at s. 9. Other uses requiring a Class B permit are the use of between 110 and 330 pounds of explosives within any 30-day period, campsite use by more than two people for between 100 and 400 person-days, petroleum storage from 800 gallons to 17597 gallons, using a petroleum container from 439 to 800 gallons and preparation of a trail or right-of-way over 4.92 feet wide but less than 9.88 acres in area.
113. C.R.C. 1978, c. 1524, s. 31(1).
114. R.S.C. 1970, c. T-6, s. 19(h).

115. *Ibid.*, at s. 3(3).
116. R.S.C. 1970, c. Y-4.
117. R.S.C. 1970, c. Y-3. For a discussion of this Act and the Yukon Quartz Mining Act, see *infra* notes 262—65, and accompanying text.
118. C.R.C. 1978, c. 1524, s. 5(c).
119. John K. Naysmith, *Land Use and Public Policy in Northern Canada* (1977), p. 64. Cabinet would still retain the right, however, to impose conditions as part of the lease or purchase contracts. R.S.C. 1970, c. T-6, s. 4.
120. Alastair R. Lucas and E.B. Peterson, "Northern Land Use Law and Policy Development: 1972—78 and the Future", *Northern Transitions Volume II* (1978), p. 67.
121. Only minimal review procedures are set forth. For a Class A permit, the Engineer is bound by time constraints (10 days for initial response, 42 days for second response if required, and 12 months for final decision), and by an informational constraint (reasons for delay or refusal decisions must be given). Appeal is limited to the Minister of Indian Affairs and Northern Development, C.R.C. 1978, c. 1524, ss. 25, 45.
122. C.R.C. 1978, c. 1237.
123. C.R.C. 1978, c. 1612.
124. A fourth Act, the Territorial Lands Act, may also be used to set aside land but the Act is generally used as only a first step in the process towards creating a National Park or National Wildlife Area. For example, an area of Bathurst Island (Polar Bear Pass) was set aside since 16 March 1978 pending a working group's decision on the final status of the area. SI/78—53 as amended. Environment Minister John Roberts subsequently announced his decision to accept the site as a National Wildlife Area under the Canada Wildlife Act to be jointly managed with the Department of Indian Affairs and Northern Development and the Government of the Northwest Territories. "IBP Ecological Site Approved," Department of Indian Affairs and Northern Development, Communique No. 1—8214 (30 July 1982). On 10 March 1983 Cabinet withdrew the area for wildlife and habitat protection. SI/83—65 substituted by SOR/84—409.
125. R.S.C. 1970, c. N-13.
126. National Park Reserves, pending settlement of native land claims, would remain open to traditional hunting, fishing and trapping by native people. An Act to Amend the National Parks Act, S.C. 1974, c. 11, s. 11(1).
127. Other zones would include Natural Environment, Outdoor Recreation and Park Services. For a complete description of the zones, see Parks Canada, *Parks Canada Policy* (1979), p. 40.
128. S.C. 1973–74, c. 21.
129. *Ibid.*, at s. 4(1).
130. Wildlife Area Regulations, C.R.C. 1978, c. 1609, s. 3(1).
131. R.S.C. 1970, c. M-12.
132. Migratory Bird Sanctuary Regulations, C.R.C. 1978, c. 1036, s. 3(1).
133. R.S.C. 1970, c. N-13, s. 6(2).
134. *Ibid.*, at s. 7(1)(k).
135. *Ibid.*, at s. 71(1) as amended by S.C. 1974, c. 11, s. 3(2).
136. Wildlife Area Regulations, C.R.C. 1978, c. 1609, s. 4 as amended by SOR/82–871.
137. Migratory Bird Sanctuary Regulations, C.R.C. 1978, c. 1036, s. 9(4)(b).
138. For a discussion of such legislative proposals, see Constance Hunt, Rusty Miller and Donna Tingley, "Legislative Alternatives for the Establishment of a Wilderness Area", *Northern Perspectives* V. 7 No. 2(1979), pp. 2–6.
139. National Parks Act, R.S.C. 1970, c. N-13, Schedule, Part I(5).

140. An Act to Amend the National Parks Act, S.C. 1974, c. 11, Schedule V, Part I.
141. *Ibid.*, at Part II.
142. *Ibid.*, at Part III.
143. S.C. 1983–84, c. 24, s. 7.
144. "IBP Ecological Site Approved", Indian and Northern Affairs Canada, Communique No. 1–8214 (30 July 1982). Lands have been withdrawn from disposal under the Territorial Lands Act. Polar Bear Pass Withdrawal Order, SOR/84–409. For a discussion of special areas protection in Canada, see P.M. Taschereau, *The Status of Ecological Reserves in Canada* (1985).
145. For a listing of the actual sites, see Migratory Bird Sanctuary Regulations, C.R.C. 1978, c. 1036, Schedule, Parts X, XI and XII.
146. Proceedings of the Special Committee of the Senate on the Northern Pipeline, 1st Sess., 32nd Parliament, Issue 37 (Appendix B - The Mandates and Programs of Environment Canada) (16 September 1982), p. 113. Parks Canada proposes to establish at least one marine park for each of eight Arctic Ocean regions. See Parks Canada, *National Marine Parks* (Third Draft, August 1983).
147. The Department could also exercise a few slivers of water control through the Territorial Land Use Regulations. The Department could prohibit any excavations below the normal high-water mark of streams, prohibit deposit of excavated material in streams, order debris removal from streams and order channel-bed restorations. C.R.C. 1978, c. 1524, ss. 10, 13. Pursuant to the Public Lands Grants Act, the Department may also control offshore activities, such as oil and gas explorations out to 12 nautical miles (the limit of the territorial sea), by placing conditions in the lease of federal lands. R.S.C. 1970, c. P-29. Such leases may be granted for a term not exceeding five years and for an annual consideration not exceeding $ 75,000. Public Lands Leasing and Licensing Regulations, SOR/82–839. The applicability of the Public Lands Grants Act to the offshore is open to question since the statutory language does not expressly mention the offshore and since at common law the territorial realm ended at the low water mark. See *R* v. *Keyn*, {1876} 2 Ex. D. 63.
148. R.S.C. 1970, c. 28 (1st Supp.).
149. R.S.C. 1970, c. 2 (1st Supp.).
150. For a detailed discussion of the law and administration surrounding inland water management, see William MacLeod, *Water Management in the Canadian North: The Administration of Inland Waters North of* 60° (1977).
151. R.S.C. 1970, c. 28 (1st Supp.), s. 5.
152. *Ibid.*, at s. 27.
153. *Ibid.*, at s. 26(d).
154. *Ibid.*, at s. 26(e).
155. *Ibid.*, at s. 7(1).
156. *Ibid.*, at s. 10(2).
157. Northern Inland Waters Regulations, C.R.C. 1978, c. 1234, s. 11(a) as amended by SOR 84–157.
158. *Ibid.*, at s. 11(b).
159. *Ibid.*, at s. 11(c). The Federal Court, Trial Division declared the original Section 11 of Northern Inland Waters Regulations as *ultra vires* (beyond the power of) the Governor in Council, since the original wording left the Comptroller too much discretion in deciding when to issue water authorizations and thus Cabinet had undertaken an unauthorized sub-delegation of legislative powers. See *Dene Nation and the Metis Association of the Northwest Territories* v. *R* (1984), 13 C.E.L.R. 139.
160. The nearest statement to a criterion is the broad policy directive to the boards to "provide the optimum benefit ... for all Canadians and for the residents of

the Yukon ... and the Northwest Territories ...", Northern Inland Waters Act, R.S.C. 1970, c. 28 (1st Supp.), s. 9.

161. Provisions cover such matters as notice of a public hearing (35 days prior to fixed date), need to file intent to intervene, confidentiality of information, production of documents, pre-hearing conferences and conduct of the hearing (for example, order of presentation). See Yukon Territory Water Board Rules of Procedures, SOR/79–838.
162. *R* v. *United Keno Hill Mines Limited* (1980), 10 C.E.L.R. 43 (Y. Ter. Ct.).
163. The maximum fine per day of offence is $ 5000. R.S.C. 1970, c. 28 (1st Supp.), s. 32. Even then, courts have been hesitant to impose the maximum. See *R* v. *United Keno Hill Mines Limited* (1980), 10 C.E.L.R. 43 (Y. Terr. Ct.) ($ 1500 fine); *R* v. *Canadian Industries Limited* (1979), 8 C.E.L.R. 121 (Y. Mag. Ct.) ($ 1000 fine); and *R* v. *Cyprus Anvil Mining Corporation* (1976), 5 Canadian Environmental Law News 145 (Y. Mag. Ct.) ($ 49,000 for nearly a month of continuous pollution). For a general discussion of mining activities in the North, see Robert F. Keith, Anne Kerr, Ray Ules, "Mining in the North", *Northern Perspectives*, V. 9, No. 2 (1981).
164. R.S.C. 1970, c. 2 (1st Supp.)
165. Section 4(1) of the Act prohibits waste deposits except as authorized by regulations. The Arctic Waters Pollution Prevention Regulations allow domestic waste disposals if they meet the requirements of the *Public Health Ordinance* of the Northwest Territories or the *Public Health Ordinance* of Yukon, whichever is applicable. C.R.C. 1978, c. 354, s. 5.
166. Arctic waters are defined to include a 100-nautical mile zone adjacent to the mainland and Arctic Islands and to include natural resource exploration or exploitation activities beyond the 100-mile zone so far as allowed by international law. R.S.C. 1970, c. 2, (1st Supp.), s. 3.
167. More specifically, waste disposals on the mainland or islands are prohibited where the wastes may enter Arctic waters. *Ibid.*, at s. 4(1).
168. Arctic Waters Pollution Prevention Regulations, C.R.C. 1978, c. 354, s. 6.
169. Work is defined broadly to include "all or part of an undertaking ... which will or is likely to result in the deposit of waste of any type in the Arctic waters...". R.S.C. 1970, c. 2 (1st Supp.), s. 10(1).
170. *Ibid.*
171. *Ibid.*, at s. 10(2).
172. *Ibid.* While the statutory provision provides for Cabinet review, the power of review has been delegated to the Minister of Indian Affairs and Northern Development. See Governor in Council Authority Delegation Order, C.R.C. 1978, c. 355, s. 2(b).
173. R.S.C. 1970, c. 2 (1st Supp.), s. 3(2).
174. The same definitional problem lingers in the *Re Offshore Mineral Rights of British Columbia* case. There the Supreme Court of Canada, while deciding offshore seabed jurisdiction, refrained from deciding the status of inland waters and failed to make clear which inland waters were being excepted. {1967} S.C.R. 792, 65 D.L.R. 2d 353. In a more recent decision, *Re Strait of Georgia*, {1984} 1 S.C.R. 388, 8 D.L.R. 4th 161, the Supreme Court of Canada specifically noted the ambiguity posed by the term "inland waters" in relation to offshore jurisdictional claims. For a discussion of the uncertainty, see Rowland J. Harrison, "Jurisdiction over the Canadian Offshore: A Sea of Confusion", *Osgoode Hall L.J.* V. 17, No. 3 (1979), pp. 469–505 and Lawrence L. Herman, "Proof of Offshore Territorial Claims in Canada", *Dalhousie L.J.* V. 7, No. 1 (1982), pp. 3–38. Since the Territorial Sea and Fishing Zones Act in Section 3(2) uses the term "internal waters" to include all sea areas to the

landward side of the territorial sea baselines, the term "inland waters" would seem to depend for definition on common law principles.

175. S.C. 1974–75–76, c. 55.
176. R.S.C. 1970, c. F-14.
177. R.S.C. 1970, c. 5 (1st Supp.).
178. S.C. 1974–75–76, c. 55, ss. 2(1), 4(1). Disposal by incineration at sea would also be covered. *Ibid.*, at s. 2(4).
179. *Ibid.*, at s. 5.
180. *Ibid.*, at s. 6.
181. *Ibid.*, at s. 7.
182. *Ibid.*, at s. 2(1).
183. *Ibid.*, at s. 8(1). An emergency dumping is defined as "dumping ... necessary to avert danger to human life at sea or to any ship or aircraft".
184. *Ibid.*, at s. 2(1). The exact language is: "{A}ny discharge that is incidental to or derived from the exploration for, exploitation of and associated off-shore processing of sea bed mineral resources".
185. A recent court decision, however, imposed a $ 150,000 fine and two years probation on a petroleum operator, Panarctic Oils Ltd, for violating the Ocean Dumping Control Act by disposing of 45-gallon oil barrels and a pickup truck into Arctic waters. The extent of the Act's applicability to oil and gas operations was apparently not raised. See, *R* v. *Panarctic Oils Ltd* (1983), 12 C.E.L.R. 78 (N.W.T. Terr. Ct.).
186. Approximately 92 per cent of the ocean dumping permits issued by the Department in 1979 were for dredging operations. *Environment Canada, A Summary of Permits Issued under the Ocean Dumping Control Act* (1979), p. 2.
187. In 1979 two permits were issued for the placement of 318 metric tonnes of coal dust on the ice in Summer's Harbour, Booth Island, NWT, to accelerate ice melting and assist the Dome drilling fleet with an early breakout. *Ibid.*
188. *Ibid.* Regulatory approval would also be needed under the Arctic Waters Pollution Prevention Act. See, for example, Arctic Waters Experimental Pollution Regulations, 1982, SOR/82–276.
189. S.C. 1974–75–76, c. 55, Schedule I, as amended by SOR/81–721. Other substances include organohalogen compounds, cadmium, persistent plastics, and other persistent synthetic materials and biological-chemical warfare agents.
190. Maximum levels are prescribed in Ocean Dumping Control Regulations, C.R.C. 1978, c. 1243, s. 5.
191. S.C. 1974–75–76, c. 55, s. 9(5).
192. *Ibid.*, at s. 10(1)(b). The Minister must consider, however, factors spelled out by Parliament in Schedule III, such as interference with fishing or navigation and availability of land disposal methods.
193. *Ibid.*, at s. 10(2).
194. *Ibid.*, at s. 12(1)(2).
195. *Ibid.*, at s. 12(3).
196. *Ibid.*, at s. 2(2)(b).
197. *Ibid.*, at s. 2(3).
198. Memorandum of Understanding Between the Department of Fisheries and Oceans and the Department of the Environment on the subject of the Administration of Section 33 of the Fisheries Act (May 1985).
199. R.S.C. 1970, c. F-14.
200. *Ibid.*, at s. 33(1).
201. *Ibid.*, at s. 33(2) as amended by R.S.C. 1970, c. 17, (1st Supp.), s. 3(1). The language covering ship disposals is more specifically "in any water where fishing is carried on". R.S.C. 1970, c. F-14, s. 33(1).

202. *Ibid.*, at s. 33(11) as amended by S.C. 1976–77, c. 35, s. 7(3). Four additional definitions are also given but all could be viewed as a subset of this omnibus provision.
203. R.S.C. 1970, c. F-14, s. 33(4)(a) as amended by S.C. 1976–77, c. 35, s. 7(1).
204. *Ibid.*, at s. 33(4)(b).
205. *Ibid.*, at s. 33(4)(a).
206. Ocean Dumping Control Regulations, C.R.C. 1978, c. 1243, s. 5.
207. See, for example, Arctic Waters Experimental Pollution Regulations, 1982, SOR/82–276.
208. Pulp and Paper Effluent Regulations, C.R.C. 1978, c. 830.
209. Petroleum Refinery Liquid Effluent Regulations, C.R.C. 1978, c. 828.
210. Potato Processing Plant Liquid Effluent Regulations, C.R.C. 1978, c. 829.
211. Metal Mining Liquid Effluent Regulations, C.R.C. 1978, c. 819.
212. Chlor-alkali Mercury Liquid Effluent Regulations, C.R.C. 1978, c. 811.
213. Meat and Poultry Products Plant Liquid Effluent Regulations, C.R.C. 1978, c. 818.
214. R.S.C. 1970, c. 5 (1st Supp.).
215. The Yukon River Basin study is being directed by the Yukon River Basin Committee with memberships from Environment Canada, Indian Affairs and Northern Development and the governments of Yukon and British Columbia. Department of Indian and Northern Affairs Communique (25 November 1980). The Mackenzie River Basin Committee, composed of representatives from the federal government, Alberta, British Columbia, Saskatchewan, Yukon and the Northwest Territories, issued a comprehensive report in 1981. See Mackenzie River Basin Committee, *Mackenzie River Basin Study Report* (1981).
216. Phosphorus Concentration Control Regulations, C.R.C. 1978, c. 393.
217. S.C. 1978–79, c. 13, Part I.
218. *Ibid.*, at s. 5(a)(iv).
219. *Ibid.*, at s. 5(a)(iii).
220. *Ibid.*, at s. 5(a)(i).
221. R.S.C. 1970, c. F-14.
222. R.S.C. 1970, c. 17 (1st Supp.), s. 3(2) as amended by S.C. 1976–77, c. 35, s. 8.
223. *Ibid.*
224. R.S.C. 1970, c. N-19.
225. R.S.C. 1970, c. S-9.
226. R.S.C. 1970, c. N-19, ss. 3(a), 5.
227. *Ibid.*, at ss. 3(b), 5.
228. Navigable Waters Works Regulations, C.R.C. 1978, c. 1232, ss. 5, 6.
229. R.S.C. 1970, c. N-19, s. 4.
230. *Ibid.*, at s. 5(2).
231. *Ibid.*, at s. 5(1)(a) (emphasis added).
232. Transport Canada, Canadian Coast Guard, *Code of Recommended Standards for the Prevention of Pollution in Marine Terminal Systems* (22 February 1977).
233. Governor in Council Authority Delegation Order, C.R.C. 1978, c. 355, s. 2(a).
234. R.S.C. 1970, c. 2 (1st Supp.), s. 4(1).
235. Arctic Shipping Pollution Prevention Regulations, C.R.C. 1978, c. 353, s. 28.
236. *Ibid.*, at s. 29.
237. Shipping Safety Control Zones Order, C.R.C. 1978, c. 356.
238. Arctic Shipping Pollution Prevention Regulations, C.R.C. 1978, c. 353, Schedule VIII as amended by SOR/79–152, Schedule H.
239. R.S.C. 1970, c. 2 (1st Supp.), s. 15(3)(a).
240. *Ibid.*, at s. 15(3)(b)(i)(ii).

241. *Ibid.*, at s. 15(3)(b)(iii).
242. Arctic Shipping Pollution Prevention Regulations, C.R.C. 1978, c. 353. s. 7(1)(a)(b).
243. *Ibid.*, at s. 7(1)(d).
244. *Ibid.*, at s. 7(1)(e)(f).
245. *Ibid.*, at s. 9.
246. *Ibid.*, at s. 21.
247. *Ibid.*, at s. 26.
248. Proceedings of the Special Committee of the Senate on the Northern Pipeline, 1st Sess., 32nd Parliament, Issue 30 (15 June 1982), p. 50.
249. R.S.C. 1970, c. 27 (2nd Supp.), s. 727(2)(b).
250. *Ibid.*, at s. 727(2)(c).
251. Fishing Zones of Canada (Zone 6) Order, C.R.C. 1978, c. 1549.
252. R.S.C. 1970, c. 27 (2nd Supp.), s. 728.
253. Oil Pollution Prevention Regulations, C.R.C. 1978, c. 1454.
254. Pollutant Substances Regulations, C.R.C. 1978, c. 1458, s. 4, Schedule I, as amended by SOR 83–347.
255. Territorial Timber Regulations, C.R.C. 1978, c. 1528.
256. Territorial Quarrying Regulations, C.R.C. 1978, c. 1527.
257. Territorial Coal Regulations, C.R.C. 1978, c. 1522.
258. For a detailed discussion of the legislation and its administration as to mining, see William MacLeod, *Water Management in the Canadian North* (1977), pp. 65–69.
259. Canada Dredging Regulations, C.R.C. 1978, c. 1523.
260. *Ibid.*, at s. 16(1).
261. *Ibid.*, at s. 16(2).
262. R.S.C. 1970, c. Y-3. Draft Yukon Placer Mining Guidelines have been co-operatively developed by the Department of Indian Affairs and Northern Development, Environment Canada and the Department of Fisheries and Oceans. The Yukon Placer Mining Guidelines Review Committee, established in 1983 to review the guidelines and hold public hearings, released a final report on 18 January 1984. The report made numerous recommendations including effluent standards for streams, financial compensation to assist operators of salmon streams to comply with new guidelines, and a requirement for financial security to ensure rehabilitation of the mining site. Government of Canada, News Release 1–8344.
263. R.S.C. 1970, c. Y-4.
264. R.S.C. 1970, c. Y-3 s. 15 and c. Y-4, s. 10.
265. R.S.C. 1970, c. Y-3, s. 77 and c. Y-4, s. 22.
266. C.R.C. 1978, c. 1516.
267. *Ibid.*, at s. 73(1).
268. C.R.C. 1978, c. 1524.
269. *Ibid.*, at s. 31.
270. For a general discussion of oil and gas regulation, see Constance D. Hunt and Alastair R. Lucas, *Environmental Regulation -Its Impact on Major Oil and Gas Projects: Oil Sands and Arctic* (1981), and Ian Townsend Gault, *Petroleum Operations on the Canadian Continental Margin: The Legal Issues in a Modern Perspective* (February 1983).
271. R.S.C. 1970, c. O-4, s. 3, re-enacted by R.S.C. 1970, c. 30 (1st Supp.), s. 3, as amended by S.C. 1980–81–82–83, c. 81, s. 76.
272. *Ibid.*
273. S.C. 1980–81–82–83, c. 81, s. 9.
274. The statutory womb of the Canada Oil and Gas Lands Administration is

section 5(5) of the Canada Oil and Gas Act which allows the Minister of Indian Affairs and Northern Development and the Minister of Energy, Mines and Resources to designate a public servant to execute the powers under the Act. Both Ministers have designated such power to the Administrator of COGLA through a memorandum of understanding. The Minister of Indian Affairs and Northern Development retained powers, however, over environmental and socio-economic matters while the Minister of Energy, Mines and Resources has retained powers over environmental assessment and national energy policy. For a copy of the memorandum, see Proceedings of the Special Committee of the Senate on the Northern Pipeline, *supra* note 79, pp. 35A:44–47.

275. S.C. 1980–81–82–83, c. 81, s. 10(2). Many exploration agreements also contain relinquishment requirements; that is, each year the petroleum operator must relinquish a portion of the total exploration area to the federal government up to a maximum percentage, such as 50 per cent.
276. *Ibid.*, at s. 10(3).
277. *Ibid.*, at s. 19(4).
278. *Ibid.*, at s. 12(1).
279. *Ibid.*, at s. 14(4).
280. *Ibid.*, at s. 12(2).
281. Proceedings of the Special Committee of the Senate on the Northern Pipeline, *supra* note 79, p. 35A:23.
282. SOR/79–82, ss. 5–79. Only a single onshore well not located on an artificial island and drilled with a conventional rig would be excepted from this requirement. *Ibid.*, at s. 5.
283. R.S.C. 1970, c. O-4, s. 3 as amended by S.C. 1980–81–82–83, c. 81, s. 76.
284. SOR/79–82, ss. 80–91.
285. S.C. 1980–81–82–83, s. 76.
286. *Ibid.*
287. *Ibid.*, at s. 80.
288. Canada Oil and Gas Act, S.C. 1980–81–82–83, c. 81, s. 17.
289. *Ibid.*, at s. 19(1)(b).
290. *Ibid.*, at s. 23.
291. *Ibid.*, at s. 26.
292. Proceedings of the Special Committee of the Senate on the Northern Pipeline, *supra* note 79, pp. 35A:28–29.
293. *Ibid.*, p. 35A:29.
294. *Ibid.*
295. *Ibid.*, p. 35A:28.
296. S.C. 1980–81–82–83, c. 81, s. 45.
297. *Ibid.*, at ss. 46, 47.
298. *Ibid.*, at s. 48(1)(a).
299. *Ibid.*, at s. 48(a)(b).
300. The Canada Oil and Gas Act has established two revolving funds, one for North of 60°, known as the Environmental Studies Revolving Fund (DIAND), and the other for South of 60°, known as the Environmental Studies Revolving Fund (EMR) under the administration of the Minister of Energy, Mines and Resources. *Ibid.*, at s. 49. A common advisory board having six members has been established to advise the Ministers. The senior environmental manager of COGLA or the senior environmental manager of DIAND would chair the meetings, depending on whether the question concerned South of 60° or North of 60°. They would be joined by two industry representatives and a member from Fisheries and Oceans and a member from Environment. Proceedings of

the Special Committee of the Senate on the Northern Pipeline, 1st Sess., 32nd Parliament, Issue 31 (11 June 1982), p. 32.

301. S.C. 1980–81–82–83, c. 81 s. 49(9)(10)(11)(13).
302. *Ibid.*, at s. 5(2).
303. *Ibid.*, at s. 6.
304. R.S.C. 1970, c. O-4, s. 12 as amended by R.S.C. 1970, c. 30 (1st Supp.), s. 6 and S.C. 1980–81–82–83, c. 81, s. 77.
305. SOR/79–82.
306. C.R.C. 1978, c. 1517.
307. C.R.C. 1978, c. 1518.
308. A.R. Lucas, D. MacLeod and R.S. Miller, "Regulation of High Arctic Development", in Symposium Proceedings, *supra* note 69, p. 115.
309. C.R.C. 1978, c. 1517, ss. 35–38.
310. SOR/79–82, s. 99.
311. *Ibid.*, at s. 130(4)(j)(k).
312. *Ibid.*, at s. 137.
313. *Ibid.*, at s. 155.
314. *Ibid.*, at s. 156.
315. *Ibid.*, at ss. 169–185.
316. *Ibid.*, at ss. 205, 210.
317. *Ibid.*, at s. 228.
318. *Ibid.*, at s. 29(1)(c).
319. *Ibid.*, at s. 130(4)(j).
320. *Ibid.*, at s. 130(4)(k).
321. *Ibid.*, at s. 137. The Oil and Gas Production and Conservation Act defines "Chief" as the public servant appointed by the Minister of Indian Affairs and Northern Development to administer the Act. S.C. 1980–81–82–83, c. 81, s. 74(1). In practice that person is the Administrator of COGLA.
322. SOR/83–149.
323. SOR/83–151.
324. SOR/83–460.
325. Proceedings of the Special Committee of the Senate on the Northern Pipeline, *supra* note 79, p. 35A:41.
326. C.R.C. 1978, c. 1052 as amended (hereinafter referred to as Gas Pipeline Regulations).
327. SOR/78–746 as amended (hereinafter referred to as Oil Pipeline Regulations).
328. R.S.C. 1970, c. N-6.
329. Gas Pipeline Regulations, *supra* note 326, at ss. 32–37 and Oil Pipeline Regulations, *supra* note 327, at ss. 37–45.
330. Gas Pipeline Regulations, *supra* note 326, at ss. 38–40 and Oil Pipeline Regulations, *supra* note 327, at ss. 46–48.
331. Gas Pipeline Regulations, *supra* note 326, at ss. 47–64 and Oil Pipeline Regulations, *supra* note 327, at ss. 63–98.
332. Gas Pipeline Regulations, *supra* note 326, at s. 24 and Oil Pipeline regulations, *supra* note 327, at s. 51.
333. Gas Pipeline Regulations, *supra* note 326, at s. 28 and Oil Pipeline Regulations, *supra* note 327, at s. 55. The Oil Regulations are actually more lenient, only requiring a minimization of water contamination.
334. Gas Pipeline Regulations, *supra* note 326, at s. 26(e) and Oil Pipeline Regulations, *supra* note 327, at s. 53(e). The Oil Regulations only require protection of spawning beds for "fish species of economic importance" in contrast to the Gas Regulations which apply to all fish spawning beds.

335. Gas Pipeline Regulations, *supra* note 326, at s. 77(7) and Oil Pipeline Regulations, *supra* note 327, at s. 112.
336. S.C. 1977–78, c. 20.
337. *Ibid.*, at s. 20.
338. The Yukon Territory Water Board would have to commence public hearings within six months of application, would have to complete hearings within sixty days and would have to render a decision within forty-five days thereafter. *Ibid.*, at s. 42.
339. *Ibid.*, at s. 14.
340. SI/80–156.
341. SI/80–128, s. 53 (emphasis added). The designated officer would be responsible to the Minister of the Northern Pipeline Agency. S.C. 1977–78, c. 20, s. 5.
342. SI/80–128, s. 65 (emphasis added).
343. SI/80–128, s. 67 (emphasis added).
344. National Energy Board Act, R.S.C. 1970, c. N-6, s. 26.
345. *Ibid.*, at s. 81 as amended by S.C. 1980–81–82–83, c. 116, s. 23.
346. *Ibid.*
347. R.S.C. 1970, c. N-6, s. 40, as amended by S.C. 1980–81–82–83, c. 116, s. 12.
348. *Ibid.*, at s. 81, as amended by S.C. 1980–81–82–83, c. 116, s. 23.
349. R.S.C. 1970, c. N-6, s. 44.
350. C.R.C. 1978, c. 1057, as amended.
351. SOR/78–926. The informational requirements for an international power line are left much more general. The applicant is merely required to file "An assessment of the probable environmental impact of the international power line, including a description of the existing environment and a statement of the measures proposed to mitigate the impact". C.R.C. 1978, c. 1057, Part III(11).
352. R.S.C. 1970, c. N-6, s. 46, as amended by S.C. 1980–81–82–83, c. 116, s. 14.
353. R.S.C. 1970, c. N-6, s. 82(1), as amended by S.C. 1980–81–82–83. c. 116, s. 24. The regulations reiterate the freedom: "Every licence for the exportation or importation of gas or oil is subject to such terms and conditions as the Board may prescribe ...". National Energy Board Part VI Regulations, C.R.C. 1978, c. 1056, s. 13. Environmental consideration may take place in licensing by such general power and by the specific provision that "except as otherwise provided", the Board's Rules of Practice and Procedure covering certifications, also cover licensing. *Ibid.*, at s. 3. The Board is specifically allowed to impose environmental conditions within power export licences. *Ibid.*, at s. 15(m).
354. R.S.C. 1970, c. N-6, s. 83(a), as amended by S.C. 1980–81–82–83, c. 116, s. 25.
355. *Ibid.*, at s. 83.
356. *Ibid.*, at s. 44 and National Energy Board Part VI Regulations, C.R.C. 1978, c. 1056, s. 10 as amended by SOR/84–722 and SOR/84–390. Board rejection, meanwhile, would be final. R.S.C. 1970, c. N-6, s. 19(1).
357. R.S.C. 1970, c. N-6, s. 87, as amended by S.C. 1980–81–82–83, c. 116, s. 29.
358. Written reasons for decisions are, however, usually issued for major applications. Minutes of Proceedings and Evidence of the Special Committee on Regulatory Reform, 1st Sess., 32nd Parliament, Issue 5 (23 September 1980), pp. 77–79. If a landowner opposes the routing of a pipeline or international power line, the Board would be required to issue reasons for a decision. R.S.C. 1970, c. N-6, as amended by S.C. 1980–81–82–83, c. 80, s. 2.
359. In practice the Board does, however, generally follow the rules of evidence as applied in courts, but the Board retains flexibility, for example by not strictly

applying the rules on hearsay evidence. Minutes of Proceedings and Evidence of the Special Committee on Regulatory Reform, 1st Sess., 32nd Parliament, Issue 5 (23 September 1980), pp. 79–80.

360. National Energy Board, Rules of Practice and Procedure, C.R.C. 1978, c. 1057, s. 3(2).
361. For a summary of the decision, see Constance D. Hunt and Alastair R. Lucas, *Canada Energy Law Service* (1982), pp. 10–4117.
362. For a summary of the decision, see *ibid*, at pp. 10–4229.
363. For a summary of the decision, see *ibid*., at pp. 10–4217.
364. *Ibid*., at pp. 10–1524.
365. (1982) 1 Fed.R. 619, 622 (F.C.A.D.). Also see *Memorial Gardens Assoc. (Canada Ltd)* v. *Colwood Cemetery Co*., {1958}1 S.C.R. 353, 358.
366. S.C. 1978–79, c. 13.
367. *Ibid*., at s. 14.
368. Federal Environmental Assessment Review Office, *Detailed Outline of Contents of the Cabinet Memoranda Establishing the Federal Environmental Assessment and Review Process* (April 1978). Cabinet accepted the recommendation on 20 December 1973.
369. Cabinet accepted the recommendation on 15 February 1977. The federal government was to accept financial responsibility for environmental baseline studies. The proponent was to pay for environmental evaluation reports and share the cost if baseline studies had to be accelerated.
370. The Department of the Environment may assist in the screening process through a Regional Screening and Coordinating Committee (one for each of five regions). Federal Environmental Assessment Review Office, *Revised Guide to the Federal Environmental Assessment and Review Process* (May 1978). The Department also has issued a screening guidebook. Federal Environmental Assessment Review Office, *Federal Environmental Assessment and Review Process: Guide for Environmental Screening* (1978).
371. The Department of the Environment may assist in the process and has issued an evaluation guidebook. Federal Environmental Assessment Review Office, *Guidelines for Preparing Initial Environmental Evaluations* (October 1976).
372. The panel might consult the public before issuing guidelines as well.
373. See, for example, J. Swaigen (ed.), *Environmental Rights in Canada* (1981), pp. 251–255.
374. *Ibid*., at 254. However, production plans have been subject to EARP revival.
375. (1981) 10 C.E.L.R. 142 (F.C.T.D.), rev'd sub nom. *Tosorantio* v. *Atomic Energy Board* (1981), 10 C.E.L.R. 146 (F.C.A.D.).
376. (1981) 10 C.E.L.R. 142, 143 (F.C.T.D.),rev'd on other grounds (1981) 10 C.E.L.R. 146, 147 (F.C.A.D.).
377. Federal Environmental Assessment Review Office, *Beaufort Sea Hydrocarbon Production Proposal* (1982).
378. Environmental Assessment and Review Process Guidelines Order, SOR/84–467.
379. Discussion of Nanisivik decision making is based largely on two extensive case studies, R. Gibson, *The Strathcona Sound Mining Project: A Case Study of Decision Making* (February 1978), and Margot J. Wojciechowski, *Eastern Arctic Case Study Series: The Nanisivik Mine* (April 1982).
380. Texas Gulf retained a 35 per cent interest in net profits after recovery of development costs. R. Gibson, *ibid*., p. 33.
381. *Ibid*., p. 39.
382. The fourteen additional actors were Transport, Industry, Trade and Commerce, Energy, Mines and Resources, Manpower and Immigration,

Environment, Central Mortgage and Housing Corporation, Finance, Treasury Board, Northwest Territories Water Board, Northern Canada Power Commission, Foreign Investment Review Agency, Communications, National Health and Welfare, and Justice. *Ibid.*, pp. 49–50.

383. For a copy of the agreement, see *Ibid.*, pp. 179–213.
384. Other interest holders were Mineral Resources International Ltd (59.5 per cent), Metallgellschaft Canada Ltd (11.25 per cent) and Billiton BV (11.25 per cent). R. Gibson and W. McLeod, "Nanisivik", *Northern Perspectives* V. 4, No. 2 (1976).
385. Minutes of Proceedings and Evidence of the Standing Committee on Indian Affairs and Northern Development, 1st Sess., 30th Parliament, Issue 54 (27 May 1976), pp. 19–20. (statement of Mr A.B. Yates, Director, Northern Policy and Program Planning Branch, Department of Indian Affairs and Northern Development).
386. For a more detailed discussion of the Committees, see Wojciechowski, *supra* note 379, pp. 2—24.
387. The abbreviated discussion of the Polaris project is drawn largely from two sources, Memoranda provided by the Department of the Environment and Katherine A. Graham, *Eastern Arctic Study Case Study Series*: *The Development of the Polaris Mine* (March 1982).
388. Cominco Ltd, *Annual Report*, 1981 (1981).
389. *APOA Review*, V. 4, No. 1 (May 1981), p. 14.
390. Francois Bregha, "Canadian Arctic Marine Energy Projects", Canadian Arctic Resources Committee (January 1982), p. 30.
391. Department of Indian Affairs and Northern Development, "Northern Hydrocarbon Developments: A Government Planning Strategy", Discussion paper (13 May 1982), p. 9.
392. *APOA Review*, V. 3, No. 1 (March 1980), p. 1.
393. *APOA Review*, V. 4, No. 1 (May 1981), p. 14.
394. Canadian Arctic Resources Committee, *Northern Perspectives* V. 2, No. 4 (1974), pp. 1–4.
395. *APOA Review*, V. 5 No. 3 (Winter 82/83), pp. 9–10.
396. COGLA News Release No. 1–8224 (November 1982).
397. Canadian Arctic Resources Committee, *Northern Decisions*, V. 1, No. 21 (15 March 1984), p. 153.
398. *Ibid.*, V. 2, No. 11 (30 September 1984), p. 69.
399. "First High Arctic Oil Production Approval", Department of Indian Affairs and Northern Development, Communique No. 1–8506.
400. Federal Environmental Assessment and Review Office, *Report of the Environmental Assessment Panel, Eastern Arctic Offshore Drilling – South Davis Strait Project* (1978), p. 8 (hereinafter referred to as EARP Panel Report).
401. *Ibid.*, p. 11.
402. See Department of Indian Affairs and Northern Development, *EAMES*: *The Achievements of an Environmental Program* (1981).
403. See EARP Panel Report, *supra* note 400, pp. 40–45.
404. *APOA Review*, V. 4, No. 1 (March 1981), pp. 16–17; *APOA Review*, V. 5, No. 2 (Fall, 1982), pp. 6–7; *APOA Review*, V. 5, No. 3 (Winter, 1982/83), p. 10.
405. Thomas R. Berger, *Northern Frontier Northern Homeland, The Report of the Mackenzie Valley Pipeline Inquiry*: *Volume One* (1977), p. xxvi.
406. While the Arctic Gas Project and the Maple Leaf Project had filed applications with the National Energy Board, March 1974 and March 1975 respectively, Alcan did not file an application until 31 August 1976.

407. Foothills Pipe Lines (Yukon) Ltd is owned equally by Nova, an Alberta Corporation (formerly known as Alberta Gas Trunk Line Company Ltd) and Westcoast Transmission Company Ltd of Vancouver, British Columbia. Six subsidiary companies were established to build the various segments - Foothills Pipe Lines (South Yukon) Ltd, Foothills Pipe Lines (North B.C.) Ltd, Foothills Pipe Lines (Alta) Ltd, Foothills Pipe Lines (South B.C.) Ltd, Foothills Pipe Lines (Sask) Ltd and Foothills Pipe Lines (North Yukon) Ltd (for a possible future Dempster Lateral). Northern Pipeline Agency, *Annual Report* 1981–1982 (1983), p. 32.
408. National Energy Board, *Reasons for Decision Northern Pipelines Volume* 1 (June 1977), pp. 1.173–1.179.
409. Also called the Lysyk Inquiry after the Chairman, Kenneth M. Lysyk.
410. Kenneth M. Lysyk, Edith E. Bohmer, Willard L. Phelps, *Alaska Highway Pipeline Inquiry* (1977), pp. xi-xvi.
411. Federal Environmental Assessment Review Office, *Alaska Highway Gas Pipeline*: *Yukon Hearings* (*March–April* 1979) (August 1979).
412. Federal Environmental Assessment Review Office, *Alaska Highway Gas Pipeline*: *Routing Alternatives Whitehorse/Ibex Region* (July 1981).
413. Federal Environmental Assessment Review Office, *Alaska Highway Gas Pipeline*: *Technical Hearings* (*June* 7–12, 1982) (September 1982).
414. Personal communication with Northern Pipelines Agency official. For a criticism of governmental shunning of the EARP process, see Don Gamble and Max Fraser, "Hearings as a Way of Life", *Northern Perspectives* V. VIII, No. 7 & 8 (1980), pp. 15–16.
415. See *supra* notes 336–343 and accompanying text.
416. Northern Pipeline Agency, *Annual Report* 1981–1982 (1983), p. 9.
417. For discussions of the financing problems, see *ibid.*, pp. 1–5, and "Alaska Highway Gas Pipeline: Financeability Questioned", *Northern Perspectives* V. VII, No. 5 (1979).
418. For a discussion of the approval process, see Francois Bregha, "Ironclad Commitments: The Selling of a Pipeline", *Northern Perspectives* V. VIII, No. 7 & 8 (1980).
419. Economic Council of Canada, *Connections*: *An Energy Strategy for the Future* (1985), p. 63.
420. Department of Indian Affairs and Northern Development, *Information Pipeline* (January 1983).
421. Proceedings of the Special Committee of the Senate on the Northern Pipeline, *supra* note 79, p. 35A:28.
422. National Energy Board, *National Energy Board* 1981 *Annual Report* (1982), pp. 16–17.
423. Federal Environmental Assessment Review Office, *Norman Wells Oilfield Development and Pipeline Project*: *Report of the Environmental Assessment Panel* (January 1981).
424. Minutes of Proceedings and Evidence of the Standing Committee on Indian Affairs and Northern Development, 1st Sess., 32nd Parliament, Issue 42 (21 April 1982), pp. 9–10.
425. "Arctic Pilot Project", Petro-Canada information brochure, (November 1981).
426. See Department of Indian Affairs and Northern Development, "Northern Hydrocarbon Developments: A Government Planning Strategy", Summary document (1982), Appendix I.
427. Federal Environmental Assessment Review Office, *Report of the Environmental Assessment Panel*, *Arctic Pilot Project* (*Northern Component*) (October 1980), p. 13.

428. *Ibid.*, pp. 12 and 17.
429. *Ibid.*, p. 93.
430. Transport Canada, *supra* note 34, pp. 29–32.
431. "Arctic Pilot Project" *supra* note 425.
432. Canadian Arctic Resources Committee, *Northern Perspectives* V. 10, No. 3 (April-May 1982), p. 10.
433. National Energy Board, "Regulatory Agenda" (March 1983), p. 4. The Board directed the project sponsors to submit their views concerning continuance or discontinuance of APP hearings by 1 December 1983. National Energy Board, *News Release* (10 June 1983).
434. Canadian Arctic Resources Committee, *Northern Decisions*, V. 2, No. 8 (15 August 1984), p. 52.
435. In 1968 and 1969, twelve independent companies filed for exploration permits; in 1970, the firms pooled their resources to form Magnorth Petroleum Ltd. See *APOA Review*, V. 5, No. 2 (Fall 1982), pp. 2–3. See also, Department of Indian Affairs and Northern Development, "Information: Lancaster Sound Regional Study" (July 1982), p. 2.
436. Federal Environmental Assessment Review Office, *Report of the Environmental Assessment Review Panel, Lancaster Sound Drilling* (1979), p. 5 (hereinafter referred to as Lancaster Sound EARP Report).
437. Peter Harrison, "Lancaster Sound: Confusion and Confrontation", *Nature Canada* V. 8, No. 2 (April-June 1979), pp. 37–42.
438. *Ibid.*, p. 38.
439. *Ibid.*, pp. 39–40.
440. Lancaster Sound EARP Report, *supra* note 436, p. 73 (emphasis in original).
441. For a review of the public review phase of the Lancaster Sound Regional study, see Peter Jacobs, *People, Resources and the Environment* (1981).
442. Peter Jacobs and Jonathan Palluq, *Public Review: Public Prospect* (1984).
443. François Bregha, *supra* note 390, p. 13. Imperial Oil has since created a wholly-owned subsidiary, responsible for the upstream activities of the company, named Esso Resources Canada.
444. "Beaufort Sea Project in Retrospect", *APOA Review*, V. 5, No. 1 (Spring-Summer 1982), p. 20. See also "Petroleum Industry in the Beaufort Sea", in *Report of the Task Force on Beaufort Sea Development*, *supra* note 61, pp. 49–56.
445. Department of Indian Affairs and Northern Development, Northern Affairs Program, *Review of the* 1981 *Beaufort Sea Drilling Program* (1982), p. 1. See critique by Beaufort Sea Project Manager, Al Milne, in *APOA Review*, V. 5, No. 1 (Spring-Summer 1982), pp. 20–21: "The Drilling Authorities granted by the Cabinet for Canmar's two initial sites in 1976 amounted to a 'clearing' all of the Beaufort Shelf for exploratory drilling." See also testimony by E.M.R. Cotterill, ADM, Northern Affairs Program, DIAND, to the House Standing Committee on Indian Affairs and Northern Development, 1st Sess., 30th Parliament, Issue 47 (6 May 1976), pp. 16–18.
446. Department of Indian Affairs and Northern Development, *Review of the Beaufort Sea Drilling Program* 1981 (1982), p. 1.
447. *Ibid.*, p. 5.
448. *Report of the Task Force on Beaufort Sea Developments*, *supra* note 61. See also Nepean Development Consultants, *Government Regulatory Capability in the Beaufort* (1982), p. 4.
449. For a brief description of the Dome, Esso and Gulf proposal for Beaufort Sea production, see *APOA Review*, V. 4, No. 3 (December 1981), pp. 4–5.
450. *Supra* note 391, at pp. 9–11.

451. *Supra* note 446.
452. Beaufort Sea Environmental Assessment Panel, "A Statement of Deficiencies on the Environmental Impact Statement for Hydrocarbon Development in the Beaufort Sea-Mackenzie Delta Region" (March 1983).
453. Federal Environmental Assessment Review Office, *Beaufort Sea Hydrocarbon Production and Transportation Final Report of the Environmental Assessment Panel* (July 1984).
454. Federal Environmental Assessment and Review Office, *Register of Panel Projects, No.* 18 (May 1982), p. 6.
455. *APOA Review*, V. 4, No. 3 (December 1981), p. 11.
456. Personal communication from the Federal Environmental Assessment Review Office (21 March 1983).
457. "Exploration Agreements Concluded", Department of Indian Affairs and Northern Development Communique No. 1–8237 (March 1983).
458. An initial plan would have divided Arctic lands and waters into six regions. Six Northern Land Use Planning Commissions would have developed and recommended plans to the Minister of Indian Affairs and Northern Development. Department of Indian Affairs and Northern Development, *Land Use Planning in Northern Canada*: *Preliminary Draft* (14 October 1982). Following a letter of agreement, signed on June 18, 1985, by the Minister of Renewable Resources, NWT and the Minister of Indian Affairs and Northern Development, a NWT Land Use Planning Commission and a Land Use Planning Policy Advisory Committee have been established, and regional land use planning commissions are in the process of being established for the Beaufort Sea and Lancaster Sound regions.
459. In December 1985 a Task Force to Review Comprehensive Claims Policy added a further voice in the call for revamping the Northern decision-making system. The Task Force, appointed by the Minister of Indian Affairs and Northern Development, David Crombie on July 1985 to review federal policy towards aboriginal claims, urged government to enter into new creative management arrangements with aboriginal groups including broader participation in offshore management and revenue sharing from sub-sea minerals and hydrocarbons. Task Force to Review Comprehensive Claims Policy, *Living Treaties*: *Lasting Agreements Report of the Task Force to Review Comprehensive Claims Policy* (1985).

7

Constitutional development in the Northwest Territories

J. F. SHERWOOD

I Introduction

Almost eighty years after responsible government was withdrawn from the Northwest Territories (NWT), current initiatives now seek its restoration through the creation of one or more Northern provinces. Dismissed for decades as merely a residual territory of the federal government, the NWT is now considered by some to merit greater stature in Canadian politics. However, a number of obstacles inhibit rapid change. For example, there is still no consensus regarding the resolution of the political conflict between theoretical entitlement to responsible government and the practical requirements of political autonomy. The North deserves the former but, at present, is ill-prepared to undertake the latter. Whereas claims for provincial status may be premature, the quest for an improved and more representative Northern government does not preclude the vigorous pursuit of intermediate goals. Specifically, a redefinition of the federal–territorial relationship and division of powers could greatly advance the Northern cause. Since this relationship is not defined by any constitutional document but merely by federal legislation,[1] it is particularly malleable and easily amended. In addition, the relationship is sufficiently ill defined that traditional federalist doctrine may not be applicable.

Being the body which will ultimately effect any constitutional changes in the NWT, the federal government will examine closely the various criteria related to development prior to acting. Questions such as the North's fiscal dependence on the federal government, and Northern claims to greater control of resources in an attempt to alleviate financial reliance are particularly important. The large federal presence in the North suggests that a devolution of total jurisdiction over natural resources is unlikely. However, other possible arrangements may accommodate both the Northern desire to acquire management control of resources, particularly offshore resources, and the federal interest in Northern resource development for the benefit of all Canadians. One such arrangement, described below, involves the transfer of some authority and jurisdiction over offshore regions of the Eastern Arctic to local authorities. This would provide a source of locally generated revenue for use by a new Northern government and would vest in Northern residents greater control over their marine environment.

This paper examines the present issues of constitutional and political development in the NWT. Several Northern proposals currently being promoted raise questions, from the constitutional law perspective, about Canadian federalism and its ability to adapt to unique circumstances. The proposal calling for a new Eastern Arctic political unit has been selected for examination in this context. Other proposals, such as provincial status for Yukon Territory[2] and those dealing with the Western Arctic,[3] are playing equally important roles in the constitutional development of Northern Canada. However, the Eastern Arctic proposal is currently commanding the most federal attention.

The Eastern Arctic proposal is best evaluated against the background of the evolutionary history of Northern government and within the context of current political realities. Certain practical obstacles stand in the way of current demands for provincehood. Conversely, some historic arguments against constitutional development may no longer be relevant to prevent its advancement in terms of a redefinition of the federal–territorial relationship.

II Background

Canadian control over most of the present NWT and other Northern areas began in 1870.[4] Almost immediately, the area began to be severed into separate political units.[5] Early administration was conducted from various locations in what are now the Prairie provinces,[6] but, following the creation of Alberta and Saskatchewan in 1905, Ottawa became the administrative centre of the North. It continued to be so until Yellowknife became the capital of the NWT in 1967 (see Figure 7.1).

When the capital was located at Fort Garry (now Winnipeg, Manitoba), the Lieutenant-Governor of the newly formed province of Manitoba was given the duties of the Lieutenant-Governor of the NWT as well. Later, when the seat of government was located within the Territories, a separate Lieutenant-Governor for the NWT was appointed. In 1888 the North-Western Territories Amendment Act[7] provided for an elected Assembly to be established and by 1897 the Territories enjoyed a fully responsible government.

When Alberta and Saskatchewan were created in 1905, however, the remaining population of the Territories was deemed inadequate to merit an *in situ* capital or an elected responsible government. Therefore, Ottawa was established as an administrative centre and Northern government reverted to a colonial form.

In 1905, the federal government appointed a Commissioner and an Advisory Council to govern the NWT from Ottawa. No elections took place in or for the North until 1951, when a minority of elected persons were included on the Council.[8] All appointed members of the Council were senior civil servants of various federal departments with interests in the North. The position of the Commissioner itself, until 1963, was part of the duties undertaken by the Deputy Minister of the Department of Northern Affairs and National Resources.

Elected representation, when inaugurated in 1951, was extended only to the Western area of the NWT, the High Arctic and Eastern regions did not enjoy elected representation until 1966.[9] By 1979 the appointment of members to the Advisory Council was abolished, and a fully elected Council was in operation. The Legislative Assembly of the Northwest Territories presently consists of twenty-two members, and a seven-person Committee, the NWT Cabinet, is chosen from Assembly members. At present, representation within the Legislative Assembly is not organized along the lines of political parties.

This gradual progression towards elected representation should not be interpreted as a phased re-establishment of responsible government in the North. The Commissioner of the NWT remains a federal appointee whose duties and powers should not be confused with those of a provincial Lieutenant-Governor. The Commissioner commands considerable power for the federal government and consequently the authority of the Legislative Assembly is quite seriously circumscribed.

The 'constituion' of the NWT is contained in the Northwest Territories Act.[10] Section 13 defines the legislative powers of the Commissioner in Council. Jurisdiction over twenty six enumerated classes of subjects is subordinate, however, to the NWT Act in general, and to any other Act of the Parliament of Canada. Therefore, despite the *prima facie* similarity between section 13 of the NWT Act and section 92 of the Constitution Act, 1867,[11] the powers of the Legislative Assembly are exercised only at the pleasure of the federal government.[12] The chief executive officer of the NWT remains a federal appointee, whose loyalties may be influenced by the nature of his or her tenure.

A particular problem with the present division of powers between the federal and territorial governments is in the allocation of land rights and

Figure 7.1. Evolution of the Northwest Territories.

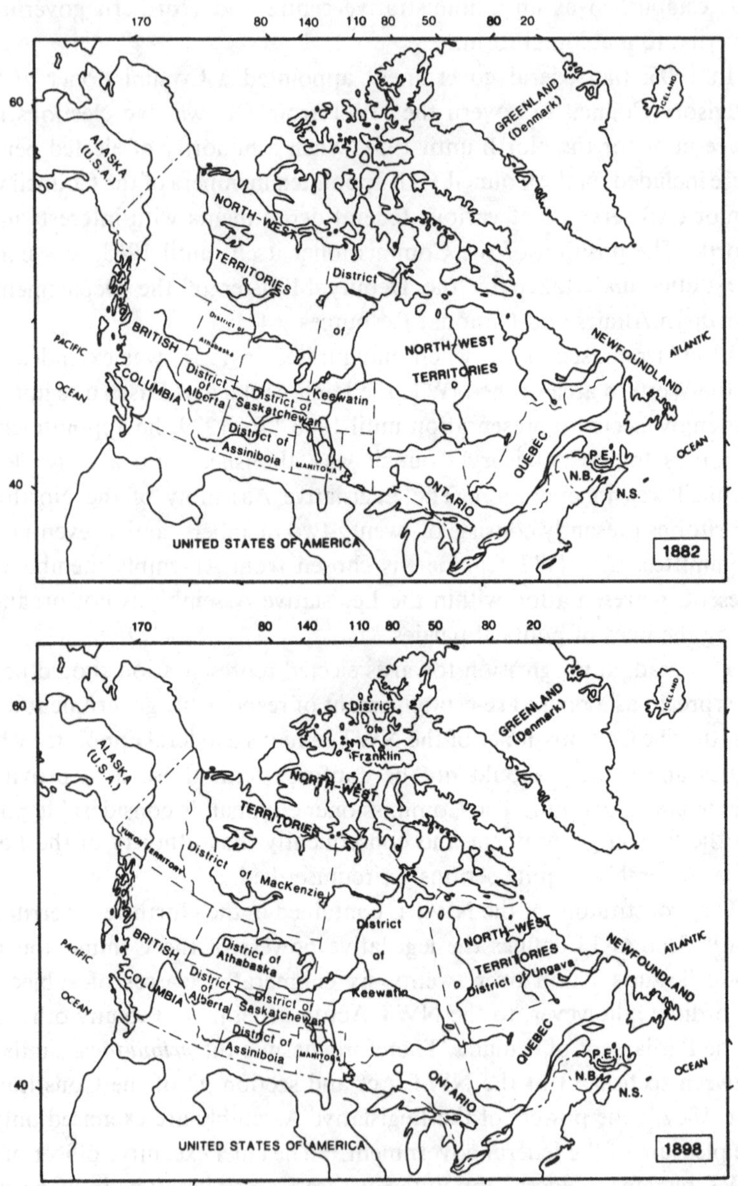

Source: Department of Energy, Mines and Resources (1969), MCR 2306.

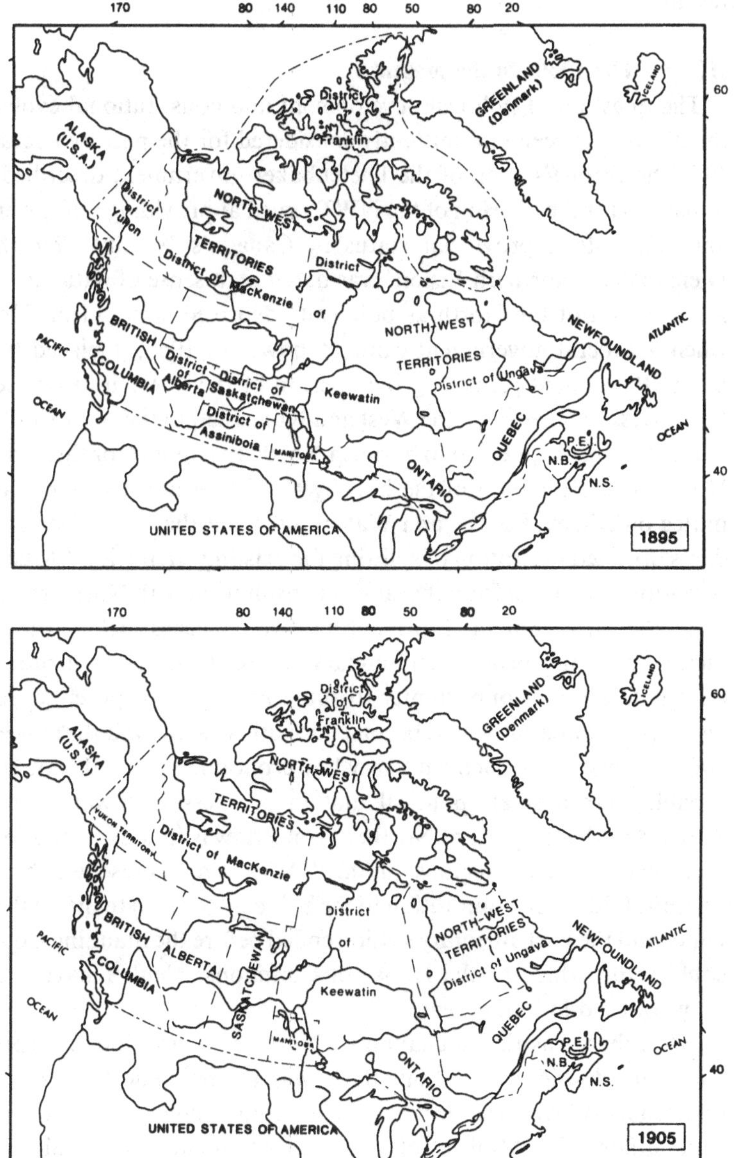
170
80
140
110
80
50
80
20
60
40
ALASKA (U.S.A.)
District of Yukon
NORTH-WEST TERRITORIES
District of MacKenzie
District of Franklin
District of
GREENLAND (Denmark)
ICELAND
NORTH-WEST TERRITORIES
District of Ungava
NEWFOUNDLAND
ATLANTIC
OCEAN
PACIFIC
OCEAN
BRITISH COLUMBIA
District of Alberta
District of Saskatchewan
District of Assiniboia
MANITOBA
Keewatin
ONTARIO
QUEBEC
P.E.I.
N.B.
N.S.
UNITED STATES OF AMERICA
1895
YUKON TERRITORY
ALBERTA
SASKATCHEWAN
1905

ownership. Section 46 of the NWT Act vests certain lands – effectively all lands – in Her Majesty in the Right of Canada.[13] This precludes territorial control of natural resources and prevents the NWT from generating revenue locally.

III The 1960s to the present

The question of political development and constitutional evolution of the North has been intermittently considered for the past two decades. In 1962 the throne speech of the Diefenbaker government described future initiatives for the division of the NWT aimed at providing self-government and, ultimately, provincial status in Canada's North.[14] Whether the Diefenbaker 'Northern Vision' was based on a sense of national pride or a realization of the North as politically deprived is not clear. However, when a Liberal government came to power in 1963, it placed two bills before Parliament proposing the division of the NWT into two separate territories: Mackenzie in the West and Nunassiaq in the North and East.[15]

Initially there appeared to be much Northern support for this legislation. However, examination of the bills by the Government's Standing Committee on Mines, Forests and Waters indicated that most Northern residents opposed the notion of splitting the existing territories.[16] Much of the opposition stemmed from the lack of consultation with Northern residents as to the bills' content. More substantive concerns included a dissatisfaction with the proposed boundary between any new territories; the anticipated loss of opportunity to share in known or potential resource development in what would become the 'other' area; and the concern that two new territories, each with very low economic potential, would be less capable than a single political unit in progressing towards provincial status. Another important Northern concern was that division should not take place until all regions of the NWT were represented by elected officials. Ultimately, opposition to the bills caused them to die on the Order Paper and the Parliamentary session ended before the Standing Committee could return them to the House of Commons. Neither have ever been re-introduced in Parliament.

While there was almost unanimous Northern opposition to the proposed territorial division legislation, there was at the same time a pervasive dissatisfaction with the degree of federal government control over territorial affairs. The Minister of Northern Affairs and National Resources therefore established an Advisory Commission to gather the views of NWT residents and other Canadians concerning the responsibility of the federal government for the Territories.

The Carrothers Commission[17] concluded that any division of the NWT would be undesirable at that time (1966 report date) or within the next decade. It suggested that many of the objectives sought by Northerners could be achieved through measures 'less drastic' than political division. Political sophistication in the Eastern Arctic was thought to be inadequate to respond to the requirements imposed by division. The Report also concluded that division would separate the East from the political expertise of the Western area of the NWT and further its position as a federal dependency.[18] The Commission did suggest, however, that conditions in the North could change and that political division was inevitable at some future date.[19]

The recommendation of the Carrothers Commission to maintain the status quo for a decade surely played a role in the abandonment of any renewed efforts to enact division legislation. Other conclusions of the report aimed at improving the existing administration were acted upon by the federal government as well; in particular, the seat of government for the Territories was moved from Ottawa to Yellowknife.

The Carrothers Commission also discussed the problems of boundary delimitation, of the tax base of the NWT being inadequate to support provincial government, and of high operating costs needed to support a divided NWT. These issues remain central to the question of division today.[20]

Constitutional development in the NWT took a complicated turn in 1976 when governmental evolution became a component of native land claims negotiations. In that year the Inuit Tapirisat of Canada, representing all Eastern and Central Arctic Inuit, submitted to the Minister of Indian Affairs and Northern Development a proposal for the creation of a new Eastern territory, to be known as Nunavut, comprised of those portions of the NWT and Yukon Territory north of the tree line.[21] Similarly, the Dene Nation formulated a proposal concerning the Mackenzie Valley area.[22] The fusion of comprehensive land claims negotiations and proposals for constitutional development complicated matters because additional consideration had to be given to division based on aboriginal right or title. However, the establishment of any new public government in the North was never meant to be in substitution for other land claims concerns, or vice versa. Ultimately, native groups made a distinction between land claims and public government objectives and pursued each through complementary but independent processes.

The linkage between native groups' land-claim assertions and proposals for public government spurred the federal government to consider constitutional development in the NWT once again. The Report of the Special Representative (of the Prime Minister) on Constitutional Development in the Northwest Territories[23] (the Drury Report) was released in 1979 after a two-year study period. Ultimately recommending that the NWT should remain unified, the Special Representative based his assessment on the perceived inadequacy of existing political structures in the Territories. The Report did, however, advocate that Northern government should become accountable and responsible to Northern residents who, in turn, should play a major role in determining the course of their political evolution.[24]

The conclusions of the Drury Report supporting a unified NWT were accompanied by recommendations to redefine the federal-territorial relationship by devolving more authority to the Government of the Northwest Territories (GNWT). It recognized that any future development should come about through a consensus of informed Northern residents and, to that end, certain recommendations were directed towards improving local and regional government to allow for greater local representation.[25]

Following the release of the Drury Report, a Special Committee, established by the NWT Legislative Assembly, reported its own findings with respect to Northern opinion on territorial government.[26] The Special Committee found that most Northerners, especially those in the Eastern Arctic, felt poorly represented by Yellowknife. Eastern residents perceived that the GNWT was more aligned with Mackenzie Valley concerns than their own. In general, the Special Committee held that the existing NWT, as a "geo-political" entity, did not engender any sense of common identity among residents, particularly native residents. Despite disparate political objectives of the several native organizations, all groups were found to be dissatisfied with the status quo. The Special Committee on Unity concluded that:

> The Northwest Territories as a geo-political jurisdiction simply does not inspire a natural sense of identity amongst many of its indigenous peoples; its government does not enjoy in the most fundamental sense the uncompromising loyalty and commitment of significant numbers of those who are now subject to it [...] it came into being without the consent of those who inhabit the area [...][27]

After considering the report of the Special Committee, the NWT Legislative Assembly transmitted the recommendations to the Minister of Indian Affairs and Northern Development and the Commissioner of the NWT. The GNWT acknowledged the need for constitutional development and endorsed the concept of territorial division, as indicated in several of its Recommendations. The first Recommendation acknowledged, for example, that successful constitutional development could proceed only if native interests were given attention. A second Recommendation characterized the present government of the NWT as an 'interim arrangement' subject to change as negotiated by its residents. Recommendation Four gave commitment-in-principle to the division of NWT into Eastern and Western sectors, subject to a plebiscite determining the desires of territorial residents.

The recent growth of support for a division of the NWT is based on two perceptions:

(a) that the Yellowknife government is subordinate in character; and
(b) that the GNWT has failed to adequately represent Eastern interests.

The plebiscite, referred to in Recommendation Four, was held on 14 April 1982. The GNWT sought a mandate from Northerners to request the federal government to divide the NWT so as to create a new territory in the Eastern Arctic along the lines of the Nunavut proposal. After a short preamble, the question posed in the plebiscite was simply:

DO YOU THINK THAT THE NORTHWEST TERRITORIES SHOULD BE DIVIDED?

——YES
——NO.[28]

All overall majority of voters (56 per cent) answered in the affirmative. Regional analysis of voting indicates, however, that support for division was very high among Eastern Arctic, Inuit-dominated communities, but less so in the West among Dene groups. Non-native communities in the West generally voted against the idea of division, with many non-natives abstaining from voting.[29]

Following the plebiscite, the NWT Legislative Assembly unanimously endorsed the division and creation of a new Eastern territory. After several months of deliberation, the Minister of Indian Affairs and Northern Development announced, in November 1982, that there was also federal agreement-in-principle for division, subject to four conditions:

(i) that native land claims are settled before division;
(ii) that agreement is reached on a division of powers interterritorially and between the Territories and the federal government;
(iii) that agreement is reached as to the boundaries between any new territories; and
(iv) that there is continued consensus among Northerners for division.[30]

All four conditions appear to be in the process of fulfilment. On 28 March, 1984 the federal Cabinet approved a comprehensive land-claim agreement, affecting 2,500 Inuvialuit in the Western Arctic, between the federal government and the Committee for Original Peoples' Entitlement (COPE). On 22–23 October, 1984 the Western Constitutional Forum held formal negotiations with representatives of the Legislative Assembly of the Northwest Territories, the Dene Nation, the Metis Association of the Northwest Territories, the Kitikmeot West area, and non-native residents, concerning principles for regional government. The Western Constitutional Forum released two reports on the effects of a new political boundary in the Northwest Territories. In the November 1983 Northwest Territories general election, a substantial number of incumbents, who advocated territorial division, were returned to the Legislative Assembly and, thus, a continued consensus among Northerners for division arguably exists.

A The Nunavut proposal

As stated earlier, the Nunavut proposal arose out of the land claims negotiations of the Inuit Tapirisat of Canada (ITC). In 1976 the ITC proposal described a territory comprising the historic homeland of Inuit and Inuvialuit over which the Inuit would have greater control of natural resources. The proposal advocated Inuit title to surface rights for 250,000 square miles of land, with such rights pertaining as well to the first 1,500 feet of stratum thereunder and any mineral wealth extracted from that stratum. This proposal was aimed at securing for Inuit real property compensation for extinguishment or reduction of their aboriginal rights. It was also meant to ensure that Northern residents would be served by a more representative government than the Yellowknife administration by creating a separate political entity in the East comprised of a majority of Inuit.

Despite attempts by proponents to dispel federal and territorial apprehensions, the two tiers of government perceived the object to be the establishment of an ethnic government since Nunavut was first proposed as part of land claims negotiations. Ethnic overtones concerned the federal government in particular, and led to a rejection of the Inuit claim, based on aboriginal title, for rights to land surface and the first 1,500 vertical feet of stratum.

As a result of this rejection, proponents of Nunavut shifted their strategy to one of promoting a new Eastern territory outside of the sphere of land claims negotiations. Land claims and territorial division did not, however, become ideologically separated. Because the legal issues of aboriginal rights were so unclear,[31] native leaders felt that a procedural separation of proposals for territorial development would enable them to progress more quickly. Nunavut is presently being promoted as a basis for 'public government',[32] without any kind of special status for Inuit except to the extent that that group would constitute the majority (80 per cent) of the resident population. Nunavut is still premised upon the uniqueness of the environment, culture and language of its majority population group.

Several differences between the East and West of the NWT may justify territorial division. Since 1967, when Yellowknife was named as the territorial capital, residents of the Eastern Arctic have progressively felt more and more alienated by the Northern administration. Because most of the commercial development in the NWT has been concentrated in the more populated District of Mackenzie with its oil, gas and mineral reserves, Easterners have perceived the GNWT as addressing Western concerns at the expense of Eastern interests. This perception is further accentuated by major cultural differences between East and West. Eastern residents are mainly Inuit, while the Western population is a mixture of non-native, Inuit, Dene and Metis peoples. The native languages spoken in the East and West are different,[33] and, perhaps more importantly, the culture and traditional economy of the Inuit are unlike any other in the North. Almost all Inuit settlements are coastal, reflecting the historic yet continuing reliance and dependence upon the marine environment for their livelihood. Inuit are primarily a maritime culture living above the tree line while other native groups and non-natives in the NWT have a predominantly land-based economy and culture. Inuit fear a lack of protection for their language and culture by the Yellowknife administration and believe that a more representative and responsible form of government should be created.

A new territory of Nunavut would have, at its inception, a political relationship similar to current NWT–federal arrangements. That is, proponents seek the creation of a new and restructured territory, and do not advocate that a 'Province of Nunavut' be formed at once. However, the long-term objective is for provincial status to accrue to the Eastern Arctic. The proposed division of powers between the federal government and a Nunavut administration is premised upon the attainment of provincehood approximately fifteen to twenty years after the territory is established. The proposed division of powers, however, goes beyond the traditional separation of jurisdiction between the existing provinces and the federal government. In particular, it suggests that the Nunavut government have some jurisdiction concerning criminal law[14] and offshore regions[35] The justification for these demands is based upon the unique needs of the Northern culture. But because these desired spheres are beyond those traditionally recognized as within provincial domain, the proposed federal–Nunavut relationship must be examined to consider whether Canadian federalism can accommodate such innovations.

B The Federal-Nunavut relationship

There is little question that Northern Canada suffers from a political deprivation that would not be tolerated in the South. While the April 1982 plebiscite indicated that a majority of Northern residents are dissatisfied with their present government, questions remain as to whether division is a practical alternative. Before initiating political change in the North, the federal government will consider several criteria, including the adequacy of the local tax base, the political sophistication of Northern residents, whether a Northern consensus continues to favour division, and, most importantly, whether it is politically advantangeous to proceed with division at this time.

The issues of the adequacy of the Northern tax base and of the level of political sophistication of Northern residents must be dealt with in the context of a small population spread over a large land area. While occupying approximately one-third of the land area of Canada, the NWT has a population of only 47000—less than one-fifth of one per cent of the population of the nation. The small population has been insufficient to establish more than a few natural resource-related, primarily extractive, industries, with most activity centred on the oil and gas reserves in the District of Mackenzie. Because of these natural resources reserves and the small population, the NWT has a per capita revenue potential higher than the Canadian average. However, in fact, the NWT receives the highest per capita expenditures of any part of Canada,[36] and is continually in a deficit situation to which the federal government must respond with significant annual transfer payments.[37] Because the NWT, unlike economically disadvantaged provinces, is unable to receive equalization payments from the federal government,[38] all funds received from the federal government come in the form of shared-cost programmes over which federal control is plenary.[39]

Northern regions will clearly require continued federal monetary and administrative support even after political development. Although Manitoba entered into confederation with only 11000 people, a population smaller than either portion of a divided NWT, the population was of a sufficient density so that the new provincial government could feasibly provide services.[40] The vast area of the North, even after the proposed division, works against all economies of scale. One commentator has said:

> The fact is that there is likely a threshold minimum population beyond which the per capita cost of providing the sorts of services

provided by provincial governments in Canada is simply too high.[41]

Questions of the level of political sophistication in the North were central to both the Carrothers Commission and the Drury Report which recommended that territorial division should not immediately proceed. The Eastern Arctic was perceived as being particularly vulnerable to long-term federal domination if division occurred. However, political expertise is a more dynamic factor than Northern ability to generate revenue, and significant developments in local political skills have occurred since 1979 making constitutional development in the 1980's more justifiable than it was in 1966 or 1979 when Carrothers and Drury reported.

Two political advances can be noted. First, fully elected representation in the NWT Legislative Assembly was achieved in 1975 with the elimination of federal appointments to the Council. Because it continues to be constrained by the supremacy of the federal government through sections 13 and 16 of the Northwest Territories Act, the Legislative Assembly is not a fully responsible government although it has become much more representative of Northern residents.

Second, politically active organizations have emerged outside of the legislative process. Native groups developed in the early 1970s[42] to address land claims issues and are now active lobbyists in Yellowknife and Ottawa, frequently appearing before House of Commons and Senate Committees to discuss a broad range of issues.[43]

Thus, in the 1970s two disjunctive political entities existed in the NWT: the Legislative Assembly which, however inadequate, was the best forum from which to press Ottawa for changes in administration and native organizations which had little political clout but identified compelling problems and formulated possible solutions. Since the Legislative Assembly endorsed the concept of division, however, it and several native groups have united in an effort to bring about division. Until February 1985, the Associate Minister for Aboriginal Rights and Constitutional Development of the Northwest Territories and the presidents of the Inuit Tapirisat of Canada and the Committee of Original Peoples' Entitlement co-chaired the Nunavut Constitutional Forum which is now the 'official' proponent of a new Eastern territory or province.[44] Such political co-operation evidences a much greater level of political sophistication and coordination in the North than in the past.

The Nunavut proposal offers a method of political evolution in the North pursuant to past Canadian tradition. Just as the Prairie provinces were created from the then North-Western Territories, new political units could arise out of the NWT through the Parliamentary process.[45] And if

Nunavut is considered as a component of Inuit land claims, then a 'Canadian' method of implementing the proposal may be more attractive than either of the two geographically proximate yet politically dissimilar examples of Alaska and Greenland.[46]

Ottawa will not favour all facets of the Nunavut proposal. Because of the present inability of the NWT to generate more than a small proportion of its fiscal requirements, potential revenues from Northern resources have been considered as a source of local income for any new territory created. The federal government's interest in retaining such revenues for itself is a major obstacle that the proponents of Nunavut must overcome.

One can understand the reluctance of Ottawa to devolve jurisdiction respecting oil and gas resources to a new territorial authority in light of the federal government's relationship with energy producing provinces. Together with Pacific and Atlantic offshore regions, Northern Canada is the last major stronghold of federal Crown land and, as such, constitutes a federal energy reserve. Any political development in the North that jeopardizes federal control over Northern 'wealth', federal officials argue, should be strongly resisted. For many years the federal position has been that Northern Canada's petroleum and mineral wealth should be for the benefit of all Canadians and therefore retained by the federal government:

> These resources [of the NWT] are held by the Government of Canada for all of the people of Canada. All Canadians, through the Government of Canada and through private enterprise, have contributed heavily to the development thus far. They will do far more generously in the future (based largely on the investment and taxation of people south of the 60th parallel) and it is the interests of Canadians, both in the North and South alike, that Canada as a whole shoulder the costs and share the benefits of the great developments which may be expected.[47]

Recent legislative initiatives such as the National Energy Board Act[48] and the Canada Oil and Gas Act[49] are restatements of the federal intention to maintain a presence in the NWT. Any proposal for development which threatens this policy will be strongly resisted by federal officials, as a result of the established bureucratic infrastructure that would require modification.

While notions of a minimum threshold population and economic base may be desirable preconditions to provincehood in a practical sense, they are not necessarily conditions precedent for entitlement to responsible government.[50] Responsible government requires provincial autonomy. The present NWT Legislative Assembly is very representative government but its subordination to the federal government denies its responsibility to the people. There is little substantive value to such representation if the government's initiatives are possible merely at the pleasure of a higher authority. To conclude that provincehood should not accrue to the NWT at this time, and that the status quo should be maintained, is to entrench a colonial form of government in the North that Southern Canada abandoned in 1867.

The debate over Northern political development does not have to be resolved by deciding between either of the two ends of the provincial status – status quo continuum. Rejection of the thesis of provincial status for the North at this time still leaves open the opportunity for a redefinition of the federal-territorial relationship and thereby an improvement on the status quo. This was a conclusion reached by the Drury Report, though most parties to the debate appear to have rejected that study. A more representative and responsible government in the NWT would result from a greater arm's length relationship between the federal government and territorial administrations.

"Inherently malleable" is an appropriate way to describe the current federal-territorial relationship because it is so ill-defined and discretionary. Modification is not difficult, and need not involve provincial consent or participation. Changes in the existing NWT government, *qua* territories, are solely the matter of the federal government.[51] The potential for a new federal-territorial relationship with respect to the division of powers is not impaired by the constraints of interdelegation[52] or the judicial interpretation of sections 91 and 92 of the Constitution Act, 1867.[53] Because of this relative simplicity, the potential for change in Northern Canadian administration is quite broad.

C Offshore resources and revenues

Reform could be achieved in the areas of the territorial tax base and of Northern participation in resource development through a restructuring of natural resource jurisdiction in the NWT. If it is conceded that an outright transfer of resource jurisdiction will not be tolerated by the federal government, there still remain other options that give Northern residents a greater role in resource policy formulation and would enable the new territories to generate revenue for themselves. Proponents of Nunavut are particularly interested in acquiring some interest in offshore Arctic areas and in related renewable and non-renewable resources.

The utilization of the Arctic offshore by would-be Nunavut residents has been central to their culture for centuries. The present distribution of settlements in the Eastern Arctic emphasizes the importance of the marine environment to traditional ways of life; all but one are located in coastal areas. Therefore, as the majority population of Nunavut, Inuit proponents of the new territory have stressed both a need and the justification for greater control over adjacent marine areas.[54] These claims are not founded solely upon aboriginal right and cultural necessity; they are also based upon the perceived needs to better protect the economic base of the majority of Northern residents and to improve the fiscal position of a newly created Eastern territory.

During the term of his Conservative government in 1979, then Prime

Figure 7.2. Inuvialuit land claims settlement area.

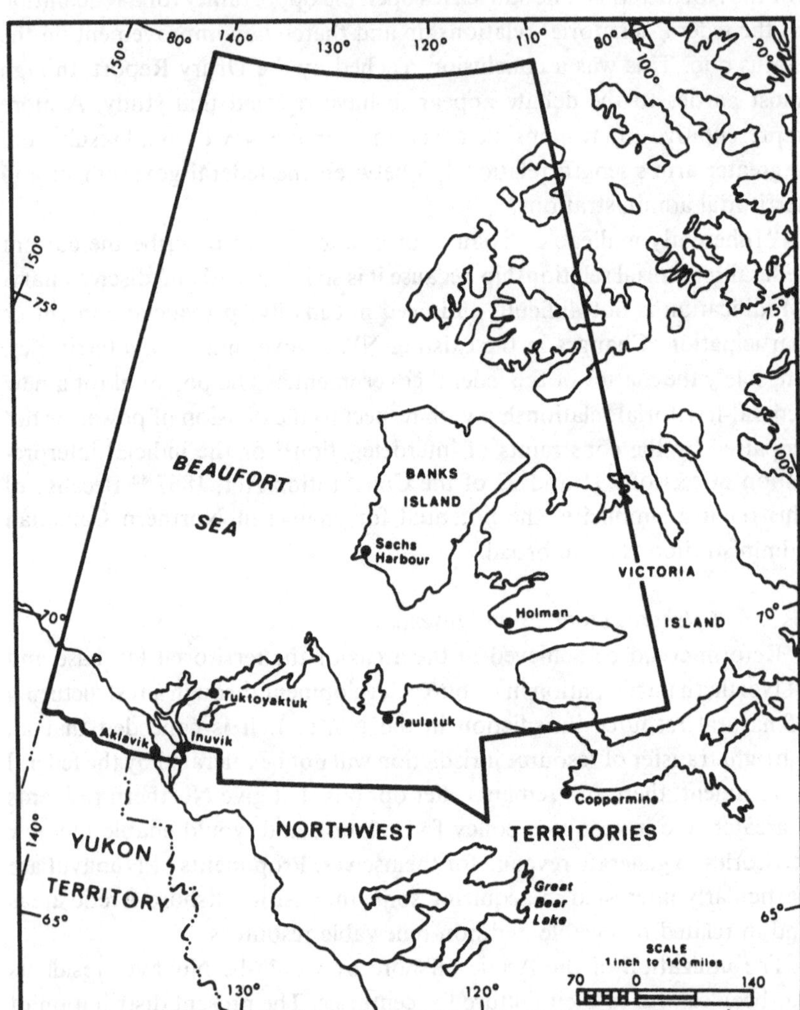

Source: *The Western Arctic Claim, The Inuvialuit Final Agreement*, Department of Indian Affairs and Northern Development, 1985, p. 38.

Minister Clark stated that coastal communities should become more involved with the decision-making process for, and recipients of, substantial revenue from offshore development. This statement was made with respect to Atlantic Canada but it should be no less applicable to Northern areas. Because most of the exploratory activity for mineral and petroleum reserves in the Eastern Arctic has been conducted offshore, it may be surmissed that the major internal source of revenue for any new territorial administration will be generated from activity in marine areas if oil and gas production was to be commenced in the region. In light of the historic criticism that Northern political development is impaired by its fiscal dependence upon Ottawa, offshore development could constitute an opportunity to alleviate that problem and at the same time vest Northerners with some degree of control over the environment upon which they depend.

In the past ten years an enormous Southern interest has developed in the NWT as a vast reserve of energy and minerals. During the review phase of proposed projects such as the Arctic Pilot Project, local and native opposition became both frequent and vociferous. From the Southern perspective in general, and certainly from the viewpoint of the proponents of major resource development projects in particular, such Northern intervention was characterized as inherently 'anti-development' – a position interpreted as a ploy to help achieve land claims objectives by holding the North 'ransom' until political demands were met. However, non-native as well as native groups have expressed concern about socio-economic and environmental effects of proposed development, and because of their weak political position, native and other groups have found appearances before Royal Commissions or environmental impact assessment panels to be their only effective opportunities for intervention. Close analysis reveals that many intervenors are not opposed to development *per se*, but are opposed to development over which they have no control or substantive input.

Possible models for Northern offshore development might be the existing agreements between Canada and Nova Scotia and Canada and Newfoundland respecting offshore oil and gas resource management.[55] The agreements provide for joint federal-provincial boards to exercise the discretions of the Minister of the Energy, Mines and Resources under the Canada Oil and Gas Act.[56] There is provision for provincial participation in both resource management and in the sharing of generated revenues. Therefore, Nova Scotia and Newfoundland have a negotiated entitlement to revenues from offshore development despite the fact that ownership of offshore areas and resources in general appears to be decisively within the federal sphere.[57] Implementation of these types of agreement would be consistent with Mr. Clark's statement referred to above. With such an arrangement, the perceived anti-development posture of Northern residents would probably change to one more accurately reflecting their disposition, *viz.* pro-development for projects over which they have some measure of control and from which they derive some fiscal benefit.

Nunavut proponents feel that given past experience, and in light of the Nova Scotia and Newfoundland examples such arrangements should be acceptable to the federal government. Various boards and agencies have already been devised to implement some sort of negotiated co-operative arrangement.[58]

One specific region of the Northern offshore is of vital interest to the proponents of Nunavut. Historically, the most widely used and occupied area of the offshore by Inuit are the zones of landfast ice; that is, areas where ocean ice is contiguous to land. The interface of this landfast ice and open sea is a productive hunting and fishing area. To the extent that Inuit have a concept of land usuage, no distinction is made between land *per se* and areas of landfast ice.[59] Both are used to gain access to the important waters at the ice edge. Therefore it has been suggested that special consideration be given to this area and that it be subject to special Northern control.[60]

This idea has inherent difficulties since the breadth of landfast ice zones varies from place to place and with climate. Thus, it would be difficult to define its boundaries and, therefore, to delimit the region of special Nunavut control and areas of more traditional federal jurisdiction beyond. In addition, it would represent a functionalist approach to the division of powers unique in Canada, even given the flexibility of the federal-territorial relationship.[61] However difficult or unique, the concept merits thorough investigation. Special consideration should be given to this issue as a vehicle constituting a form of concurrent jurisdiction, giving attention both to particular 'territorial interests' and to the general federal interest in the offshore.

Such a form of local administration for the management of coastal land and ice zones would go far in satifying Inuit and other Northerners who presently feel they are ignored when decisions are made regarding their environment. It would engender greater representative government in the East by enabling residents to become integral participants in a management scheme for which the Yellowknife administration seems to have an inadequate interest. While expediting valuable resource development and revenue generation objectives, recognition of concurrent jurisdictional responsibilities would also constitute a step towards more sophisticated regional government in the Eastern Arctic.

Significant political development in the North will only occur when the region acquires some measure of economic independence. Until certain changes are made regarding the channelling of future funds generated from offshore regions, there can be little hope for fiscal responsibility in the North. In contrast to the recent Canadian constitutional exercise where constitutional patriation and amendment were not fundamental to economic vitality, the constitutional restructuring of Northern Canada, as well as furthering representative and responsible government, is a condition precedent to lessening its financial dependence upon Ottawa.

On 5 February 1985, the Minister of Indian Affairs and Northern Development indicated an openness to consider avenues for devolving financial and management powers. He agreed to consider the Resource Management and Revenue Sharing Proposal prepared by the Energy, Mining and Resources Secretariat of the GNWT. The proposal calls for negotiation of a revenue-sharing regime and direct taxation powers for the territorial government.[62]

IV Other issues and conclusion

The federal government's condition that parties in the North agree to the boundaries between Nunavut and the remaining Western territories may be a problematical requirement for division. When reference is made to a boundary that will run "generally north of the tree line",[63] debate over a large area remains unresolved. One of the concerns of Northerners in 1963 was that the Mackenzie and Nuanassiaq bills would alienate some people from the potential resources of the 'other' territory after division. This concern remains no less today than twenty years ago. Further, because the Inuvialuit in the Mackenzie Delta, represented by COPE, also inhabit northern Yukon Territory, inclusion of that group in the Nunavut territory (as is planned subject to change by ongoing federal government-COPE negotiations) would necessitate a boundary agreement with Yukon which may not be eager to give up its access to the Beaufort Sea.

Another possible boundary of a Nunavut territory requiring settlement will be the status of the High Arctic Islands, a major area of energy exploration activity. Because no Inuit permanently inhabit these islands, some federal officials consider them as inappropriate inclusions to the Nunavut territory and argue that they should remain under full federal stewardship. A requirement to identify which islands would or would not fall within a new Eastern territory would further increase the Nunavut boundary delimitation problem. The task of establishing boundaries for Nunavut acceptable to all parties will probably require extensive research and lengthy negotiations. Resolution of boundary conflicts among Northerners, as a federal pre-condition to territorial division, could prove to be more difficult to achieve than agreement as to the division of powers with the federal government. Some lower court pronouncements have suggested that the boundaries of the present NWT include certain offshore areas.[64]

Should the federal government consider granting provincial status to Nunavut and the transfer of jurisdiction over natural resources, it will need to determine whether a future Nunavut government would be entitled to an accounting for resource revenues accruing to the federal government before full provincial status is attained. The situation of the reserved jurisdiction over natural resources in the Prairie provinces until 1930[65] is similar, but in some ways different, to the case of Nunavut. *In re Transfer of Natural Resources to the Province of Saskatchewan*[66] established that the period over which an accounting must take place is from the date provincial status is granted until resource ownership is fully transferred from the federal government to the provinces. It should be noted, however, that Alberta and Saskatchewan were created in 1905 as full provinces but the federal government retained natural resources on provincial lands to serve as security for railway construction.[67] In the case of Nunavut, the progression towards provincial status is designed to be incremental and leading to full provincial autonomy after fifteen or twenty years. It could be argued that if full Nunavut provincehood is projected, an accounting would be required from the time that the incremental progression begins. To the extent that certain offshore areas could be claimed as within Nunavut, 'federal' revenue generated therefrom would be held in trust for the future provincial government from the time the territory of Nunavut is created.

Given this possibility, it is unlikely that the federal government will endorse or agree to any provincial status for the NWT until, if ever, it is prepared to fully transfer all traditional provincial jurisdiction that such recognition entails.

Notes

1. The principle governing statute for the NWT is the Northwest Territories Act R.S.C. 1970, c. N-22.
2. In several respects – population base, population density and local revenue generation capabilities – Yukon Territory is a better position to argue for provincial status than is the Northwest Territories. As well, the presence of a Member of Parliament from Yukon in a senior position within the Progressive Conservative party should bode well for the Territory's future. Yet Yukon still relies on the federal government for a substantial portion of its operating budget ($38 million of $122 million in 1979–80). See Proceedings, *Eighth National Northern Development Conference 1979* (Edmonton, 1979), p. 30. For the position of the Yukon Legislative Assembly on territorial constitutional

development see "Second Report of the Standing Committee on Constitutional Development in Yukon", in Robert F. Keith and Janet B. Wright (eds), *Northern Transitions*, V.2 (1978), pp. 269–74. An analysis of the Yukon constitutional development issue is given in Glen B. Toner, *Constitutional Development in the Yukon* (M.A. Thesis, University of Alberta) (1978).

3. Dene Nation and the Metis Association of the NWT, *Public Government for the People of the North* (1981).
4. S.C. 1869, c. III. The Arctic Islands north of the mainland were admitted into the Confederation by an Order in Council (U.K.) of 31 July 1880, reproduced in R.S.C. 1970 (Appendices) No. 14.
5. The province of Manitoba was created through S.C. 1870, c. III, and later expanded by S.C. 1877, c. 6 and S.C. 1881, c. 14. The District of Keewatin was made independent of the NWT by S.C. 1876, c. 21. Other contractions of NWT included the creation of a separate Yukon Territory, S.C. 1898, c. 6, and the provinces of Alberta, S.C. 1905, c. 5, and Saskatchewan, S.C. 1905, c. 42.
6. For Garry [Manitoba], 1870–1875; Fort Livingston [Saskatchewan], 1875–1877; Battleford [Saskatchewan], 1877–1882; Regina [Saskatchewan], 1882–1905.
7. S.C. 1888, c. 10.
8. S.C. 1950–51, c. 21.
9. S.C. 1966–67, c. 22.
10. R.S.C. 1970, c. N-22.
11. 30–31 Vict., c. 3 (U.K.), as amended.
12. *Supra* note 1, s. 16(2).
13. Of the more than 1.3 million square miles in the NWT, the Crown in the Right of Canada holds all but twelve square miles. Nunavut Constitutional Forum, *Nunavut and Land Claims: Options for a Public Land Regime*. Working Paper No. 2 (not dated), p. 2.
14. Debates of the House of Commons, 1st Sess., 27th Parliament (27 September 1962), p. 7.
15. Bill C-83 and Bill C-84, 8 July 1963.
16. See Minutes of Proceedings and Evidence of the Standing Committee on Mines, Forests and Waters, 1st Sess., 26th Parliament, Issue 3 (11 December 1963).
17. A. W. R. Carrothers, Jean Beetz and J. M. Parker, *Report of the Advisory Commission on the Development of Government in the Northwest Territories*, 3 vols (1966). [Hereinafter referred to as The Carrothers Commission].
18. *Ibid.*, V. 1, pp. 147–148.
19. *Ibid.*, p. 147.
20. See, for example, "$200 million cost seen to divide NWT in two", (Toronto) *Globe and Mail* (28 November 1983), p. 10.
21. The initial Nunavut proposal was to include all lands traditionally occupied by Inuit and by Inuvialiut. Such lands included:

   "that part of the existing Yukon Territory north of the height of land between the Arctic Ocean and the Yukon River basin [...and] that part [of the NWT] generally north of the treeline [...]', [including all harbours, bays estuaries and other similar inland waters].

   See, Inuit Tapirisat of Canada, *Nunavut, a Proposal for the Settlement of Inuit Lands in the Northwest Territories* (1976), p. 14.

   Although the proposed geographic area has remained constant, the political claims and players have varied. In February 1982 the Constitutional Alliance of the Northwest Territories, consisting of native and non-native members of the Legislative Assembly and representatives of native organizations in the NWT, was formed to foster territorial division through two subcommittees. The Nunavut Constitutional Forum and the Western Constitutional Forum are conducting research and holding public meetings to facilitate the eventual division of the NWT into an Eastern and a Western territory. *NWT Data Book 84/85* (1984), p. 50. For the most recent Nunavut proposal, see Nunavut Constitutional Forum, *Building Nunavut: a Discussion Paper Containing Proposals for an Arctic Constitution* (1983).

22. *Supra* note 3.
23. C. M. Drury, *Constitutional Development in the Northwest Territories* (1979). [Hereinafter referred to as the Drury Report].
24. *Ibid.*, p. 13.
25. *Ibid.*, pp. 42–68.
26. *Report of the Special Committee on Unity to the 3rd Session of the 9th Assembly at Frobisher Bay, 22 October 1980*. NWT Debates, 3rd Sess., 9th Assembly, Appendix B.
27. *Ibid.*, p. 2.
28. This question was prefaced by the preamble:

    In response to a proposal to create a new territory in the eastern part of the Northwest Territories, the Legislative Assembly agreed to hold a plebiscite.

    If a majority of the voters agree that the Northwest Territories should be divided, the Legislative Assembly will request the Government of Canada to divide the Northwest Territories and create a new territory in the eastern part of the Northwest Territories.

    If the Government of Canada agrees to divide the Northwest Territories, the Legislative Assembly will also request that a federal boundaries commission be appointed to consult with the people of the Northwest Territories and to recommend the exact boundaries of the new territory.

    Three years ordinary residence in the NWT was required to be eligible to vote in the plebiscite. This requirement was challenged on the basis that it infringed upon the mobility rights provided for in the *Canadian Charter of Rights and Freedoms*, but upheld in *Re Allman et al. and Commissioner of the Northwest Territories* (1983), 144 D.L.R. (3d) 467 (NWT S.C.) *affirmed* [1984] N.W.T.R. 65 (NWT C.A.); Issue to appeal denied, S.C.C. (17 May 1984).
29. NWT Bureau of Statistics, *Plebicite on the Division of the Northwest Territories – Voter Choice, Participation and Eligibility* (1982).
30. Sheilagh M. Dunn, *The Year in Review 1982: Intergovernmental Relations in Canada* (1982), p. 57.
31. The concept of aboriginal title in Canada has received little judicial consideration. Important cases dealing with the question have served more to indicate what it is not, rather than to supply a positive description. See *Calder* v. *A.G. of B.C.*, [1973] S.C.R. 313, and *Baker Lake* v. *Minister of Indian Affairs and Northern Development* (1978), 87 D.L.R. (3d) 342 (F.C.T.D.).
32. This point is emphasized in *Building Nunavut*, *supra* note 21, p. 1.
33. The language of all Inuit is Inuktitut. In the District of Mackenzie, other languages all quite distinct from Inuktitut are spoken.
34. Greater ability to incorporate native values and perceptions into criminal law is desired. To some extent this presently occurs in the sentencing process. Input into criminal legislation is the proposed objective.
35. See Nunavut Constitutional Forum, *Nunavut: the Division of Power*. Working Paper No. 1 (not dated), pp. 115–116.
36. Drury, *supra* note 23, p. 111. Provincial per capita expenditures for the 1978–79 fiscal year ranged from $1764 in Ontario to $2548 in Alberta. Yukon Territory and NWT per capita expenditures for the same period were $4742 and $5933 respectively.

37. *Ibid.*, p. 110. Of the $291 million NWT budget for the fiscal year 1979–80, the federal government contributed $202.7 million.
38. *Constitution Act, 1982*, being Schedule B to the *Canada Act 1982*, (U.K.) 1982, c. 11, at s. 36.
39. J. W. Pickersgill, "Responsible Government in a Federal State", *Canadian Public Administration* V. 15. No. 4 (1972), p. 520.
40. In the 1970s [and today] a "small" province was considered to be one having a population of less than 1.5 million people. See P.J.T. O'Hearn, "From Representation by Population to the Pursuit of Elegance", *Dalhousie L. J.* V. 7 No. 3 (1983), p. 779. See also Robert M. Bone, "Population Changes in the Northwest Territories and Some Geopolitical Consequences", *Journal of Canadian Studies* V. 16 No. 2 (1981), p. 81.
41. Michael S. Whittington, "Canada's North in the Eighties", in Michael S. Whittington and Glen Williams (ed), *Canadian Politics in the 1980s* (1981), p. 64. In contrast, consider the following statement concerning an objective of Confederation:

    That economic disparities between the different provinces and regions of Canada should be eliminated, or at least reduced so that a decent basic minimum of income level, social services and welfare is available in all provinces and regions. In W. R. Lederman, *Report of the Proceedings of the Twenty-First Tax Conference* (1964), p. 4.
42. The Inuit Tapirisat of Canada (ITC) and the Committee for Original Peoples' Entitlement (COPE) are two examples. Other Northern organizations include the Baffin Regional Inuit Association and the Inuit Committee on National Issues, both of which work closely with the ITC, the Dene Nation, the Metis Association of the NWT and the Council of Yukon Indians.
43. See for example, Proceedings of the Special Committee of the Senate on the Northern Pipeline, 1st Sess., 32nd Parliament, Issue 26 (18 May 1982) (Presentation of the Baffin Regional Inuit Association) and Minutes of Proceedings and Evidence of the Special Joint Committees of the Senate and of the House of Commons on the Constitution of Canada, 1st Sess., 32nd Parliament, Issue 16 (1 December 1980) (Presentation of the Inuit Committee on National Issues).
44. On 25 February 1985, Dennis Patterson resigned as chairman of the Nunavut Constitutional Forum because colleagues did not agree with a boundary agreement which would have placed Beaufort Sea Inuit communities in the Western territory. (Toronto) *Globe and Mail* (26 February 1985), p. 4.
45. The *Constitution Act, 1871*, 34–35 Vict., c. 28 (U.K.) describes the process of provincial establishment (s. 2), provincial boundary amendments (s. 3), and of enacting legislation for the 'peace, order and good government' of the territories (s. 4). Paragraph 42(1)(f) of the *Constitution Act, 1982* may repeal the 1871 legislation, but the latter nonetheless reflects Canadian custom.
46. See Kenneth Lysyk, "Approaches to Settlement of Indian Title Claims: the Alaskan Model", *U.B.C. Law Rev.* V. 8 No. 2 (1973), p. 321, and Isi Foighel, "Home Rule in Greenland", *Meddelser om Grønland, Man and Society* No. 1 (1980), p. 3. The Special Committee of the House of Commons on Indian Self-Government concluded that: "No international models [for self government] were found that would be readily transferable to Canada." However in a footnote to that comment the Committee stated that: "Greenland does, however, offer some possibilities as a model for the Eastern Arctic." See Minutes of Proceedings of the Special Committee on Indian Self-Government, 1st Sess., 32nd Parliament, Issue No. 40 (12 and 20 October 1983), p. 42.

47. Statement by the Honourable Arthur Laing, then Minister of Indian Affairs and Northern Development, at Yellowknife, NWT 18 January 1967, quoted in A. R. Thompson, "Ownership of Natural Resources in the Northwest Territories", *Alta. L. Rev.* V. 5 (1967), p. 304.
48. R.S.C. 1970, c. N-6, as amended.
49. S.C. 1980–81–82–83, c. 81.
50. "Every citizen of Canada has a claim to participate in the institutions of responsible government under the Canadian constitution; [...]", in The Carrothers Commission, *supra* note 17, V 1, p. 128.
51. Section 42(1) of the Constitution Act, 1982 requires provincial participation in the establishment of new provinces or the alteration of boundaries of existing ones. Any restructuring of the territories themselves does not require provincial participation or even that of the GNWT.
52. Under Canada's constitutional structure, Parliament and the provincial legislatures are each vested with areas of legislative authority. Section 91 of the *Constitution Act* 1867 describes federal powers and s. 92 enumerates the provincial authority. The doctrine of delegation prohibits the transfer of such legislative powers from one level of government to the other. See *A.G. for N.S.* v. *A.G. for Canada*, [1951] S.C.R. 31. Delegation of administrative authority is possible however; see *P.E.I. Potato Marketing Board* v. *H. B. Willis Inc.*, [1952] 4 D.L.R. 146 (S.C.C.). *Ibid.*
53. See Peter Jull and Nigel Bankes, "Inuit Interest in the Arctic Offshore". *National and Regional Interests in the North.* Canadian Arctic Resources Committee, (Proceedings of the Third National Workshop on People, Resources and the Environment North of 60°) (1984), pp. 557–86.
54. *Ibid.*, pp. 557–86.
55. The Nova Scotia agreement has been implemented by both federal and provincial legislation. See Canadian Nova Scotia Oil and Gas Agreement Act, S.C. 1983–84, c. 29 and Canada – Nova Scotia Oil and Gas Agreement (Nova Scotia) Act, S.N.S. 1984, c.2. For the Newfoundland agreement, see the Atlantic Accord, memorandum of agreement between the Government of Newfoundland and Labrador on offshore oil and gas resource management and revenue sharing (11 February 1985).
56. S.C. 1980–81–82–83, c. 81.
57. *Re Offshore Mineral Rights of British Columbia*, [1967] S.C.R. 792; 65 D.L.R. (2d) 353, *Reference re Property in and Legislative Jurisdiction over the Seabed and Subsoil of the Continental Shelf Offshore Newfoundland* [1984] 1.S.C.R. 86, 5 D.L.R. (4th) 385. But *contra*, see Reference re Ownership of the Bed of the Strait of Georgia and Related Areas, [1984] I.S.C.R. 338, 8 D.L.R. (4th) 161. *R.* v. *Tootaiik E4-321* (1970), 71 W.W.R. 435; 9 C.R. N.S. 92 (NWTS.C.), and *BP Exploration Company* (*Lybia*) *Limited* v. *Hunt* (1980), 23 A.R. 271: 16 C.P.C. 168 (NWT S.C.).
58. Jull and Bankes, *supra* note 53, describe proposed Inuit participation in a Wildlife Management Board and other planning and review authorities dealing with land and water use.
59. Whittington, *supra* note 41, p. 52 and *R.* v. *Tootalik E4-321*, supra note 57.
60. Nunavut Constitutional Forum, *supra* note 35.
61. Perhaps because of its novelty, this proposal has not been discussed by those outside the Nunavut negotiations. See James A. Dobbin, Michele H. Lemay and Nancy E. Dobbin, "Lancaster Sound Regional Study: Coastal and Marine Resource Planning in the Canadian High Arctic", *Coastal Zone Management Journal* V. 11 No. 2 (1983), p. 71.

62. Canadian Arctic Resources Committee, *Northern Decisions*, V.2 No. 20 (28 February 1985) 136–7.
63. *Supra* note 21.
64. Two cases from the NWT Supreme Court suggest this: *R.* v. *Tootalik E4-321* and *B.P. Exploration Co.* (*Lybia*) *Ltd.* v. *Hunt*, *supra* note 57. In the *Tootalik* case the jurisdiction of the Court to try the defendant for unlawfully killing a bear was disputed as the alleged offence took place on ice covering a bay. In interpreting the NWT Act, the Court held that the boundaries of the NWT did not stop at land's end but extended over Northern waters. In *B.P. Exploration*, the same conclusion was reached after comparing the relevant provisions of the NWT Act and the *Yukon Act*, R.S.C. 1970, c. Y-2. In a Schedule to the *Yukon Act* the Northern boundary of that Territory is described to be "that part of the Arctic Ocean called the Beaufort Sea". However, Yukon territory includes all islands within twenty statute miles from the shores of the Beaufort Sea. The statutory geographic decription of the NWT is in part as follows: "all that part of Canada north of the Sixtieth Parallel of North Latitude, except the portions thereof that are within the Yukon Territory, the Province of Quebec or the Province of Newfoundland".
65. See Thompson, *supra* note 47.
66. [1932] A.C. 28.
67. The three Canadian Prairie provinces, Alberta, Saskatchewan and Manitoba, each entered Confederation without jurisdiction over their natural resources; such jurisdiction was reserved by the federal government in order to finance railway construction in those provinces. In 1930 the federal government decided that resource jurisdiction would be transferred to the respective provinces, and that the federal government should produce an accounting for all resource revenues it received during its "trusteeship" of the resources from the time the provinces were established (Alberta and Saskatchewan – 1905, Manitoba – 1870) until the 1930 transfer. The province of Saskatchewan argued, however, that the accounting period should commence not at 1905 but rather at 1870 when the era of the present province became, then as part of Rupert's Land, under the control of Parliament (S.C. 1869, c. III). The Saskatchewan claim was based on the thesis that the resources within the provincial territory should have been administered for the benefit of the people in the area at all times, regardless of the attainment of provincehood. The Privy Council ruled against Saskatchewan's submissions. Note, however, that United Nations Resolution 1803 (XVII) of 14 December 1962 declared that permanent sovereignty to natural resources is to be an interest held by "peoples and nations', suggesting sovereignty to be independent of political or constitutional status.

Part II: Paradigms and prospects

8

The designing of a transit management system

DOUGLAS M. JOHNSTON

I The Concept

A Function

The Canadian Northern Waters Project of Dalhousie Ocean Studies Programme is based on a fundamental assumption: that all major industrial and governmental activities involving the use or transit of Canada's Arctic waters should be brought under a single, comprehensive system of regulation and management. The envisaged system would not only constitute an entire transportation system,[1] but would also be designed to affect the regulation and management of related matters, such as Arctic science, resource development, environmental protection, community concerns, industrial strategy, and international affairs. Since the primary purpose of such a system would be to maintain safe, efficient, and harmless passage of cargoes or products through, over or under these waters, it is referred to as a "transit-management system". The object of this preliminary collection of essays has been to review some of the problems associated with designing a Northern waters transit-management system and to consider some of the basic or strategic options for decision-makers in the 1980s.

The concept of a Northern waters transit-management system arises, of course, from the growing temptation to open up the Northwest Passage for the surface and-or subsurface transportation of a variety of cargoes. For over a decade, shipments of ores have been transported by surface craft out of the Eastern Arctic to overseas markets, and it seems certain that the extraction and transportation of hard minerals from the Arctic region will continue on a larger scale in the years ahead. In recent years, however, public attention has been focused more sharply on the prospective extraction and transportation of Arctic oil and gas. Canadians look to the development of petroleum on the islands and in the offshore areas of the Eastern and High Arctic; Americans contemplate the shipment eastwards of onshore oil from the North Slope of Alaska; and both Canadians and Americans anticipate the shipment eastwards of offshore oil from the Beaufort Sea. Others, both in North America and beyond, foresee the emergence of a safe, efficient and harmless transportation system for general cargoes proceeding in both directions through the Northwest Passage.

Much of this "conventional" forecasting assumes the primacy of surface craft of one kind or another within a Northern-waters transit-management system. But even a non-prophetic extrapolation of current trends suggests that an entire Arctic transportation system *for all possible cargoes in the early twenty-first century* will consist of several modes of transit: surface water, subsurface water, overland, and airspace. Eventually, the surface water vessel may be a minor, rather than major, component of an Arctic transit system.[2]

Moreover, the effective management of sophisticated modes of transit through the Arctic will require an array of supportive technologies of the sort reviewed by Professor Frankel.[3] What must be envisaged is an unprecedented network of technologies and administrative agencies, requiring a high degree of co-operation among governments, industrial users and the communities affected by expanded uses of Northern waters.

No present system offers an adequate model. The Soviet Union has, of course, preceded Canada in managing transit through its Northern waters for a mixture of industrial, commercial, military, and other purposes. But, although instructive, the Soviet experience is unlikely to afford an appropriate model for Canadian policy making.[4] The Canadian problem may be analogized with other megaprojects of comparable magnitude elsewhere,[5] but the Arctic setting poses unique challenges to science and technology, and arguably also to the human spirit.

After taking account of ideas and experience evolving elsewhere, Canada will have to accept responsibility for designing its own transit-management system for Northern waters. Such an ambitious and sophisticated undertaking will have to be a truly Canadian endeavour, calling for an impressive deployment of national will, imagination, and talent.

B Scope

Given the functional parameters of a transit-management system, as envisaged here, one must assume that the system would be designed to have operational effect not only in the Northwest Passage (however defined),[6] but also, in some degree, in the "approaches" on each side of the Passage. Accordingly, the system would extend, in part, to the western approaches at least to the limit of Canadian jurisdiction in the Beaufort Sea, and to the eastern approaches at least to the limit of Canadian jurisdiction in Davis Strait.

If the system is conceived entirely as a Canadian operation, then the fixing of its spatial limits on the western side would be contingent on a resolution of the current boundary dispute between Canada and the United States in the Beaufort Sea.[7] On the eastern side, boundary issues in the Davis Strait area have been largely resolved through successful negotiations between Canada and Denmark-Greenland.[8] The southern limit of the system in Davis Strait might be fixed, somewhat arbitrarily, at 60° North latitude,[9] but because of community and environmental concerns in the Labrador Sea region, a more appropriate boundary might be set at 52°N.

Ideally a system, though managed by Canada and operated mostly within Canadian waters, would be approved and accepted by all users, and especially by Canada's Arctic neighbours to the immediate west and east of the Northwest Passage. Accordingly, one would prefer to design the transit-management system in such a way as to be operationally compatible with appropriate regional ocean development and management arrangements between Canada and the United States in the Beaufort Sea[10] and between Canada and Denmark-Greenland in Davis Strait.[11]

II Alternative approaches, frameworks and theories

A Introduction

The task of designing a Northern waters transit-management system will require an extremely sophisticated and unprecedented co-operative effort by the Canadian research community. Obviously, no one traditional discipline has more than a modest contribution to make to such an effort. Nor is a conventional multidisciplinary approach, employing a number of traditional unidisciplinary techniques, likely to be adequate to the task. The decisions that must be made will require *integrative thinking* based on *integrative research*.

The first problem, then, is to select or to design an appropriate approach to this integrative research task. Ideally, one would wish to discover a single, but sufficiently encompassing, framework of analysis possessing a solid theoretical base and proven methodologies. Such a framework would raise fundamental hypotheses and crucial questions and point out the most relevant areas of investigation, so as to permit rational decision making in light of the best available knowledge and ideas. But contemporary scholarship reveals a bewildering diversity of approaches to integrative thinking.

This section identifies some of the more salient modes of integrative research available as models for the design of a transit-management system. Some of these models might be regarded as a potential theoretical or conceptual framework, others merely as a set of techniques or modes of analysis.

B Systems theory

It seems natural, indeed inevitable, to refer to the required transit-management network as a "system". What is envisaged is a complicated and interlocked set of institutional arrangements among government, industry and the affected communities, with a view to harnessing appropriate technologies for the common purpose of transit management in Canadian Northern waters. One tries to imagine a pulling together of selected "components" from the political, social, economic, legal, and administrative "systems" of Canada. In so doing one seeks to develop an appropriate "technological system", one which will have minimal adverse effect on the natural "ecosystem" of the Arctic environment and on the human environment of Canadian Arctic communities. The concept of a system is all-pervasive, and one reaches out, almost instinctively today, for *systems theory* and the techniques of *systems analysis*.[12]

In designing a system one considers the best mix of characteristics in terms of various (and sometimes conflicting) concepts, such as stability, flexibility, authority, control, limits, and equilibrium. Theorists are concerned with the relationship between the "structure" and "functioning" (and sometimes the "state") of a system. Accordingly, systems theory may be defined as "an intellectual tool for studying the relation between the structure of a system and its functioning"[13] and system design (or "systems synthesis") is "an instrument of social planning, which answers the question, 'if we want a system that functions in a given way, what are the possible structures that would result in this manner of functioning?'".[14]

Both systems analysis and its cognate, *operations research*, developed during the Second World War, have offered important practical applications in engineering and military planning respectively. Both methods of analysis are frequently used today in government and industry to help decision-makers analyse the potential effectiveness of proposed policies or programmes.[15] But because of an intensely rational and linear approach, these methods tend to be inadequate in contexts (such as Arctic transit management) where the complexity and irrationality of social systems are an important aspect of the problem being studied.[16] This inadequacy of systems analysis may be likened to that of *cost-benefit analysis*, which seems too sharply focused on a single value (economic efficiency) and on a particular form of rationality to be the sole method of evaluation of a social problem.[17]

In summary, systems theory would make a valuable conceptual contribution to the design of a transit-management system, but the rigorous (often mathematically precise) techniques of systems analysis developed in recent years,[18] are not sufficient as a research strategy.

C *Planning theory*

Although strongly influenced by the concepts of systems theory, planners have been required by the social nature of their responsibilities to develop their own frameworks and techniques. The range of considerations injected into most large-scale planning problems tends to force the planner to engage in synthesis rather than analysis.

1 *Three models of planning*

Although some theorists have challenged the assumption,[19] most planners believe that planning is a rational undertaking. Rational planning, however, can be undertaken in at least three different ways.

(*a*) *Comprehensive planning*: This model assumes that the goal of planning is to bring together all relevant areas of knowledge, as systematically and comprehensively as possible, so as to ensure the most rational plan. Particularly in the case of multiple-objective planning at the national or regional level, the rationalist planner tends to set an ambitious range of tasks before purporting to present a plan to decision-makers. For example, a planner following the comprehensive model might undertake seven tasks in the following sequence:

(i) to identify the whole complex of problems, and the interrelationship of these problems;
(ii) to determine the goals and objectives for each of these problems;
(iii) to appraise existing policies;
(iv) to formulate alternatives for the future;
(v) to evaluate alternatives;
(vi) to select one alternative; and
(vii) to develop this alternative in the form of a plan.[20]

Sometimes the comprehensive planning model leads the planner into "meta-planning", which involves the devising of major reforms in existing institutions.[21] This approach is more likely to succeed under conditions of authoritarianism than under the more constraining conditions of an open society.[22] Even if meta-planning were necessary for the designing of a transit-management system in Canada, that approach would almost certainly result in the delay of any final decision. Yet it may be argued that it is better to have a delayed but properly planned outcome than one which is prompt but improperly planned. In any event, the comprehensive planning model involves the planner, more or less consciously, in questions of *values*. It has been said that human growth is the aim of planning,[23] but a much subtler balancing of ideals would be necessary under a comprehensive planning approach to transit management in the Canadian Arctic.[24]

(*b*) *Disjointed incrementalism*: As a counter-argument against the trend to comprehensive planning, it has been contended by Charles E. Lindblom and associates[25] that planning strategies must be adapted to the limited cognitive capacities of decision-makers. A much more selective approach to planning, usually referred to as "disjointed incrementalism", is advocated, so as to make the scope and cost of information collection and processing manageable and affordable. The advantages of this model have been summarized as follows:

(i) rather than attempting a comprehensive survey and evaluation of all alternatives, the decision-maker focuses only on those policies which differ incrementally from existing policies;
(ii) only a small number of policy alternatives are considered;
(iii) for each policy alternative, only a restricted number of 'important' consequences are evaluated;
(iv) the problem confronting the decision-maker is continually redefined. Incrementalism allows for countless ends-means and means-ends adjustments which, in effect, make the problem more manageable;
(v) thus, there is no one decision or "right" solution, but a "never-ending series of attacks" on issues at hand through serial analysis and evaluation;
(vi) as such, incremental decision-making or planning is described as remedially geared, as it is more to the alleviation of present concrete social imperfections than to the promotion of future goals.[26]

The incremental approach is held out as "a more realistic description of what actually goes on when a decision-maker is faced with a problem he must solve more or less quickly; he moves incrementally, making adjustments to existing policies at the margin".[27]

In the context of Canadian Northern waters transit management, it is accepted that decisions will have to be made in the absence of complete information. There are obviously some advantages for an analyst to work within a framework recognizing the political, administrative and corporate limitations of real-world decision making. It is unfair to accuse incremental planning of creating chaos or arbitrary decisions. As a matter of necessity, rather than choice, a certain degree of coordination follows from incremental decision-making, even if each incremental decision-maker pursues policies from a personal, and often narrow, perspective.[28]

(*c*) *Mixed scanning*: Planners today commonly reject the incrementalist model in its extreme form. The truth is that government, and often industry, is required to make fundamental decisions, in the form of long-range, multiple-objective plans for large numbers of people affected by complex institutions. Accordingly, a third intermediate model known as the "mixed scanning" approach has been introduced.[29] One of the purposes of this model is to distinguish between fundamental and incremental decisions, and to apply modes of analysis appropriate to each.[30]

Mixed scanning seeks a balanced strategy between all-encompassing and sectoral planning. Incrementalism may offer the decision-maker insufficient information, while comprehensive planning may offer an excessive database. Design of a transit-management system must go beyond the actual limits of any particular decision, not least in order to provide an adequate basis for evaluating the actual limits. The researcher or planner must look to the subsequent problems of policy implementation as much as to the earlier problems of policy making, to the processes engaged as well as the product, to questions of procedure as well as substance.[31]

2 *Four sectors of planning*

The field of planning is so large that different areas of specialized planning can be distinguished, each with its own emphasis and its own set of characteristics. For the purposes of transit management, four sectors of planning are potentially relevant.

(*a*) *Technological planning*: The designing of a transit-management system for the Canadian North is, in large part, a matter of technological planning: the assessment of present and projected technological capabilities and the forecasting of technological impacts. The concept of "technology" has, of course, become emotive for those most deeply concerned with its impact on individuals and communities,[32] but since the 1960s a growing number of technical experts have attempted to view *technological forecasting and assessment* more neutrally as a tool of professional planning.[33] One such writer has attempted to develop the concept of a "socio-technical system" and, in the effort to design an appropriate framework, has called for the integration of system, organization and management theory.[34] A transit-management system for Canada's Northern waters would be nothing if not a socio-technical system.

Another writer on technological forecasting has distinguished "intuitive" techniques (for example, individual opinion, polls, the Delphi technique, scenario-writing and cross-impact analysis) from "analytical" techniques. The latter, mostly of a mathematical variety, are divided into three modes: trend extrapolation (for example, straight-line and fitted-curve analysis), network construction (for example, matrices, contextual mapping, morphological analysis, functional array, graphic models, relevance trees, decision trees, mission networks and functional analysis, systems analysis and operations research) and modelling.[35]

An approach, more amenable to non-mathematicians, would be to integrate two or more fields which have something to contribute to governmental and industrial strategy for *technological innovation*, such as the fields of science policy and economic policy.[36] In the context of Canadian Arctic waters transit management, the first step in technological planning might be to establish a *technology-assessment system* comprising the appropriate sectors of government and industry, and possibly the academic community, as was done in the early years of planning offshore development on the East Coast of Canada.[37]

(*b*) *Transport planning*: Designing a transit-management system for Canadian Northern waters is, in significant measure, a transport-planning problem. Narrowing the planning focus to a manageable context, such as transport policy, allows a reasonable degree of specificity without surrendering the advantages of "integrative thinking" addressed to multiple objectives. Normally the level of analysis in this sector of planning is national. At the national level every transport-policy decision is linked with policy choices in other, non-transport sectors. National transport plans are often required to serve a variety of purposes, not least as "a rally point for the co-operation and support of the electorate and a generator of common hopes and expectations".[38] A national (or macro-regional) transport-planning approach to the problems of transit management in Canadian Arctic waters would be invaluable by emphasizing the symbolic as well as the substantive importance of national planning and by accentuating the overall economic significance of a transit-management system as an integral component of macro-regional economic planning for Canada's North.[39]

In the context of transport policy, one can be quite specific about the diversity of interests affected by the project under consideration, and thus come closer to suggesting the kinds of trade-offs or compromises that decision-makers must address. For example, in the context of Arctic Ocean transit management, one can explicitly distinguish *user interests*, such as costs, time, accessibility, convenience-comfort, and safety, from *community interests*, such as pollution, noise, preservation of wilderness areas or areas of special significance, community welfare, income distribution, resource utilization, and fiscal efficiency.[40]

(*c*) *Developmental planning*: In the case of an industrially underdeveloped or developing region, such as the Canadian North, it is normal to assume that any proposed project of considerable magnitude must be placed within an overall framework of national developmental planning. In the Canadian Arctic region, anything as potentially large and pervasive as an entire transit-management system will be required to pass the "developmental test" of the day, if it is to have any chance of receiving political approval. If the matter were solely one for technical consideration, it might be dealt with primarily as a problem of *public investment decision making* within such a developmental framework.[41]

But the Canadian Arctic is a special case: a highly "underdeveloped" region within a highly developed nation; a region characterized by small, isolated communities, whose lifestyle seems unusually vulnerable to insensitive technologies imported from outside; and a region consisting of a unique and susceptible natural environment. A developmental planning approach must be "sensitized" in two important respects: first, in a way that avoids uncritical assumptions inherent in developmental planning for the totally different conditions of the industrially advanced southern regions of Canada; second, in a way that takes full account of the probable impact of projects like a transit-management system on the human and natural environment of the region. An appropriate Northern model of *eco-development* should be followed in developmental planning.[42]

It must be acknowledged, however, that any developmental planning approach to transit management in the Canadian North would be naive and self-defeating unless it were firmly based on the realities of the market-place. Hard-headed appraisal requires that characteristics of such a system, for example, scale, complexity, time-frame, regulatory stringency, and formality, must conform with the demand for resource and transportation opportunities of the kinds anticipated within such a system.[43]

Subject to the conditions referred to above, a developmental planning approach should also be grounded on considerations of *industrial strategy* with a view to the nation's actual and potential place in the international economy.

(*d*) *Regional planning*: Regional planning, also of potential relevance to the design of a transit-management system for Canada's North, owes much to systems analysis.[44] But regional planning, like operational research planning, is more complex than systems analysis. Regional planning almost always consists of general, multiple-objective planning, often with a physical planning component.[45] As a result of increasing informational complexity, regional planning is forced to take account of a new grouping of disciplines (especially, economics, geography, and sociology) under the heading of *regional science*.[46]

As applied to the problems of transit management in Canadian Northern waters, the regional-planning and regional-science approaches would be useful in accentuating the regional character of the "environment" to be planned, not only as a region of Canada but also as an international region. Even if the system is to be designed as a Canadian management system, it would be unwise to ignore the fact that the Northwest Passage and adjacent Canadian waters "belong" also to a larger region, in an ethnic as well as a strictly physical sense, if not yet in the full political sense.[47] The chief merit of this approach would be to emphasize the need for linkages between a Northwest Passage transit system on one hand, and regional management arrangements in the Beaufort Sea to the west and the Davis Strait region to the east on the other.

Even within a single, sufficiently large framework for regional planning it would be possible to approach the problems of transit management at several different levels. In the last decade the Canadian government has already shown its interest in developing *land-use planning* for the North at the macro-regional level, where the Northwest Passage would be a special-use zone within a much larger regional setting.[48] The federal government has also published a green paper, which focuses on the problems of regional planning at the micro-regional level within the Lancaster Sound sector of the Northwest Passage.[49] This latter study explicitly includes the development of a year-round shipping route through the Northwest Passage as one of six suggested "resource-use options".[50] Within the last 12 years Canadian planners have begun the task of trying to reconcile land-use trends with transportation requirements within a Canadian setting.[51]

While citizens usually react critically, specialists still tend to advocate the comprehensive planning approach to regional resource-use problems. A recent study by US planners concluded that, despite legitimate criticism, "the comprehensive land-use plan is the best instrument to use, if it is objectively developed on the basis of accurate and complete data".[52] The authors qualified their conclusions with these wise words:

> Like any other tool, it still must be used with caution. It is an error to accept a plan with a completion date that does not allow time for modifications. A weakness of some plans is that they are too complicated for local people to implement. In other cases, they require a much higher level of funding or technology than the local area can afford. Still another problem with a comprehensive land-use plan approach is the possibility of creating a situation where the planning agency is without the power to execute the plan. In this case, the development of a planning programme is an exercise in futility. Finally, the land-use plan should have self-correcting steps or it can become a self-fulfilling prophecy.[53]

D *Organizational theory*

This field of "integrative thinking" has evolved in the twentieth century as a result of two kinds of nineteenth-century theories: the theory of scientific management and the theory of bureaucracy. The former originated at least as early as 1832 with the seminal writing of Charles Babbage, but it was not developed as a theoretical framework until the end of the century with the work of Frederick W. Taylor.[54] The classical theory of bureaucracy evolved from Hegel and Marx to Weber.[55] To the extent that both kinds of theorizing focus on "complex organizations", they can be said to constitute the field of organizational theory, and thus have something to contribute to the designing of a governmental-industrial transit-management system.

Organizational theory rests upon the theory of action. Three predominant theories of action can be distinguished. Under the first, action is seen as "purposive, boundary or intendedly rational, and prospective or goal-directed".[56] The second envisages action as "externally constrained or situationally determined".[57] The third portrays action as "somewhat more random and dependent on an emergent, unfolding process".[58] Under the second and third theories rationality is constructed after the fact to make sense of behavioural trends after they have been evidenced.[59] Although either of these two theories of action may appeal to our sense of reality, their "passive" orientation tends to limit them to explanatory purposes. It is the first theory, based on the notion of rational choice —

which may or may not be illusory – which has dominated organizational theory, and it is chiefly in that branch of theory that we must hope to find assistance in the design and planning of a transit-management system. Yet, to some extent, the requirements of the external situation can and must be incorporated into the design of such a system, and the structure of the system should perhaps be left sufficiently flexible to permit adaptation to chance as well as directed events.

Since a transit-management system for the Northwest Passage will primarily be the responsibility of government rather than industry, one may have to turn to *public administration theory* for a proper balancing of the three theories of action. The original "classic bureaucratic" model of public administration, based on the values of efficiency, economy, and effectiveness,[60] has been succeeded by other models: for example, the "neo-bureaucratic", the "institutional", the "human-relations", and the "public choice"[61] models. In recent years some theorists have begun to advocate a "new public administration", which would "foster humanistic, decentralized, democratic organizations that distribute public service equitably"[62] and produce organizations that further cognate values such as responsiveness, worker and citizen participation in decision making, social equity, citizen choice, and administrative responsibility for programme effectiveness.[63] Within the parameters of this new theory of public administration – based on a combination of all three theories of action, but with special emphasis on the second ("situational") theory – there emerges an approach to the designing of a Canadian Arctic transit-management system that would take full account of the need to involve Northerners in the planning process. Only with a proper balancing between the first and second theories, and among the various values associated with them, would it be possible to design a transit-management system with an appropriate blend of Northern and Southern characteristics.[64]

E *Impact assessment studies*

A more recent phenomenon in the world of system planning and design is the emergence of impact assessment as a prerequisite to political approval of an industrial project having potential adverse impacts on the economy, the natural environment, or human society. Today, economic, environmental, or social-impact assessment studies are frequently required by law, but the procedures involved may vary enormously.[65] Whatever form impact assessment requirements take, they invariably involve an *interdisciplinary* evaluation of the possible consequences of a proposed action, policy, or project.

1 *Environmental impact assessment*

Apparently the basic concept of environmental "impact" originated in the work, published in 1909, of a Scottish city planner,[66] but both the theory and practice of modern environmental impact assessment are derived chiefly from the growth of environmental consciousness in North America. The National Environmental Policy Act (NEPA) was enacted in 1969 by the US Congress in terms that ensured that environmental impact assessment would become an integral feature of federal plan making.[67] Specifically, section 102(2)(c) of NEPA required every federal agency making recommendations on a planned major initiative to provide a detailed statement on the anticipated environmental impact of the proposed action. Such a statement was termed an environmental impact statement (EIS).[68] Since then various EIS guidelines have been developed by the US Council on Environmental Quality (CEQ), the Environmental Protection Agency and other agencies.[69]

In Canada somewhat different, yet comparable, environmental impact-assessment requirements have been developed, either under the guidelines of the Federal Environmental Assessment Review Office (FEARO) of Environment Canada[70] or, in the case of energy-related projects, under the aegis of the National Energy Board (NEB).[71] Under present Canadian law, any major federal initiative such as a proposal for a transit-management system in the Northwest Passage would be subject to extensive review by various federal agencies.[72]

The scientific validity of these procedures has recently been subjected to critical scrutiny.[73] The authors of a recent study discern a gradual drifting apart of governmental and industrial approaches to science:

> On the one hand are the administrators and their scientific advisors, who are responsible for establishing the terms of reference for particular assessments and judging the adequacy of the resulting studies. In contrast are the project proponents and their environmental consultants, who must translate the terms of reference into a study programme, but are seldom sure of the scientific standards which the reviewers will finally adopt.[74]

Concerned about the lack of a meaningful scientific basis for environmental impact assessment, the authors identify five major constraints:

(i) the need for a common standard;
(ii) the need for early agreement on the basic approach to be adopted;
(iii) the need for continuity of study;
(iv) the need for information transfer; and
(v) the need for better communications.

Special emphasis is placed on the need to "break out of the EIS syndrome", since "the rationale for baseline studies and predictions of impact becomes rather tenuous without some follow-up monitoring" after the project has been initiated.[75] Monitoring, they recommend, should be formally recognized as an integral component of the assessment process.[76] Despite these deficiencies, the authors also recognize that "the decisions resulting from environmental impact assessment may be based as much on subjective judgements involving values, feelings and beliefs, as on the results of scientific studies", and that the ultimate concern in practice can almost always be traced to human values.[77]

Another Canadian scientist, also challenging the assumptions underlying environmental assessment practice, has recently articulated eight "myths" of environmental assessment, namely:

(i) that environmental assessment should consider *all* possible impacts of the proposed development;
(ii) that each new assessment is unique;
(iii) that comprehensive "state of the system" surveys are a necessary step in environmental assessment;
(iv) that detailed descriptive studies of the present condition of system parts can be integrated by systems analysis to provide overall understanding and predictions of systems impacts;
(v) that any good scientific study contributes to better decision making;
(vi) that physical boundaries based on watershed areas or political jurisdictions can provide sensible limits for impact investigations;
(vii) that systems analysis will allow effective selection of the best alternative from several proposed plans and programmes; and
(viii) that ecological evaluation and impact assessment aim to eliminate uncertainty regarding the consequences of proposed developments.[78]

Despite these criticisms of current practices, environmental assessment, if modified according to recommended improvements, may contribute greatly to the design of a transit-management system for the Northwest Passage.

2 *Socio-economic impact assessment*

The emergence of social or socio-economic impact assessment can be traced to the same modern origins as environmental impact assessment. In the United States, sections 102(A) and 107 of NEPA require the use of the social sciences and mandate "the consideration of unquantified values by means of a systematic interdisciplinary approach".[79] The NEPA requirements are extended and made more specific by CEQ guidelines and by the internal guidelines of many US federal agencies.[80] Socio-economic impact-assessment requirements have also been developed in Canada.[81]

Certainly the quality of many socio-economic impact assessments has been questioned.[82] But given the saliency of such predictions within the political system, it seems safe to assume that pressures will force critical refinements to social and socio-economic assessments. For example, the North Sea offshore-inshore planning studies of the early 1970s have been reevaluated.[83]

Given the unique vulnerability of human settlements in the Arctic, and the extreme riskiness of major industrial initiatives in the ice-bound region, social and economic impact assessment will be crucial to any systematic effort to design a transit-management system for the Northwest Passage.

F *Management system design*

Several management approaches might be combined to produce an appropriate transit-management system design. The narrowest and most specific, as well as the most closely related to the transportation function of transit management, is the *vessel traffic management* approach.[84] This approach requires the designer to create a practical subsystem of administrative rules and procedures around a set of navigation and communications technologies, and the relevant operational and maintenance training requirements.[85]

Also familiar is the *coastal management* framework for planning, and the central concept of *coastal zone management*.[86] The Canadian government system has not enthusiastically embraced this concept, originating in the United States,[87] for the Atlantic, Pacific, and Great Lakes shoreline areas.[88] However, the concept might be accepted as relevant for planning in the North.

The last, and least specific, of these approaches is that of *environmental management*, or, as applied to Canadian Northern waters, that of "marine environmental management". The most conspicuous efforts to develop this kind of planning framework have been made within the United Nations system. The UN Environment Programme (UNEP) has designated ten "regional seas"[89] where Regional Action Plans[90] will facilitate the development of marine environmental management. Certain elements incorporated into these Regional Action Plans may be valuable for the design of a transit-management system for the Northwest Passage, as well as for the neighbouring marine regions to the west (the Beaufort Sea) and to the east (Davis Strait).[91] Environment Canada, Transport Canada, the Department of Fisheries and Oceans and DIAND should consider the design of a transit-management system for Canadian Northern waters within the framework of a Canadian Arctic Ocean Regional Plan.[92]

III The question of values

A Challenge to a nation

Now that several alternative approaches for designing a transit-management system have been reviewed, there follows the problem of choice. Which of these alternatives should be used in the research and planning? All, most, or only a few? Which, if any, should be dominant or central within the overall framework of research and planning strategy? What kinds of data, ideas and considerations should be given the heaviest weighting? What is the appropriate level, kind or degree of sophistication to be achieved as a reliable basis for this kind of decision making?

All of these questions, directly or indirectly, involve judgement about the *scale* of such an endeavour, which is, finally, a question of *values*. How important is it, to Canada in general and to the Canadian North in particular, to bring *excellence* to bear in the designing of a transit-management system for the Northwest Passage? Do the values which are at stake justify a quest for excellence, and the costs involved in such a quest?

This last question cannot be answered in monetary terms alone. Such a question is essentially addressed to the nation at large, and although the answer can only be given by leadership, it must be evinced from the various expressions of national will intuited from the mood of the people. It might even be said that the question constitutes a challenge to the nation which requires a cultural response.

If we assume that the best outcome is worth working for, even if a long-term time-frame is required, then we can afford to envisage the designing of a transit-management system for the Northwest Passage as a *heroic endeavour*. It may, then, be useful, in dealing with these difficult questions, to recall other heroic endeavours in history requiring a comparably massive deployment of national will, energy and imagination.

In Canada's own rather short history, there has been no lack of willingness to undertake heroic endeavours of two sorts: *megasystems* and *megaprojects*. The Distant Early Warning (DEW) line and the North American Air Defence (NORAD) system are impressive examples of megasystems, both admittedly joint undertakings with the United States, with much of the will, energy, imagination and money coming from south of the border. The building of the Canadian Pacific Railway, the first Canadian megaproject, across the face of a wilderness nation, is certainly a better and more soul-stirring example of a truly Canadian endeavour. Less stirring, but impressive in scale, was the recent construction of the national Trans-Canada Highway.

In the category of joint (Canadian–US) megaprojects belong the St Lawrence Seaway of the early 1950s and the Alaska Highway of the same decade. Even more spectacular are the Churchill Falls and James Bay hydroelectric power projects, among the largest megaprojects in North America. Today Canadians are faced with major decisions on other megaprojects, such as the Athabasca Tar Sands project and the Fundy Tidal Power project. If Canada holds back indefinitely from the opening up of the Northwest Passage on a heroic scale, it is unlikely to be due to national diffidence.

Outside Canada innumerable examples can be given of heroic endeavours of a similar magnitude. The opening of the Suez and Panama Canals probably comes closest to exemplifying the scale of effort required to implement a transit-management system for the Northwest Passage. It may seem unfair to refer to the vast investments of energy, talent and money in the United States and Soviet space programmes, but even scaled down these efforts still represent a striking expression of national faith and an impressive example of national will. Certainly Canadians can identify with the Zuider Zee reclamation megaproject in The Netherlands, or the projected English Channel tunnel project.

In short, the planning and construction of a full-scale transit-management system for the Northwest Passage is an appropriate endeavour for the Canadian people under the conditions of the late twentieth century. It is a challenge the nation should be asked to meet.

B The values at stake

It is assumed, then, that the opening up of the Northwest Passage in the near future is necessary for a growing variety of transit purposes; that these future activities must be brought under a single, comprehensive system of regulation and management; that the design of such a system should be carefully and rationally planned within a large-scale and sophisticated framework; and that this heroic undertaking deserves the most sophisticated, integrative thinking and research that Canadians can generate for such a challenging task. But the *direction* of the appropriate thinking and research should be determined only after a weighing of the values at stake.

In seeking the right direction, we should remind ourselves that what is envisaged is not merely a project, nor even a megaproject, but a system. Perhaps a "megasystem". If we were merely planning a project for the extraction and transportation of non-renewable resources, we would be bound to yield to considerations of economy and limit the framework of planning and design to sensible proportions. The short-lived and local nature of such a project tends to dictate that special weighting be given to short-term prospects of economic exploitation and to social and environmental impacts in the immediate vicinity of the extraction site.

A transit-management system such as the one proposed will be *permanent* and *pervasive* throughout an extensive area of the Canadian North. The designing of such a system must take account of *all* the values at stake, directly and indirectly, in the long term as well as the short. The framework for this task of system design must be sufficiently broad to ensure that the researchers, planners, decision-makers, administrators and managers involved—and eventually the users too—will be, and remain, as sensitive as possible to the legitimacy of all these values.

The apparent values at stake may be discussed under six headings: social, economic, administrative, ethical, scientific, and aesthetic.

1 *Social values*

The opening up of the Northwest Passage has been described as a "challenge to the nation" in terms of national culture. This aspect must be stressed because it is essentially a matter of *social investment* by the Canadian nation as a whole. But social investment centred at the national level shifts the focus to *social impact* at the regional or subregional level. The most important, most direct, and most lasting social impacts will be felt most directly by the remote and isolated Northern communities of the Northwest Territories. The prospect of a transit-management system in the Canadian North complicates the difficult problems of community (and individual) welfare, which exist in all other contexts of Northern planning and development. On one matter, at least, most planners today are likely to agree: the Northern perception and understanding of the North must be accepted as an indispensable input into the design of a transit-management system for the Northwest Passage. Even though Northerners may be divided in their opinions about this projected system, Northern participation both in the planning and in the operation of such a system is crucial, ethically desirable and politically advisable.

The designers of the system must also be concerned with the welfare of incomers whose services will be essential for certain operations, even after Northerners wishing employment within the system complete training programmes. Whether, or to what extent, these incomers should be confined to special communities of their own, or integrated with local personnel in new communities, is a difficult question of social values that will have to be addressed.

2 *Economic values*

Planning, in general, is devoted to the value of human growth, and it is difficult to deny the importance of economic development in human affairs. But a crassly growth-orientated strategy for the Canadian North would clash with other values, and it may be necessary to define some kind of "steady-state" economy for the region. A serious effort should be made to incorporate the development of a transit-management system into a Northern economic plan which is not unduly influenced by Southern Canadian economic thought and which represents a strategy of *eco-development* or *socio-ecological economic development.*

In some degree the problems of economic planning for Canadian Northern waters have been eased, or at least clarified, by the introduction of the 200-nautical-mile exclusive economic zone (EEZ) regime in the international law of the sea. Although not all legal issues in the Arctic have been resolved by the new UN Convention on the Law of the Sea, the advent of this quasi-monopolistic resource regime for the benefit of the coastal state enables Canada to make fairly bold assumptions about its exclusive entitlement to establish and maintain a transit-management system in Northern waters. With legal control essentially in Canadian hands, the overall cost of running such a system is significantly reduced and revenues derived from the system can defray operating costs. But a Northwest Passage transit system, like any other large-scale transit system, must also be designed with a view to extracting the economic advantages of co-operation: that is, of co-operation between Canada, as the managing state, and the various user and neighbouring states whose interests must be considered in the design of an efficient as well as equitable system.

3 *Administrative values*

Efficiency is not the sole preserve of the economist. It is also one of the major value concerns of the public administrator and of the corporate manager. Most observers and participants are likely to accept the need for a high degree of administrative efficiency in the operation of a transit-management system for the Northwest Passage. But a wide difference may exist between governmental and corporate concepts of efficiency, since government is primarily concerned with providing public services at reasonable cost, and private business is primarily concerned with making a profit. Whose concept of efficiency should prevail?

The bureaucratic model of efficiency invokes the goal of *interorganizational co-operation and coordination*, but consideration should also be given to two alternative schemes: a corporate structure (on the model of a federal crown corporation) or contributions of equity and other resources from both private and public sectors. Whatever form the operational authority of the system might assume, it will have to interact in some degree both with the governmental system of regulatory bodies and with the industrial system of shippers, cargo owners, bankers and insurers.

Another kind of administrative efficiency, that of intergovernmental co-operation, may be more difficult to attain. Efficient operation of a system designed for managerial effectiveness will depend on smooth and harmonious relationships between federal and territorial agencies with overlapping responsibilities. Such an overlap of responsibilities may increase in the future when the governments of the Northwest Territories (or its successors) and Yukon will have a good deal of involvement, if not autonomy, in certain sectors of Northern development and management. Administrative efficiency will also require *international intergovernmental co-operation*, particularly with the United States in the Beaufort Sea and with Denmark-Greenland in Davis Strait.

4 *Ethical values*

The principle of *democracy* requires the design of a transit-management system to be based on consultations with those most likely to be directly affected. The ethos of democracy is particularly persuasive in the Northern context, since the Arctic region is not fully represented at the leadership level of our political system. Meaningful Northern participation in the design and management of the system is part of the desired response to this value demand, and the raising of the political status of the Territories (or their successors) may be another. Indeed, Northerners should be full participants in deciding whether and when Arctic shipping should increase. However, the ethical challenge goes further, beyond institutions, to other *values* which are not effectively represented in the interest-directed process of decision making. Planners of a transit-management system for the Northwest Passage cannot ethically disregard the legitimacy of non-social, non-economic, and non-administrative values such as those of science and aesthetics, and even broader *humanistic* values which might be defined out of these two categories. Neither a scientist nor an aesthete feels compelled to consider the preservation of all species, or of wildlife in particular, but a humanist does. To a humanist of the Julian Huxley tradition it is hardly enough, or even necessary, to agonize over abstract questions of social justice. The Arctic context projects special humanistic or ethical implications. Arguably nowhere is the demand for humanistic ethics more compelling than in the Arctic, precisely because it is there, more than anywhere else, that the benevolence of human impact can be effectively monitored.

5 *Scientific values*

A strictly scientific set of values must also be respected in the design of any large-scale system or project in the North. Any substantial investment of public funds in scientific investigation should be justified to some extent by reference to "basic" as well as "applied" areas of knowledge. At least in a modern and wealthy state like Canada, government has a commitment to the advancement of knowledge. Public policy requires that, wherever possible, a publicly funded research programme should be designed with sufficient latitude and imagination to permit the satisfaction of curiosity as well as the attainment of designated goals in project-relevant areas of inquiry. Especially in a largely unknown region like the Arctic, science has claims of its own. When government or industry is prepared to spend large sums on research for planning a project or system in the Canadian North, the Canadian scientific community is entitled to expect a balanced and well-considered Arctic science policy, which will ensure that scientific effort is not entirely devoted to short-term and politically attractive considerations.

In short, the "interest" of science prescribes that designing of a transit-management system for the Northwest Passage should be done in the light of professionally approved criteria of Arctic science policy, and that priorities should be set to ensure that investigations specific to projects or systems should be balanced against other kinds of more fundamental research.[93] If the Arctic states are not prepared to concede this to their own scientific community, they should not be surprised or resentful if the demand of science in the Arctic is taken up and asserted vigorously by the international science community.

6 *Aesthetic values*

Finally, the beauty of the Arctic environment must be considered. Those who have visited the Arctic can scarcely fail to be impressed by its lonely and mysterious grandeur. Natural beauty is a value unto itself, even when it remains unseen for most of a normal year. As long as mystery and wonder are a valued part of the human experience, all citizens of the world are entitled to know that the beauty of the Arctic wilderness is not being thoughtlessly ravaged. Ultimately, in a spiritual sense, the Arctic belongs to us all.

IV Some preliminary conclusions

It is premature to offer bold conclusions on the complicated questions arising from the task of designing a comprehensive transit-management system for the Northwest Passage. But a few suggestions might be offered with a view to avoiding some early mistakes.

First, the federal government of Canada should invest in a *national programme of research into the design of Arctic systems*, so as to take advantage of most, if not all, of the appropriate kinds of "integrative thinking" available in the academic and research communities of this country. It is timely, in the late-1980s, to carve out of the general field of "Northern studies" a more clearly demarcated area of "transit-management studies" for the Northwest Passage, as well as the cognate area of regional (or subregional) "ocean development and management studies" for the neighbouring waters of the Beaufort Sea and Davis Strait. It is not enough merely to designate these areas as "areas of national significance" under the grant systems of agencies such as the Social Sciences and Humanities Research Council of Canada (SSHRCC), and then to sit back and wait for separate applications for individual studies. The programme proposed here should be *organized* under criteria which have been carefully vetted by different kinds of specialists and institutions familiar with the intellectual problems of designing complex systems. The problem here is simply too difficult and too important to be left to the random processes of the academic or consultant market-place. Something like a five-year time-frame should be allowed for this round of research and conceptualization, and during that period the research teams engaged should be brought together in various groupings in workshop settings to encourage and accelerate the process of cross-fertilization of ideas. The recent efforts of the Environmental Studies Revolving Funds might be an appropriate point of departure in the design and structuring of such a national programme of preliminary research.

Second, the federal government of Canada should launch *a series of technical conferences, workshops and symposia* devoted to the technological problems involved in the design and construction of the transportation and communications components of a transit-management system for the Northwest Passage. This proposed series would bring together all the best technical expertise in Canada — and, if necessary, elsewhere — to ensure that the system will be built at the highest level of technological capability in existence.

Third, the federal government of Canada should initiate *a series of scientific conferences and workshops* designed to create productive synergistic interactions between Arctic scientists and administrators. The purpose of these meetings would be to construct a complete picture of Arctic science requirements for the remainder of the twentieth century, so as to provide an appropriate basis for the setting of Arctic research priorities. With the setting of national Arctic science-policy priorities, the research requirements for proposed systems and projects in the North will be amenable to more intelligent scrutiny and appraisal.

Fourth, the federal government of Canada, the governments of the Northwest Territories and Yukon, and the appropriate sector of Canadian industry should jointly sponsor *meetings and workshops of government officials and industry representatives* concerned with the technical and operational aspects of design of a transit-management system for the Northwest Passage. These meetings would be devoted primarily to the identification of stresses likely to be encountered between government and industry both in the design and in the operation of such a system, so that some of the more serious differences can be reconciled within the design. These meetings might also present an opportunity for government, industry and public interest groups to work out certain financial problems such as those of insurance and alternative forms of compensation in the event of loss or liability.

Fifth, the federal government of Canada and the territorial governments should hold *a series of workshops to investigate institutional options for integrating local communities into the decision-making process.* A transit-management system will require many value choices such as selecting socially and environmentally acceptable vessel routings and perhaps closing certain sensitive areas to human imposition altogether. Institutional mechanisms must be developed to allow Northerners full participation in the setting of value priorities.

Sixth, and last, the federal government of Canada, the governments of the Northwest Territories and Yukon, and the appropriate sectors of Canadian industry should jointly sponsor *a national Arctic training programme* with a view to preparing Canadians to undertake government and industry jobs in the North associated with such undertakings as a transit-management system for the Northwest Passage. Such a training programme might utilize existing facilities at Thebacha College in Fort Smith, Frobisher Bay's Gordon Robertson Education Centre, or the Vocational and Technical Training Centre in Whitehorse.[94] For Southerners, one component of the training programme might include a period of residency in a Northern community to enhance their understanding of the North through firsthand experience and observation. Northern trainees might benefit from a residency period in an administrative centre in the South such as Ottawa or in a provincial capital city.[95]

All of these preliminary steps could be initiated, and in most cases completed, within a five-year period of initial planning, before any of the critical questions of design and construction must be answered by decision-makers. Anything less than a sensible and systematic approach to this difficult task will be less than Canada deserves.

Notes

1. Rather surprisingly, no treatise deals comprehensively and systematically with transportation systems in general. The transportation literature tends to break down into sectors such as urban, intercity, rail, air, ocean, and so forth. For some representative general or theoretical works, see Alan Hay, *Transport for the Space Economy: A Geographical Study* (1973); W.R. Blunden, *The Land-Use/Transport System* (1973); and Edward J. Taaffe and Howard L. Ganthier, *Geography of Transportation* (1973). For an overview of the literature on transportation systems and issues, see Evert J. Visser (ed.), *Transport Decisions in an Age of Uncertainty* (1977) (Proceedings of the Third World Conference on Transport Research, Rotterdam, 26—28 April 1977).
2. See Lawrence R. Jacobsen and James J. Murphy, "Submarine Transportation of Hydrocarbons from the Arctic", in Louis Rey (ed.), *Arctic Energy Resources* (1983) (Proceedings of the Comité Arctique International Conference on Arctic Energy Resources, Oslo, 22–24 September 1982), pp. 273–283; and J.L. Courtney, "Arctic Resource Transportation: Present System and Future Development", in P.J. Amaria, A.A. Bruneau and P.A. Lapp (eds), *Arctic Systems* (1977) (Proceedings of a Conference on Arctic Systems, St John's, 18–22 August 1975), pp. 373–400.
3. See Chapter 4, *supra* Ernst Frankel, "Arctic Marine Transport and Ancillary Technologies".
4. Despite the differences between the political systems of Canada and the Soviet Union, a case study of Soviet experience in opening up its Northern Sea route may be a valuable contribution to the designing of a Canadian transit-management system for the Northwest Passage. See Terence Armstrong, "The Northern Sea Route Today", in Rey, *supra* note 2, pp. 251–257. For a comparison of the Northern Sea Route and the Northwest Passage, see Howard Hume, "A Comparative Study of the Northern Sea Route and the Northwest Passage: With Special Reference to the Future Development of Marine Transportation Along the Latter Route" (M. Phil thesis, Scott Polar Research Institute, Cambridge) (June 1984).
5. See, for example, *A Report by the Major Projects Task Force on Major Capital Projects in Canada to the Year* 2000 (June 1981).
6. For a general discussion of the alternative Northwest Passage routes, see Donat Pharand, *The Waters of the Canadian Arctic Archipelago in International Law* (in press).
7. Canada has consistently maintained that the 141st meridian, the land boundary between Alaska and Yukon, should also be the offshore line of demarcation. Consistent with such a position, Canada has claimed pollution-control jurisdiction and oil and gas exploration rights out to the 141st meridian as the western boundary. The United States, meanwhile, has argued for an equidistant line due to the lack of special circumstances in the area, and in accordance with equitable principles. For general discussions of the boundary dispute, see Ken Beauchamp, "An Overview of Current International Legal Issues in the Arctic Waters" (Paper presented at the Third National Workshop on People, Resources and the Environment North of 60°, Yellowknife, 1–3 June 1983), pp. 10–11; and Ted L. McDorman and Susan Rolston, "Maritime Boundary Delimitation in the Arctic Region", in Douglas M. Johnston and Phillip

Saunders (eds), *Maritime Boundary Delimitation: Regional Issues and Developments* (in press).

8. The Agreement between the Government of Canada and the Government of the Kingdom of Denmark Relating to the Delimitation of the Continental Shelf between Greenland and Canada was signed in Ottawa on 17 December 1973, and entered into force 13 March 1974. *Canada Treaty Series* 9 (1974).
9. The latitude of 60° North marks the southern limit of the jurisdiction of the Department of Indian Affairs and Northern Development (DIAND). See Order-in-Council, P.C. 1968—1574, dated 14 August 1968 as an amendment to the Canada Oil and Gas Lands Regulations.
   It is assumed that DIAND will play an important part in the design and operation of a transit-management system for the Northwest Passage and in the making and maintenance of ocean management arrangements in the Davis Strait region and the Beaufort Sea.
10. For a general discussion of US–Canadian bilateral relations concerning the Beaufort Sea, see John E. Carroll, *Environmental Diplomacy: An Examination and a Prospective on Canadian-US Transboundary Environmental Relations* (1983), pp. 79–85.
11. For example, Canada and Denmark have been co-operating for several years in the study of marine environmental factors affecting the waters adjacent to their coastlines. Exploration for oil and gas, and prospects for development in both Northern Canada and Greenland prompted several binational initiatives, including the Interim Canada-Denmark Marine Pollution Contingency Plan (1977), the Canada-Denmark Marine Pollution Plan (1979) and the Marine Environment Co-operation Agreement (1983) which supersedes the earlier agreements.
12. On the emergence of systems theory and systems analysis, see Robert Lilienfield, *The Rise of Systems Theory: An Ideological Analysis* (1978), pp. 7–34.
13. Fernando Cortes, Adam Pizeworski, and John Sprague, *Systems Analysis for Social Scientists* (1974), p. 5.
14. *Ibid.*
15. Systems analysis has been defined as "a systematic approach to helping a decision-maker choose a course of action by investigating his full problem, searching out objectives and alternatives, and comparing them in the light of their consequences, using an appropriate framework". V.G. Gates, "Technology and Public Policy: The Process of Technology Assessment in the Federal Government", in *Program of Policy Studies in Science and Technology* (George Washington University, 1972), V. 1, p. I-59, cited in M. Gibbons and R. Voyer, *A Technology Assessment System: A Case study of East Coast Offshore Petroleum Exploration* (1974), p. 28.
16. *Ibid.*
17. The deficiencies of cost-benefit analysis have prompted critics in several areas, especially that of environmental studies, to develop alternative techniques of analysis, such as risk-benefit analysis.
18. See, for example, Hartmut Bossel, Salòmon Klaozko, and Norbert Miller (eds), *Systems Theory in the Social Sciences* (1976).
19. See, for example, Mario Cambis, *Planning Theory and Philosophy* (1979). This author distinguishes three approaches in planning theory: rationalist idealist, irrational idealist, and materialist.
20. Peter Hall, *The Theory and Practice of Regional Planning* (1970), pp. 10—11.
21. Andreas Faludi, *Planning Theory* (1973), p. 11.
22. See, for example, Judith Pallot and Denis J.B. Shaw, *Planning in the Soviet Union* (1981).

23. Faludi, *supra* note 21, pp. 139–48.
24. For a distinction between "ideals", "objectives", and "policies" in transport planning, see Morris Hill, *Planning for Multiple Objectives: An Approach to the Evaluation of Transportation Plans* (1973), p. 24.
25. Charles E. Lindblom, *The Intelligence of Democracy: Decision Making through Mutual Adjustment* (1965); David Braybrooke and Charles E. Lindblom, *A Strategy of Decision: Policy Evaluation as a Social Process* (1963); and Robert H. Dahl and Charles E. Lindblom, *Politics, Economics and Welfare: Planning and Political-Economic Systems Resolved into Basic Social Processes* (1983).
26. Gibbons and Voyer, *supra* note 15, p. 31.
27. *Ibid.*
28. *Ibid.*, p. 87.
29. A. Etzioni, "Mixed Scanning: A 'Third' Approach to Decision Making", *Public Administration Review* V. 27, No. 5 (December 1967), pp. 385–392.
30. See Robert F. Adie and Paul G. Thomas, "Theories of Decision Making in Government", *Canadian Public Administration: Problematical Perspectives* (1982), pp. 96–100.
31. Cambis, *supra* note 19, pp. 1–7.
32. One scholar has divided these reactions into three categories: utopian, dystopian and socialist. Bernard Gendron, *Technology and the Human Condition* (1977).
33. Erich Jantsch, *Technological Planning and Social Futures* (1972); Robert U. Ayres, *Technological Forecasting and Long-Range Planning* (1969); and Marvin J. Centron, *Technological Forecasting: A Practical Approach* (1969).
34. Kenyon B. DeGreene, *Socio-technical Systems: Factors in Analysis, Design and Management* (1973).
35. H.W. Langford, *Technological Forecasting Methodologies: A Synthesis* (1969).
36. Arthur Gerstenfeld and Robert Brainard (eds), *Technological Innovation: Government/Industry Co-operation* (1979).
37. In 1973 a study commissioned by the Science Council of Canada concluded that "the technology assessment system, operating in the mixed scanning mode, has brought together into a loose aggregation the majority of actors who have some interest in the development of an offshore petroleum recovery capability". Gibbons and Voyer, *supra* note 15, p. 88. Ten years later, a coalition of interested institutions from the governmental, industrial and academic sectors of Canada, the United Kingdom, and Norway established an international network, the Programme for Atlantic Co-operative Offshore-Onshore Development (PACOD), to facilitate offshore research, training, and related services for the Northwest Atlantic Ocean. See PACOD Draft Memorandum of Understanding (28 September 1983).
38. Raymond Vernon, "Comprehensive Model-Building in the Planning Process: The Case of the Less Developed Economies", *The Economic Journal* V. 76, No. 301 (March 1966), p. 59, quoted in Edwin T. Haefele, "Transport Planning and National Goals", in Edwin T. Haefele (ed.), *Transport and National Goals* (1969), pp. 177–193, p. 178.
39. "The control on promotion of the expansion of transport facilities is perhaps the most powerful geographically-specific instrument that government can use to guide economic development." Patrick O'Sullivan, Gary D. Holtzelaw, and Gerald Barber, *Transport Network Planning* (1979), p. 9. For recent international perspectives on transportation policy issues, see Visser, *supra* note 1; and David Bannister and Peter Hall (eds), *Transport and Public Policy Planning* (1981).
40. Hill, *supra* note 24, p. 24.

41. Kenneth J. Arrow and Mordecai Kurz, *Public Investment, the Rate of Return, and Optimal Fiscal Policy* (1980); Stephen A. Marglin, *Public Investment Criteria: Benefit-Cost Analysis for Planned Economic Growth* (1967).
42. See Chapter 3, *supra* Carlyle L. Mitchell, "The Development of Northern Ocean Industries".
43. Preston P. Le Breton and Dale A. Henning, *Planning Theory* (1961), pp. 338–342.
44. See, for example, J. Brian McLoughlin, *Urban and Regional Planning: A System Approach* (1969).
45. Hall, *supra* note 20, *passim*.
46. Despite this blending of concepts, regional science has become dominated in recent years by econometric models and mathematical techniques, as reflected in the *Journal of Regional Science* and the papers of the Regional Science Association.
47. Douglas M. Johnston, *Arctic Ocean Issues in the* 1980*s* (1982) (Report on Workshop held at Mackinac Island, Michigan, in June 1981 under joint auspices of the Law of the Sea Institute and Dalhousie Ocean Studies Programme), pp. 2–4.
48. For proposals concerning federal land-use planning in the North, see John K. Naysmith, *Land Use and Public Policy in Northern Canada* (1976); and Department of Indian Affairs and Northern Development, *Land-Use Planning in Northern Canada: Preliminary Draft* (14 October 1982).
49. Department of Indian Affairs and Northern Development, *The Lancaster Sound Region*: 1980–2000 (January 1982).
50. *Ibid.*, pp. 6–7.
51. See, for example, W.R. Blunden, *The Land-Use Transport System: Analysis and Synthesis* (1971); and Eric J. Miller and Adil Cubukgil, *Land-Use Trends and Transportation Demand Forecasting: An Empirical and Theoretical Investigation* (1981).
52. Edward A. Fernald, John F. Lounsbury and Laurence M. Sommers, "The Future and the Need for Rational Land-Use Decisions" in Lounsbury, Sommers and Fernald (eds) *Land Use: A Spatial Approach* (1981), pp. 223–230, p. 227.
53. *Ibid.*, pp. 227–228.
54. Rolf E. Rogers, *Organizational Theory* (1975), pp. 19–31.
55. *Ibid.*, pp. 3–18.
56. Jeffrey Pfeffer, *Organizations and Organization Theory* (1982), p. 5.
57. *Ibid.*
58. *Ibid.*
59. *Ibid.*
60. H. George Frederickson, "The Lineage of New Public Administration", in Carl J. Bellone (ed), *Organization Theory and the New Public Administration* (1980), pp. 33–51, p. 43.
61. *Ibid.*, pp. 36–43.
62. *Ibid.*, p. 43.
63. *Ibid.*, p. 47.
64. For a general discussion of the practical and theoretical problems of interorganizational relations, see David F. Gillespie and Dennis S. Mileti, *Technostructures and Interorganizational Relations* (1979).
65. For a summary of environmental assessment approaches in Canada, see William J. Couch (ed.), *Environmental Assessment in Canada*: 1982 *Summary of Current Practice* (January 1982).

66. Patrick Geddes argued that the physical system and facilities of a city should be planned so as to harmonize with the natural environment. Patrick Geddes, *Cities in Evolution* (1909).
67. Joseph L. Ridges, Jr., *Environmental Impact Assessment, Growth Management, and the Comprehensive Plan* (1976), pp. 29–43; and Ruthann Corwin and Patrick H. Heffernan (eds), *Environmental Impact Assessment* (1975), pp. 18–33.
68. For a detailed description of this and related requirements, see Michael Remy, "The Law of Environmental Impact Assessment", in Corwin and Heffernan, *supra* note 67, pp. 170–196.
69. On the conceptual and methodological implications of the EIS system, see Sherman J. Rosen, *Manual for Environmental Impact Evaluation* (1976), pp. 36–46; and Paul N. Cheremisinoff and Angelo C. Morresi, *Environmental Assessment and Impact Statement Handbook* (1977), pp. 75–92.
70. Environmental Assessment and Review Process Guidelines Order, SOR/84–467.
71. See, for example, National Energy Board Rules of Practice and Procedure, amendment SOR/78–926.
72. For a general discussion of the administrative mesh, see Chapter 6, *supra* David L. VanderZwaag and Cynthia Lamson, "Northern Decision Making: A Drifting Net in Restless Sea".
73. Gordon E. Beanlands and Peter N. Duinker, *An Ecological Framework for Environmental Impact Assessment in Canada* (1983).
74. *Ibid.*, p. 1.
75. *Ibid.*, p. 2.
76. *Ibid.*, p. 10.
77. *Ibid.*, pp. 2–3.
78. C.S. Holling, *Adaptive Environmental Assessment and Management* (1978), pp. 3–5.
79. Robert A. Johnston, "Assessing Social and Economic Impacts", in Corwin and Heffernan, *supra* note 67, pp. 113–57, p. 113.
80. For a discussion of the evolution of social assessment approaches see Gregory Daneke and Jerry Priscoli, "Social Assessment and Resource Policy: Lessons from Water Planning", *Natural Resources Journal* V. 19, No. 2 (1979), p. 359.
81. See Federal Environmental Assessment Review Office, *Federal Environmental Assessment and Review Process: Guide for Environmental Screening* (1978), p. 5. Since 1 August 1978, all new regulations or proposed amendments to existing regulations relating to health, safety or fairness (HSF) must be subjected to socio-economic impact analysis (SEIA). The departments and agencies affected are: Agriculture Canada, Consumer and Corporate Affairs Canada, Energy, Mines and Resources, Fisheries and Oceans, Indian Affairs and Northern Development, Health and Welfare Canada, Labour Canada, Transport Canada, Canadian Transport Commission, Atomic Energy Control Board, National Energy Board and Canada Mortgage and Housing Corporation. See Treasury Board, *Administrative Policy Manual* (1982), c. 490.
82. See Reg Lang and Audrey Armour, *The Assessment and Review of Social Impacts* (1981).
83. See, for example, G.A. Mackay and Anne C. Moir, *North Sea Oil and the Aberdeen Economy* (1980); D.I. MacKay and G.A. Mackay, *The Political Economy of North Sea Oil* (1975); and Adrian Hamilton, *North Sea Impact: Offshore Oil and the British Economy* (1975).
84. See M.A. Turner, "Arctic Traffic Management, Navigation and Communication System", in Amaria, Bruneau and Lapp, *supra* note 2, pp. 19–47.

85. *Ibid.*
86. Bostwick H. Ketchum (ed.), *The Water's Edge: Critical Problems of the Coastal Zone* (1972). For a recent overview, see Jens C. Svensen, Scott T. McCreary, and Mare J. Hershman, *Institutional Arrangements for Management of Coastal Resources* (Research Planning Institute, Inc, 1984).
87. The concept of coastal zone management originated in the recommendations of the Stratton Commission (The Marine Sciences Commission) in its Report, *Our Nation and the Sea* (1969), which led to the enactment of the federal US Coastal Zone Management Act in 1972. A special planning approach to coastal areas can be traced to an earlier period in the United Kingdom, and perhaps elsewhere, but the CZM approach has assumed a distinctive character in the United States.
88. See, for example, Douglas M. Johnston, Paul Pross, and Ian McDougall (with the special assistance of Norman G. Dale), *Coastal Zone: Framework for Management in Atlantic Canada* (1975); and Douglas M. Johnston, "Coastal Zone Management in Canada: Purposes and Prospects", *Canadian Public Administration* V. 20, No. 1 (1977), p. 140. For a recent review of Canadian coastal planning problems, see special issue on "Coastal Management in Canada", *Coastal Zone Management Journal* V. 11, Nos. 1–2 (1983).
89. The ten "regional seas" are the: Mediterranean, Red Sea and Gulf of Aden, Arabian-Persian Gulf, West Africa, Caribbean Sea, East Asian Seas, Southwest Pacific, Southeast Pacific, Southwest Atlantic and East Africa.
90. Douglas M. Johnston and Lawrence M.G. Enomoto, "Regional Approaches to the Protection and Conservation of the Marine Environment" in Douglas M. Johnston (ed.), *The Environmental Law of the Sea* (1981), pp. 285–385, pp. 324–337.
91. For a detailed analysis of the Regional Action Plan approach to environmental management of the South China Sea, see Douglas M. Johnston, *Environmental Management in the South China Sea: Legal and Institutional Developments* (East-West Environment and Policy Institute, Research Report No. 10, 1982).
92. For planning purposes, the Northwest Passage might be divided into various subregions, for example: Beaufort Sea Region, Amundsen Gulf-Prince of Wales Strait Region, Viscount Melville Sound Region, Barrow Strait Region, Lancaster Sound Region and Baffin Bay-Davis Strait Region. Within each subregion, institutional mechanisms would have to be created which integrate various political interests – local communities, territorial government(s), the federal government, Alaska-US government (at the western extremity) and Greenland-Denmark (at the eastern extremity) – into co-operative decision-making units. Many other subregional classifications are, of course, possible. For example, Parks Canada has divided the Arctic Ocean into eight marine subregions: Beaufort Sea, Viscount Melville Sound, Northern Arctic, Queen Maud Gulf, Lancaster Sound, Eastern Baffin Island Shelf, Foxe Basin and Hudson Bay. See Parks Canada, *National Marine Parks Draft Policy* (Third Draft, August 1983).
93. See, for example, Association of Canadian Universities for Northern Studies, *Ethical Principles for the Conduct of Research in the North* (1982).
94. It should be noted that a similar recommendation was among resolutions passed at the *Ninth National Northern Development Conference*:

    Resolution 13

    EDUCATION

    WHEREAS northerners aspire to acquire an education north of 60 which will enable freedom of choice with respect to access to jobs throughout Canada, while guaranteeing the survival of northern native language and culture, BE IT RESOLVED that Thebacha College, currently operating in Fort Smith, be expanded by adding satellite facilities throughout the North to provide post-secondary education in a balanced manner by offering courses of direct northern relevance as well as courses of more universal application. See *Ninth National Northern Development Conference* 1982 (Proceedings, Edmonton, 27—29 October 1982), p. 165.

The Special Committee on Education of the Northwest Territories tabled a report to the NWT Legislative Assembly in 1982 and concluded that an Arctic College with regional campuses, each under regional control, would best serve the needs of Northerners. J.D. Jacobs and M.E.Sanderson, "Toward University Education in the North", *Northline* V. 4, No. 1 (January 1984), p. 4.

95. The new National Indigenous Development Program (NIDP) which provides native people with training opportunities in managerial and advisory skills, might be incorporated into the proposed national Arctic training programme. The Donner Canadian Foundation has funded a three-year study of Northern native employment training programmes by the University of Calgary. See *APOA Review* V. 6, No. 1 (Spring/Summer 1983), p. 4. It should be noted that previous proposals have been put forward to promote coordinated research and to facilitate communications between Southern scientists and Northerners. For example, the Department of Indian Affairs and Northern Development has suggested the establishment of a Northwest Territories Science Institute, and the Association of Canadian Universities for Northern Studies has proposed a Foundation for Northern Research. See "Co-operative Research", in Milton Freeman (ed.), *Proceedings: First International Symposium on Renewable Resources and the Economy of the North* (Banff, May 1981), pp. 221—25, p. 223.

Appendix

Statement in the House of Commons by the Secretary of State for External Affairs, the Right Honourable Joe Clark, on Canadian Sovereignty (10 September 1985).

Mr Speaker,

Sovereignty can arouse deep emotion in this country. That is to be expected, for sovereignty speaks to the very identity and character of a people. We Canadians want to be ourselves. We want to control our own affairs and take charge of our own destiny. At the same time, we want to look beyond ourselves and to play a constructive part in a world community that grows more interdependent every year. We have something to offer and something to gain in so doing.

The sovereignty question has concerned this government since we were first sworn in. We have built national unity, we have strengthened the national economy, because unity and strength are hallmarks of sovereignty, as they are hallmarks of this government's policy and achievements.

In unity and strength, we have taken action to increase Canadian ownership of the Canadian petroleum industry. We have declared a Canadian ownership policy in respect of foreign investment in the publishing industry. We have made our own Canadian decisions on controversial issues of foreign policy – such as Nicaragua and South Africa. We have passed the *Foreign Extraterritorial Measures Act* to block unacceptable claims of jurisdiction by foreign governments or courts seeking to extend their writ to Canada. We have arrested foreign trawlers poaching in our fishing zones. We have taken important steps to improve Canada's defences, notably in bolstering Canadian forces in Europe and in putting into place a new North Warning System to protect Canadian sovereignty over our northern airspace. And we have reconstructed relations with traditional friends and allies, who have welcomed our renewed unity and strength and the confidence they generate.

In domestic policy, in foreign policy, and in defence policy, this government has given Canadian sovereignty a new impetus within a new maturity. But much remains to be done. The voyage of the *Polar Sea* demonstrated that Canada, in the past, had not developed the means to ensure our sovereignty over time. During that voyage, Canada's legal claim was fully protected, but when we looked for tangible ways to exercise our sovereignty, we found that our cupboard was nearly bare. We obtained from the United States a formal and explicit assurance that the voyage of the *Polar Sea* was without prejudice to Canada's legal position. That is an assurance which the government of the day, in 1969, did not receive for the voyage of the *Manhatten* and of the two United States Coast Guard icebreakers. For the future, non-prejudicial arrangements will not be enough.

The voyage of the *Polar Sea* has left no trace on Canada's Arctic waters and no mark on Canada's Arctic sovereignty. It is behind us, and our concern must be with what lies ahead.

Many countries, including the United States and the Federal Republic of Germany, are actively preparing for commerical navigation in Arctic waters. Developments are accerating in ice science, ice technology, and tanker design. Several major Japanese firms are moving to capture the market for icebreaking tankers once polar oil and gas come on stream. Soviet submarines are being deployed under the Arctic ice pack, and the United States Navy in turn has identified a need to gain Arctic operational experience to counter new Soviet deployments.

Mr Speaker

The implications for Canada are clear. As the Western country with by far the greatest frontage on the Arctic, we must come up to speed in a range of marine operations that bear on our capacity to exercise effective control over the Northwest Passage and our other Arctic waters.

To this end, I wish to declare to the House the policy of this government in respect of Canadian sovereignty in Arctic waters, and to make a number of announcements as to how we propose to give expression to that policy.

Canada is an Arctic nation. The international community has long recognized that the Arctic mainland and islands are a part of Canada like any other. But the Arctic is not only a part of Canada. It is part of Canada's greatness.

The policy of this government is to preserve that greatness undiminished.

Canada's sovereignty in the Arctic is indivisible. It embraces land, sea, and ice. It extends without interruption to the seaward-facing coasts of the Arctic islands. These islands are joined and not divided by the waters between them. They are bridged for most of the year by ice. From time immemorial Canada's Inuit people have used and occupied the ice as they have used and occupied the land.

The policy of this government is to maintain the natural unity of the Canadian Arctic archipelago, and to preserve Canada's sovereignty over land, sea, and ice undiminished and undivided.

That sovereignty has long been upheld by Canada. No previous government, however, has defined its precise limits or delineated Canada's internal waters and territorial sea in the Arctic. This government proposes to do so. An order in council establishing straight baselines around the outer perimeter of the Canadian Arctic archipelago has been signed today, and will come into effect on January 1, 1986. These baselines define the outer limit of Canada's historic internal waters. Canada's territorial waters extend 12 miles seaward of the baselines. While the *Territorial Sea and Fishing Zones Act* requires 60 days' notice only for the establishment of fisheries limits, we consider that prior notice should also be given for this important step of establishing straight baselines.

Canada enjoys the same undisputed jurisdiction over its continental margin and 200-mile fishing zone in the Arctic as elsewhere. To protect the unique ecological balance of the region, Canada also exercises jurisdiction over a 100-mile pollution prevention zone in the Arctic waters. This too has been recognized by the international community, through a special provision in the United Nations Convention on the Law of the Sea.

No previous government, however, has extended the application of Canadian civil and criminal law to offshore areas, in the Arctic and elsewhere. This government will do so. To this end, we shall give priority to the early adoption of a *Canadian Laws Offshore Application Act*.

The exercise of functional jurisdiction in Arctic waters is essential to Canadian interests. But it can never serve as a substitute for the exercise of Canada's full sovereignty over the waters of the Arctic archipelago. Only full sovereignty protects the full range of Canada's interests. This full sovereignty is vital to Canada's security. It is vital to Canada's Inuit people. And it is vital even to Canada's nationhood.

The policy of this government is to exercise Canada's full sovereignty in and over the waters of the Arctic archipelago. We will accept no substitutes.

The policy of this government is also to encourage the development of navigation in Canada's Arctic waters. Our goal is to make the Northwest Passage a reality for Canadian and foreign shipping, as a Canadian waterway. Navigation, however, will be subject to the controls and other measures required for Canada's security, for the preservation of the environment, and for the welfare of the Inuit and other inhabitants of the Canadian Arctic.

In due course, the government will announce the further steps it is taking to implement these policies, and especially to provide more extensive marine support services, to strengthen regulatory structures, and to reinforce the necessary means of control. I am announcing today that the government has decided to construct a Polar Class 8 icebreaker. The Ministers of National Defence and Transport will shortly bring to Cabinet recommendations with regard to design and construction plans. The costs are very high, in the order of half a billion dollars. But this government is not about to conclude that Canada cannot afford the Arctic. Meanwhile, we are taking immediate steps to increase surveillance overflights of our Arctic waters by Canadian Forces aircraft. In addition, we are now making plans for naval activity in eastern Arctic waters in 1986.

Canada is a strong and responsible member of the international community. Our strength and our responsibility make us all the more aware of the need for cooperation with other countries, and especially with our friends and allies. Cooperation is necessary not only in defence to our own interests but in defence of the common interests of the international community. Cooperation adds to our strength and in no way diminishes our sovereignty.

The policy of this government is to offer its cooperation to its friends and allies, and to seek their cooperation in return.

We are prepared to explore with the United States all means of cooperation that might promote the respective interests of both countries, as Arctic friends, neighbours, and allies, in the Arctic waters of Canada and Alaska. The United States has been made aware that Canada wishes to open talks on this matter in the near future. Any cooperation with the United States, or with other Arctic nations, shall only be on the basis of full respect for Canada's sovereignty. That too has been made clear.

In 1970, the government of the day barred the International Court of Justice from hearing disputes that might arise concerning the jurisdiction exercised by Canada for the prevention of pollution in Arctic waters.

This government will remove that bar. Indeed, we have today notified the Secretary General of the United Nations that Canada is withdrawing the 1970 reservation to its acceptance of the compulsory jurisdiction of the World Court.

The Arctic is a heritage for the people of Canada. They are determined to keep their heritage entire.

The policy of this government is to give full expression to that determination.

We challenge no established rights, for none have been established except by Canada. We set no precedent for other areas, for no other area compares with the Canadian Arctic archipelago. We are confident in our position. We believe in the rule of law in international relations. We shall act in accordance with our confidence and belief, as we are doing today in withdrawing the 1970 reservation to Canada's acceptance of the compulsory jurisdiction of the World Court. We are prepared to uphold our position in that Court, if necessary, and to have it freely and fully judged there.

In summary, Mr Speaker, these are the measures we are announcing today:

1. immediate adoption of an order in council establishing straight baselines around the Arctic archipelago, to be effective January 1, 1986;
2. immediate adoption of a *Canadian Laws Offshore Application Act*;
3. immediate talks with the United States on cooperation in Arctic waters, on the basis of full respect for Canadian sovereignty;
4. an immediate increase of surveillance overflights of our Arctic waters by aircraft of the Canadian Forces, and immediate planning for Canadian naval activity in the Eastern Arctic in 1986;
5. the immediate withdrawal of the 1970 reservation to Canada's acceptance of the compulsory jurisdiction of the International Court of Justice; and
6. construction of a Polar Class 8 icebreaker and urgent consideration of other means of exercising more effective control over our Arctic waters.

These are the measures we can take immediately. We know, however, that a long-term commitment is required. We are making that commitment today.

Canadian Jurisdictional Zones in the Arctic

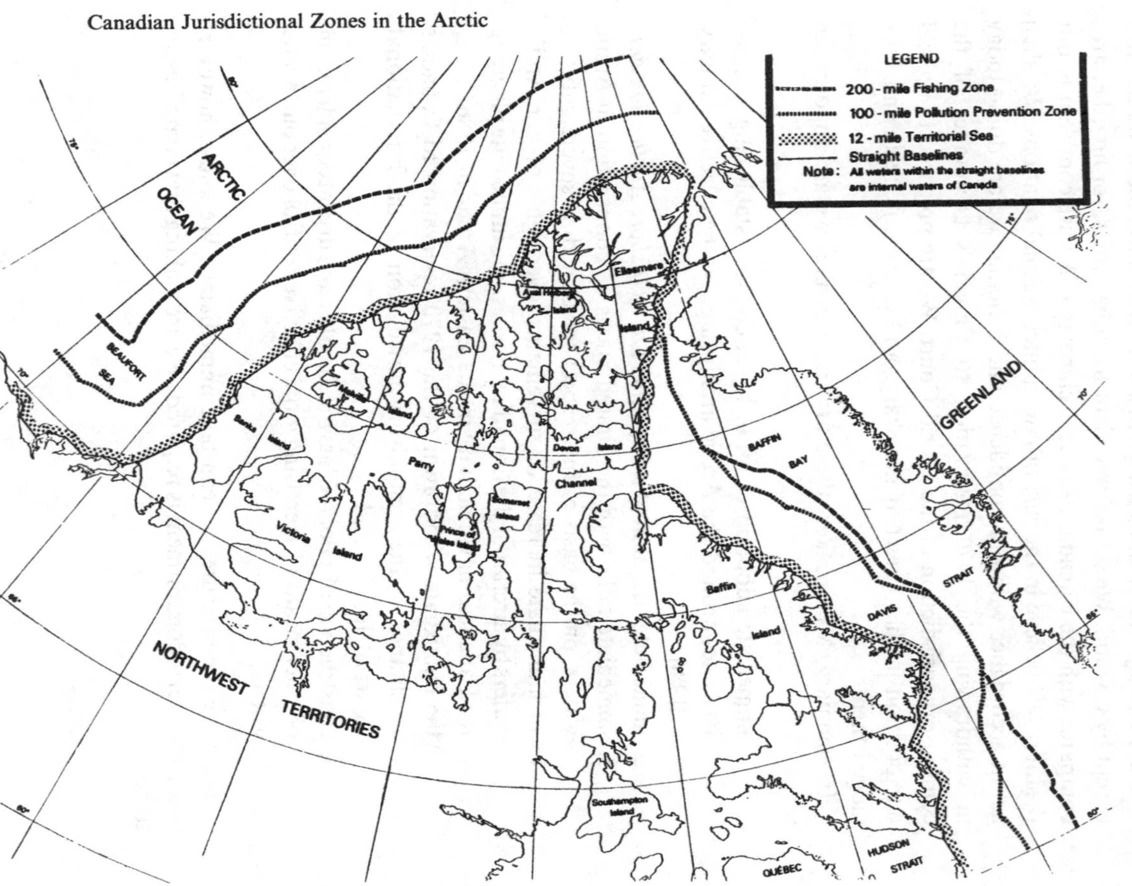

Contributors

Hal Mills is the Executive Secretary of the Northwest Territories Land Use Planning Commission. He served with the Royal Canadian Air Force and holds degrees from the University of Western Ontario. Prior to joining the government of the Northwest Territories, he worked with Environment Canada, the Department of Fisheries and Oceans, and as an independent marine policy consultant.

Carlyle L. Mitchell is a resource economist and international development consultant. He holds degrees from St Francis Xavier University, the University of Alberta, and the University of Ottawa. He has worked with the Canadian Department of Fisheries and Oceans, Environment Canada, and is President of North-South Intermedium Ltd.

Ernest G. Frankel is an international consultant specializing in ship design and engineering, shipping policy and operations management. He is Professor of Ocean Systems at the Massachusetts Institute of Technology, and frequently advises the World Bank as well as the US and foreign governments on shipping related matters. He holds degrees from the University of London, Boston University and MIT.

William J. H. Stuart served with the Royal Navy before transferring to the Royal Canadian Navy for work with the Canadian Hydrographic Service. In 1969 he was appointed Chief of Marine Emergencies, Canadian Goast Guard, and in 1973 he assumed the duties of Director Canadian Coast Guard, with responsibilities for fleet activities. Captain Stuart was a consultant to the Arctic Pilot Project and continues to make Arctic voyages aboard the cruise ship *World Discover* as Ice Master.

David VanderZwaag is an Assistant Professor at Dalhousie Law School and a Research Associate with the Dalhousie Ocean Studies Programme. His educational background includes a BA from Calvin College, M Div from Princeton Theological Seminary, the JD from the University of Arkansas School of Law and the LLM in international law of the sea from Dalhousie Law School. He is author of *The Fish Feud: The US and Canadian Boundary Dispute*.

Cynthia Lamson is a Research Associate with the Dalhousie Ocean Studies Programme. She is a graduate of Skidmore College, and holds advanced degrees from Simmons College, Memorial University of Newfoundland, and Dalhousie University. She is author of a book about the Canadian sealing controversy, *Bloody Decks and a Bumper Crop*, and is editor of *Atlantic Fisheries and Coastal Communities: Fisheries Decision-Making Case Studies*.

J. Fielding Sherwood was a graduate of the University of Toronto, and held an LLB degree from the University of British Columbia and an LLM from Dalhousie University. He was a barrister with the Halifax firm, Daley, Black and Moreira prior to his unexpected death in April, 1986.

Douglas M. Johnston is a Professor of Law and Director of the Marine Environmental Law Programme at Dalhousie University. He is a graduate of St Andrews University and McGill University, and earned his LLM and JSD degrees from Yale University. He is the author of numerous books and articles including *The International Law of the Fisheries: A Framework for Policy-Oriented Inquiries*, *Conservation and the New Law of the Sea*, and *Arctic Ocean Issues in the 1980s*.